Springer Complexity

Springer Complexity is an interdisciplinary program publishing the best research and academic-level teaching on both fundamental and applied aspects of complex systems – cutting across all traditional disciplines of the natural and life sciences, engineering, economics, medicine, neuroscience, social and computer science.

Complex Systems are systems that comprise many interacting parts with the ability to generate a new quality of macroscopic collective behavior the manifestations of which are the spontaneous formation of distinctive temporal, spatial or functional structures. Models of such systems can be successfully mapped onto quite diverse "real-life" situations like the climate, the coherent emission of light from lasers, chemical reaction-diffusion systems, biological cellular networks, the dynamics of stock markets and of the internet, earthquake statistics and prediction, freeway traffic, the human brain, or the formation of opinions in social systems, to name just some of the popular applications.

Although their scope and methodologies overlap somewhat, one can distinguish the following main concepts and tools: self-organization, nonlinear dynamics, synergetics, turbulence, dynamical systems, catastrophes, instabilities, stochastic processes, chaos, graphs and networks, cellular automata, adaptive systems, genetic algorithms and computational intelligence.

The two major book publication platforms of the Springer Complexity program are the monograph series "Understanding Complex Systems" focusing on the various applications of complexity, and the "Springer Series in Synergetics", which is devoted to the quantitative theoretical and methodological foundations. In addition to the books in these two core series, the program also incorporates individual titles ranging from textbooks to major reference works.

Springer Series in Synergetics

Founding Editor: H. Haken

The Springer Series in Synergetics was founded by Herman Haken in 1977. Since then, the series has evolved into a substantial reference library for the quantitative, theoretical and methodological foundations of the science of complex systems.

Through many enduring classic texts, such as Haken's *Synergetics and Information and Self-Organization*, Gardiner's *Handbook of Stochastic Methods*, Risken's *The Fokker Planck-Equation* or Haake's *Quantum Signatures of Chaos*, the series has made, and continues to make, important contributions to shaping the foundations of the field.

The series publishes monographs and graduate-level textbooks of broad and general interest, with a pronounced emphasis on the physico-mathematical approach.

For further volumes:
http://www.springer.com/series/712

M. Lakshmanan · D.V. Senthilkumar

Dynamics of Nonlinear
Time-Delay Systems

 Springer

Prof. Dr. M. Lakshmanan
Centre for Nonlinear Dynamics
Bharathidasan University
Palkalaiperur Campus
Tiruchirapalli-620 024
India
lakshman.cnld@gmail.com

Dr. D.V. Senthilkumar
Transdisciplinary Concepts
and Methods
Potsdam Institute for Climate
Impact Research
Telegrafenberg
14412 Potsdam
Germany
skumarusnld@gmail.com

ISSN 0172-7389
ISBN 978-3-642-14937-5   e-ISBN 978-3-642-14938-2
DOI 10.1007/978-3-642-14938-2
Springer Heidelberg Dordrecht London New York

Library of Congress Control Number: 2010936895

Cover design: Integra Software Services Pvt. Ltd., Pondicherry

Printed on acid-free paper.

Springer is part of Springer Science+Business Media (www.springer.com)

To our parents

Preface

Numerical works of Edward Lorenz on the fluid convection model of atmospheric weather and of Norman Zabusky and Martin Kruskal on the initial value problem of the Korteweg-de-Veries (KdV) equation had paved the way for identifying two of the important basic concepts, namely *Chaos* and *Soliton*. These developments indeed triggered the golden era of modern nonlinear dynamics in the early 1960s. Since then, nonlinear science has emerged as a highly interdisciplinary subject having its roots in every branch of science and technology and even extending its principles and concepts to sociology, humanities, etc.

Chaos synchronization, a patently nonlinear phenomenon, has also emerged as a highly active interdisciplinary research topic from early 1990s after the works of Pecora and Carroll, and the earlier works of Fujisaka and Yamada. The possibility of chaos synchronization has been demonstrated by introducing appropriate coupling between identical chaotically evolving dynamical systems. The phenomenon is of interest not only from a theoretical point of view but also has potential applications in diverse subjects such as biology, neuroscience, laser physics, chemical, electrical and fluid mechanical systems as well as in secure communication, cryptography and so on. Subsequently, several generalizations and interesting applications have been developed. The emergence of coherent behavior from the collective dynamics of systems ranging from just two to an ensemble/network of interacting oscillators can be explained by this phenomenon. Also synchronizability of hyperchaotic systems with just a single scalar coupling motivates further intensive research in synchronizing high-dimensional systems with appropriate applications.

In this connection, time-delay systems described by delay differential equations, whose discussion forms the main theme of this monograph, are essentially infinite dimensional in nature and they can admit hyperchaotic attractors with large number of positive Lyapunov exponents for suitable nonlinearities. Even a simple first order scalar time-delay system with appropriate nonlinearity can exhibit hyperchaotic attractors for suitable values of parameters. Time-delay is also ubiquitous in many physical systems due to finite switching speeds of amplifiers, finite lengths of vehicles in traffic flows, finite signal propagation time in biological networks and circuits, memory effects and so on. Therefore the study of chaos synchronization in coupled time-delay systems and the effect of time-delay feedback, which give rise to plethora of new collective dynamics, have become a center of attraction in nonlinear dynamics.

This monograph aims to present the basic materials on chaotic time-delay systems and their synchronization to research students of interdisciplinary study. The basics on the dynamics of time-delay systems and on the synchronization can even serve as a tutorial for master-level students to provide further insight into their knowledge on advanced topics in nonlinear science. It is also intended as an introduction to both theoreticians and experimentalists, particularly in electronics, as circuit details on time-delay systems and their PSPICE simulations are provided at appropriate places. Further, to kindle the interest of researchers, we have tried to provide a detailed discussion on the effect of time-delay feedback which gives rise to new collective dynamics and on chaos synchronization in networks of time-delay systems.

Specifically, the first four chapters are devoted to the basics of time-delay systems including different types of delays along with their applications in different areas of science and technology, salient features of time-delay systems, detailed stability and bifurcation analysis, numerical simulation of time-delay systems, collection of chaotic time-delay systems, electronic circuits of a few time-delay systems, their PSPICE simulations and applications of time-delay systems. Special attention has been given to explain the phenomenon of amplitude death due to delay, time-delay induced bifurcations, chimera states and networks with time-delay feedback/coupling in Chaps. 5 and 6, which are of current research topics in the literature. Different types of synchronizations (complete, generalized, lag, anticipatory and phase) and their transitions in coupled time-delay systems with constant delay are discussed in Chaps. 7, 8, 9 and 10. Synchronization with delay time modulation is discussed in Chap. 11. Exact solution for certain time-delay systems (car following models) are provided in Chap. 12. This monograph also contains four appendices: The first one is on calculating Lyapunov exponents of time-delay systems, while the second one deals with different types of synchronizations and their respective characterizations. The third appendix provides a brief discussion on recurrence analysis used for data analyses which has also been used for identifying different types of synchronization transitions. The final appendix contains a list of time-delay systems of practical interest.

We hope this monograph bridges a gap in the literature in providing basic materials on chaotic time-delay systems and their synchronization along with some experimental realizations using electronic circuits. Although several books are available on synchronization in chaotic systems, some of them are focused on specific applications and others without delay systems or delay coupling. Further, the available books on time-delay systems have also been focused either on complete mathematical description or on applications. However, in the present monograph, we have endeavoured to provide special attention on scalar chaotic/hyperchaotic time-delay systems including some higher order ones occurring in different branches of science and technology and also on the synchronization of their coupled versions.

In the course of our studies on synchronization and the preparation of this monograph, we have received considerable support from many colleagues, students and friends. In particular, we are very thankful to Dr. J. H. Sheeba for reading the manuscript and helping to organize Chap. 6 on networks. Mr. R. Suresh has helped us intensively in the preparation of the manuscript to whom we express our special thanks. We are also thankful to Ms. A. Durga Devi and Mr. B. Subash for their assistance in the preparation of the manuscript. However, the authors are solely responsible for any shortcomings, errors or misconceptions that remain.

We would like to express our special gratitude to Prof. Jürgen Kurths with whom we enjoyed discussions and collaborations and for his interest in this monograph and advice.

A few of the illustrations are reproduced from other sources and appropriate references are give at the relevant places. We sincerely thank the respective authors and publishers for granting us permission to use those figures. We would also like to record our thanks to the Department of Science and Technology, Government of India for providing support under various research projects, particularly IRHPA and Ramanna Fellowship, which enabled us to undertake this task. DVS is also especially thankful to the Alexander von Humboldt Foundation for its support in the form of a Research Fellowship to work at the University of Potsdam, Germany during which period a major portion of the book was completed. DVS also acknowledges the support from the PIK, Potsdam and the project PHOCUS (EU FET-Open grant number: 240763). Finally, we thank our family members for their constant support and encouragement during the course of this project.

Tiruchirapalli   M. Lakshmanan
Postdam   D.V. Senthilkumar
October 2010

Contents

Appendix

Chapter 1
Delay Differential Equations

1.1 Introduction

Dynamical systems with delay (which we simply designate hereafter as delay dynamical systems or delay systems) are abundant in nature. They occur in a wide variety of physical, chemical, engineering, economic and biological systems and their networks. One can cite many examples where delay plays an important role.

1. Consider spectators sitting in a cricket or football stadium. Eventhough they are all clapping their hands in synchrony, those who are sitting in opposite directions do not hear the clapping in synchrony with their own due to propagation delay from one end to the other [1].
2. In an auditorium echoes arise due to the bouncing of sound waves again and again at the microphone after reflection from the walls, where again propagation delay is the underlying reason.
3. Chirping of crickets causes finite time-delay in the propogation of their sound and this is measured for example to be 10 ms between two crickets that are 3 m apart [1–3]. Similarly delay arises in crocking of frogs after rain, chirping of insects, etc.
4. Another physically observed phenomenon is the El Niño/southern oscillation (ENSO) [4, 5], where delayed feedback represents the effect of oceanic waves. What happens here is that the westward propagating Rossby waves on the ocean thermocline which after getting reflected from the western boundary become eastward propagating Kelvin waves that reenter the coupled ocean atmosphere system after a time delay equal to their transit time.
5. In the case of control systems, a controller monitors the state of the system and makes adjustments to the system based on its observations. Since these adjustments can never be made instantaneously, a delay arises between the observation and the control action [6].
6. Neural activity of central nervous system is a cooperative process of neurons and the information flow among the neurons is not generally instantaneous and hence there exists finite delay in the flow of information. For example, the speed of signal conduction through unmyelinated axonal fibers is of the order of 1 m/s resulting in time delays up to 80 ms for propagation through the cortical

M. Lakshmanan, D.V. Senthilkumar, *Dynamics of Nonlinear Time-Delay Systems*,
Springer Series in Synergetics, DOI 10.1007/978-3-642-14938-2_1,
© Springer-Verlag Berlin Heidelberg 2010

network [7, 8]. Significant delays of more than 4% of the characteristic period of the 40 Hz frequency oscillations of the brain neurons occur during the nerve conduction between neurons less then 1 mm apart [9, 10].

7. Connection delays in the networks of coupled dynamical systems due to delayed flow of information among them can lead to a plethora of new behaviours like phase flip bifurcation, Neimark-Sacker-type bifurcation, etc., and the connection delays have been proved to improve the synchronizabilty of networks [8, 11].

8. Time-delay is also ubiquitous in many physical systems due to finite switching speeds of amplifiers, finite signal propagation time in biological networks and circuits, memory effects and so on [12, 13].

The mathematical description of delay dynamical systems will naturally involve the delay parameter in some specified way. This can be in the form of differential equations with delay or difference equations with delay or differential-difference equations with delay or even might include integral forms (integro-differential equations). A differential equation with delay describing a dynamical system belongs to the class of retarded functional differential equations (also sometimes called retarded differential-difference equations) [14]. One can also consider other classes of delay differential equations (DDE), namely neutral DDEs and advanced DDEs [14]. If the evolution of a DDE depends on the past rates of changes in addition to its present and past values, then the corresponding DDE is referred to as a neutral DDE. An advanced type DDE is the one in which the evolution depends on its present and future values [15]. For example, consider the simple case of a linear scalar first order equation

$$a_0 \frac{dx(t)}{dt} + a_1 \frac{dx(t-\tau)}{dt} + b_0 x(t) + b_1 x(t-\tau) = f(t), \tag{1.1}$$

where a_0, a_1, b_0 and b_1 are arbitrary constants and $f(t)$ is a forcing function. The above equation is said to be a DDE of retarded type if $a_0 \neq 0$ and $a_1 = 0$; it is said to be of neutral type if $a_0 \neq 0$ and $a_1 \neq 0$, and of advanced type if $a_0 = 0$ and $a_1 \neq 0$.

In particular, the evolution of a dynamical variable corresponding to a retarded DDE depends not only on its present value, $x(t)$, but also on its values at earlier times, $x(t')$, $t' \in (-\tau, 0)$, where $\tau > 0$ is the lag time (or delay time). As a consequence, a time-dependent solution of a system of DDEs is not uniquely determined by its initial state at a given moment alone. Instead, the solution profile (initial function) on an interval of length equal to the maximal delay prior to the time $t = 0$ has to be prescribed. That is, we need to define a set of infinite (but continuous) number of initial conditions for $-\tau < t < 0$ and hence DDEs are effectively infinite-dimensional systems, even if we have only a single scalar delay differential equation.

The most common type of infinite-dimensional dynamical systems involve the evolution of functions in time. For instance, if we want to study the evolution of chemical concentrations in time and space, we can phrase the problem as the change in time of the spatial distribution of chemicals. This distribution can be represented by a function of the spatial variables, that is, $C = C(\mathbf{r})$. This is also one of the reasons for increasing interest of physics community for DDEs as they provide a natural link with space extended systems by means of the two variable representation of the time $t = \chi + \theta\tau$, where $\chi \in (0, \tau)$ is the continuous space variable, and $\theta \in \mathbb{N}$ is a discrete temporal variable [16–18].

Another way to think about the infinite dimensionality of the function space is that its *basis set* is infinite [6]. We can represent the coordinates of a point in an n-dimensional space as a superposition of n basis vectors, typically the unit vectors lying along the coordinate axes. On the other hand, we cannot represent a general function, or even a function satisfying a few special conditions, using a finite superposition of some simple set of functions. Rather, the expansion of a general function requires an infinite basis set. For example,

1. Taylor's theorem implies that continuous, infinitely differentiable functions can be expanded in terms of the basis set $\{1, x, x^2, \cdots\}$. Only in exceptional cases (polynomials) does this expansion terminate.
2. The set of functions which are zero at $x = 0$ and at $x = 1$ can be represented as a superposition of the functions $\{s_1, s_2, s_3, \cdots\}$, where $s_n = \sin(n\pi x)$. Most functions are of course not representable exactly as a finite superposition of sine waves.

In general, the DDEs (retarded type) of our concern can be represented as

$$\dot{X} = F(t, X(t), X(t - \tau_i)), \; X = (x_1(t), x_2(t), \cdots, x_n(t))^T, \qquad \left(\dot{} = \frac{d}{dt}\right) \; (1.2)$$

where the quantities $\tau_i > 0, i = 1, 2, ...,$ are lag times or delay times and F is a vector valued smooth continuous function. One may further distinguish the types of DDEs as characterized by

1. a single constant delay $\tau = \tau_i, i = 1$,
2. discrete delays $\tau_i, i = 1, 2, ...,$
3. distributed delays (the right hand side of the differential equation is a weighted integral over the past states, see Eq. (1.4) below),
4. state-dependent delays (τ_i's depend on $X(t)$) and
5. time-dependent (modulated) delays where τ_i's depend on t.

In the following subsections, we will describe briefly the details of the different kinds of DDE's mentioned above. A list of specific interesting DDEs are given at the end of this monograph, in Appendix D.

1.1.1 DDE with Single Constant Delay

A delay differential equation with a single constant delay can be represented in general as

$$\dot{X} = F(t, X(t), X(t - \tau)), \tag{1.3}$$

where τ is a positive constant. Several dynamical systems in biology [19, 20], optics [21–23], economics [24, 25], ecology [26, 27], etc., can be described by DDEs with a single constant delay. Examples include the following:

1. The Mackey-Glass DDE [19, 20] was introduced as a model for blood production in patients with leukemia, where the time interval taken for the maturation of red blood cells (RBCs) after their production in the bone marrow is considered as the delay time τ and their respective concentration of the RBCs are considered as dynamical variables. A detailed analysis of the Mackey-Glass delay differential equation will be presented later in Chap. 4.
2. The Ikeda system [21–23] was introduced to describe the dynamics of an optical bistable resonator, and the time taken for the round trip of the light across the resonator is considered as the delay time τ. More details are presented in Chap. 4.
3. The Kaldor-Kalecki model of business cycle [24, 25] was introduced as a second order DDE to study the dynamics of economic fluctuations due to investments. This model is characterized by time delay in an investment function.
4. The dynamics of a predator-prey system has also been described using delay models with single constant delay [27]. Time-delay in this model enters through the temporal evolution of prey's growth and predator's death.

1.1.2 DDE with Discrete Delays

Delay differential equations with multiple delays are represented by Eq. (1.2) with more than one positive delay constants τ_i, $i = 1, 2, \cdots$. These delay times are also called discrete delays. Dynamical systems with multiple discrete delays are abundant in biology (neurology) [28–30], control theory [31–35], economics [36, 37], population dynamics [38] and so on. Some examples are the following.

1. Since neuronal activity is the cooperative process of several neurons, the state of each neuron depends on the history of all the other neurons situated at distinct locations in the neuronal assembly. Consequently, the ith neuron receives information from the other neurons with different delay times and hence there appears the necessity of multiple discrete delays [28]. Dynamics of different neuronal assemblies with multiple delays have been studied in [29, 30].
2. Similarly, multiple delays inevitably occur in networks of coupled dynamical systems with different architectures in analogy with the consideration of weighted networks, where different weights are considered at the couplings to account for the different degree of interactions between the various dynamical units in the network [39, 40].
3. Multiple delays were included in a chemostat model in order to study sustained oscillations in a yeast population [41]. Multiple delays were also included in the study of the intrinsic growth rate of microorganisms in a chemostat [42].

4. Several control systems, for example in a huge machinery, require past knowledge of various processes to maintain their proper functioning and hence different variables with different delays have to be considered [31–35].
5. Whenever multiple investors, multiple predators and preys, multiple species in migration/population dynamics are studied, dynamical systems with multiple delays are considered to be good approximations [36–38].

1.1.3 DDE with Distributed Delay

Delay differential equations with distributed or continuous delay can be represented in general as

$$\dot{X} = F(t, X(t), \int_{0}^{\infty} \mu(\tau)X(t - \tau)d\tau). \tag{1.4}$$

Models based on distributed delays have been proposed as early as the time of Volterra [43] and used in areas such as biology [44], ecology [45, 46], neurology [47], viscoelasticity [48] and economics [49]. It has also been pointed out in biological sciences that distributed delay leads to more realistic models [50]. Recently it has also been pointed out that distributed delay facilitates amplitude death of coupled oscillators [51]. Typical examples are as follows.

1. The presence of parallel pathways with a variety of axon sizes and length in neural networks usually have widespread spatial distribution and there will be a distribution of conduction velocities along these pathways and a distribution of propagation delays. Under such a situation, the signal propagation is not instantaneous and cannot be described with discrete delays alone and hence one needs to introduce a continuously distributed delay [52].
2. In populations of spatially separated neurons, the synaptic communications between them depend on the propagation of action potentials over appreciable distances and involve distributed delays [53].
3. In a chemostat model of a single species feeding on a limited nutrient supplied at a constant rate, distributed delay is included since the growth of the species depends on the past concentration of the nutrients [54].
4. In textile engineering, drafting of textile silver represents one of the most important stages of the textile production chain. Unlike metal wire, textile silver is composed of many discrete fibres which do not change their lengths during the process and only their positions with respect to each other and the number of fibres in cross sections are changed due to different speeds of the rollers. The lengths of individual fibers are random variables varying between a minimum and a maximum and this distribution of fibre lengths include distributed delay [55].

1.1.4 DDE with State-Dependent Delay

In population dynamics and epidemic problems the delay time has also been considered as a function of the state variable itself [see for example [14] and references therein]. Delay differential equations with state dependent delay can be represented in general as

$$\dot{X} = F(t, X(t), X(t - \tau(t, X(t)))). \tag{1.5}$$

State-dependent delay appears in processes such as

1. The time delay for turning processes in the milling operations is not only determined by the rotation of the workpiece but is also affected by the current and the delayed position of the tool. This results in a DDE with state-dependent delay, where the delay depends on the present state and also on a delayed one [56, 57].
2. More realistic and interesting models of species growth have considered DDEs with state dependent delay and have taken into account the seasonality of the changing environment which in turn depends on past seasons [58].

1.1.5 DDE with Time-Dependent Delay

In the case of time-dependent DDEs, the delay time $\tau(t)$ is considered to be dependent on time t explicitly. DDEs with time dependent delay can be represented in general as

$$\dot{X} = F(t, X(t), X(t - \tau(t))). \tag{1.6}$$

Time-dependent delays have been introduced as a stochastic process in describing the dynamics of neural networks and internet [59, 60]. Recently, it has also been demonstrated that the time-dependent delay with stochastic or chaotic modulation [61] or even simple rectified sinusiodal delay time modulation [62] increases the complexity of chaotic/hyperchaotic attractors of time-delay systems with constant delays so that the reconstructed phase trajectory does not collapse to a simple manifold, a property different from that of delayed systems with fixed delays. For example, Kye et al. [61] considered the time-dependent delay of the form

$$\tau(t) = \tau_0 + \int_0^t \xi(s)ds, \tag{1.7}$$

where $\xi(s)$ is considered to be a stochastic process. A typical example of time-dependent delay is the following.

An impact fuze with a time-delay between the impact of a flying object such as a rocket or projectile at the target and the detonation thereof renders possible penetration of the rocket or projectile into the target before detonation. The flight velocity or speed of a projectile decreases with increasing flying time. In order to ensure for a penetration depth which nevertheless is sufficient, the impact time-delay must increase with decreasing flight velocity of the rocket or projectile. For this purpose, a time-delay counter is set by means of a self-destruction counter as a function of the flight velocity (see for example, www.freepatentsonline/4455939.html).

It is thus clear that the DDEs of retarded type are of considerable relevance in nonlinear dynamics and occur in a wide range of physical, chemical, biological and engineering problems. Consequently, they are receiving considerable attention recently.

1.2 Constructing the Solution for DDEs with Single Constant Delay

Let us for the moment specialize on DDEs with a single constant delay, that is, $\tau = \tau_i =$ constant with $i = 1$ in Eq. (1.2) or Eq. (1.3), as most of the studies on DDEs considered in the literature are only with a single constant delay. This is because even simple scalar DDEs are hard to analyse for the underlying dynamics, including stability and bifurcation analysis, and to arrive at a global picture. So, to start with, we confine ourselves to DDEs with a single constant delay only.

In order to generate solutions for time $t > 0$ for the DDEs with a single constant delay, Eq. (1.3), one has to define the initial function $X(t)$ over the interval $(-\tau, 0)$. A simple way to interpret the solution of this DDE (1.3) is to consider it as a mapping of the functions from the interval $(t - \tau, t)$ to the next interval $(t, t + \tau)$ and then to the next interval $(t + \tau, t + 2\tau)$, and continue the process recursively. Alternatively, the solution of the delay dynamical system (1.3) can be thought of as a sequence of functions $f_0(t), f_1(t), f_2(t), \cdots$ defined over contiguous time intervals of length τ. The mapping of the function $f_0(t)$ defined over the interval $t \in (-\tau, 0)$ onto a solution curve in the interval $(0, \tau)$ represented by the function $f_1(t)$ is illustrated in Fig. 1.1, see [6] for more details. The action of the evolution operator ϕ^τ in

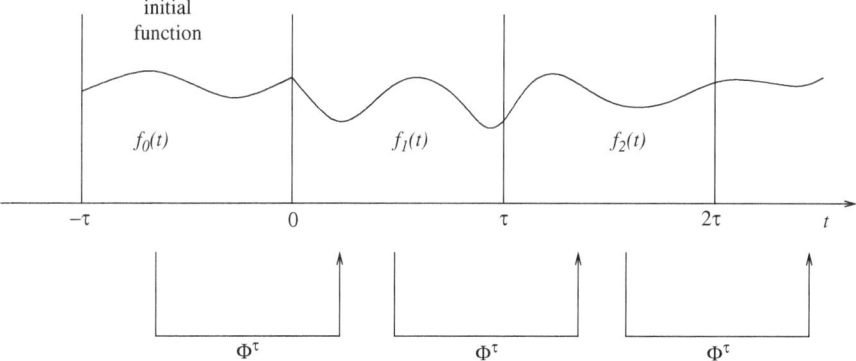

Fig. 1.1 The action of the evolution operator ϕ^τ for a delay differential equation of the form (1.3) is to take a function defined over a time interval of length τ and to map it into another function defined over a similar subsequent time interval [6]

Fig. 1.1 for a DDE of the form (1.3) is to take a function defined over a time interval of length τ and to map it into another function defined over a subsequent similar time interval. In some simple cases, one can work out this mapping analytically as discussed in the next subsection for the case of a linear delay differential equation (LDDE).

1.2.1 Linear Delay Differential Equation

The introduction of a constant delay even in a linear system will have a profound effect on the nature of solutions, for example to the rate equations in chemical kinetics. The prototype linear rate equation without delay and with rate constant k,

$$\dot{x} = -kx(t), \qquad k > 0, \tag{1.8}$$

will have only a decaying solution $x(t) = ae^{-kt}$, where a is an arbitrary constant. On the other hand, the same equation with a constant delay of the form

$$\dot{x} = -kx(t - \tau), \qquad \tau > 0, \tag{1.9}$$

will have multiple solutions ranging from monotonic decay to damped oscillations, stable periodic oscillations and undamped growing oscillations for a range of delay values [63–66]. In fact, a particular solution for the above linear rate equation with constant delay can be obtained as

$$x(t) = b\sin(\omega t) - \frac{b\sin(\omega\tau)}{1 - \cos(\omega\tau)}\cos(\omega t), \quad \omega = k, \tag{1.10}$$

where b is an arbitrary constant. However, note that the general solution to (1.9) is not known.

Now, the solution for the LDDE (1.9) can be analytically mapped from the initial function in the interval $(-\tau, 0)$ in a heuristic way as follows. Suppose we have the function $x(t) = f_{i-1}(t)$, defined in the interval (t_{i-1}, t_i), where $i = 0, 1, 2, \cdots$ corresponds to the number of integration steps of length τ. Then, the solution in the interval $(t_i, t = t_i + \tau)$ is given by a separation of variables of (1.9) as

$$\int_{f_{i-1}(t_i)}^{f_{i-1}(t)} dx' = -k \int_{t_i}^{t} f_{i-1}(t' - \tau)dt'. \tag{1.11}$$

Consequently, we can write

$$x(t) \equiv f_i(t) = f_{i-1}(t_i) - k \int_{t_i}^{t} f_{i-1}(t' - \tau)dt'. \tag{1.12}$$

This approach is known as the method of steps [6]. We can demonstrate it to obtain the solution for the LDDE (1.9) for a given set of initial values, where the function $x(t)$ is defined over the interval $(-\tau, 0)$.

To make this task simpler, we choose the value of the rate constant $k = 1$, the delay time $\tau = 1$ and suppose that

$$x(t) = 1 \qquad\qquad \text{for } t \in (-1, 0). \tag{1.13}$$

In the first interval $(0, 1)$, we have from Eq. (1.12) that

$$x(t) = 1 - \int_0^t dt' = 1 - \left[t'\right]_0^t = (1 - t). \tag{1.14}$$

In the second interval $(1, 2)$, we have

$$x(t) = 0 - \int_1^t (1 - (t' - 1))dt',$$

$$= -\left[2t' - \frac{t'^2}{2}\right]_1^t$$

$$= -2t + \frac{1}{2}t^2 - \frac{3}{2}. \tag{1.15}$$

In the third interval $(2, 3)$, we have

$$x(t) = -\frac{1}{2} - \int_2^t \left[-2(t' - 1) + \frac{1}{2}(t' - 1)^2 + \frac{3}{2}\right]dt',$$

$$= -\frac{1}{2} - \left[-(t' - 1)^2 + \frac{1}{6}(t' - 1)^3 + \frac{3}{2}t'\right]_2^t$$

$$= \frac{5}{3} + (t - 1)^2 - \frac{1}{6}(t - 1)^3 - \frac{3}{2}t. \tag{1.16}$$

This procedure can be continued further to any order. It can be summarized as follows. With the knowledge of the function $x(t) = f_{i-1}(t_i)$, $i = 1, 2, \cdots$ in the interval (t_{i-1}, t_i), one can obtain the unknown function $x(t) = f_i(t)$ in the subsequent interval (t_i, t_{i+1}) of length τ using Eq. (1.12). To illustrate the procedure further, we make the specific choice $k = 1$, $\tau = 1$ in Eq. (1.9) or (1.12). Then, to start with, using our knowledge of the function $x(t) = f_{i-1}(t_i) = 1$ in the first interval $(-1, 0)$, we can calculate the function $x(t) = f_i(t)$ in the next interval $(0, 1)$ of length $\tau = 1$ by substituting $x(t) = f_{i-1}(t_i) = 1$ which is known in the previous interval into the next interval $(0, 1)$ using Eq. (1.12). This new function $x(t)$, $t \in (0, 1)$, can then be used to find $x(t)$ now for $t \in (1, 2)$ again using the relation

(1.12). This procedure can be repeated for subsequent intervals each of length τ for as large a value of t as required and hence the solution can be calculated recursively as the evolution time t increases. It can be realized that even for this simple LDDE (1.9), evaluating the solution by this procedure becomes tedious for larger values of t and it is a difficult and time consuming task to do calculations analytically. However, this method of steps can be coded as a program and then the solutions for any t can be obtained easily using simulation in a computer.

It is to be noted that for the ease of analytical calculation and understanding we have taken the integration step Δt to be of length $\tau = 1$ and hence the solution curve seems to be constant over successive intervals of length τ. But in an actual simulation, the integration step size Δt is taken as a small but optimal value such that the prescribed initial function $x(t)$ over the interval $(-\tau, 0)$ itself is sampled into N samples, where N is sufficiently large. More details on numerical simulations will be discussed below.

1.2.2 Numerical Simulation of DDEs

To simulate the behavior of infinite-dimensional systems on a computer it is necessary to approximate the continuous evolution of an infinite-dimensional system by a finite number of elements whose values change at discrete time steps. Hence to calculate the solution $x(t)$ of a DDE of the form (1.3) for times greater than t, a function $x(t)$ over the interval $(t, t - \tau)$ must be given. This function can be approximated by N samples taken at intervals $\Delta t = \tau/(N - 1)$. These N samples can equivalently be thought of as the N variables of an N-dimensional discrete mapping [67],

$$(x_1, ..., x_{N-1}, x_N) = f(x(t - (N - 1)\Delta t), ..., x(t - \Delta t), x(t)). \qquad (1.17)$$

In particular, consider a DDE of the form

$$\dot{X} = F(t, X(t), X(t - \tau)),$$

and this can be approximated in terms of N variables of an N-dimensional discrete mapping (1.17) as pointed out above. Choosing any suitable integration scheme, for example, Euler integration (or more reliable fourth order Runge-Kutta method)

$$x(t + \Delta t) = x(t) + F(x, x_\tau)\Delta t, \qquad x_\tau = x(t - \tau), \qquad (1.18)$$

N samples and equivalently N variables of Eq. (1.17) can be reduced to an N-dimensional iterated map, $X(k + 1) = G(X(k))$ (k labels the kth iteration and $k + 1$ to its next iteration). Each iteration of the map G corresponds to N time steps Δt of the continuous equations, that is, each iteration of G moves the system forward by time Δt. Using Euler integration, the map G is defined as follows [67]:

$$x_1(k+1) = x_N(k) + F(x_N(k), x_1(k))\Delta t,$$
$$x_2(k+1) = x_1(k+1) + F(x_1(k+1), x_2(k))\Delta t, \qquad (1.19)$$
$$\vdots$$
$$x_N(k+1) = x_{N-1}(k+1) + F(x_{N-1}(k+1), x_N(k))\Delta t.$$

In this way a continuous infinite dimensional dynamical system is replaced by a finite-dimensional iterated map.

Once the numerical algorithm for obtaining the solution of the DDE is fixed, then the Lyapunov exponents for the N-dimensional discrete map can be calculated using the orthonormalization procedure which can be done, for example, using the well known Wolf algorithm [67, 68]. More discussion and details on calculating Lyapunov exponents of a DDE are provided in Appendix A. It is also to be noted that DDEs in general exhibit enormous transient effects and hence a large amount of transients should be left out before the system settles into a steady state solution. The number of transients depends largely on the integration step size (in other words sampling interval) Δt, the number of coupled equations N and even on the nature of nonlinear function $f(x)$. More details on the transient effects of DDEs will be discussed later in Chap. 3.

The nature of the solution obtained numerically for the LDDE (1.9) for different values of the delay time τ for fixed values of rate constant $k = 1$ along with the initial condition $x(t) = 0.1$ in the range $t \in (-\tau, 0)$ is shown in Fig. 1.2. Monotonically decaying solution is obtained for a delay time $\tau = 0.3$ as seen in Fig. 1.2a. Damped oscillatory solution is seen for $\tau = 1.4$ in Fig. 1.2b, whereas periodic oscillation is obtained for $\tau = 1.565$ (Fig. 1.2c) and the Eq. (1.9) exhibits undamped growing oscillation for $\tau = 1.7$ as shown in Fig. 1.2d.

1.2.3 Nonlinear Delay Differential Equations

It is a well known fact that for nonlinear dynamical systems without delay, represented as a set of coupled nonautonomous first order ODEs, at least three dimensions are required to exhibit chaotic behavior. In other words, it requires a third-order autonomous continuous nonlinear system or a second-order nonlinear nonautonomous system to have the possibility of chaotic behavior. What is the situation in the case of delay dynamical systems? A first order *linear* DDE cannot produce chaos even for a large delay, as we have seen in the previous section. On the other hand, a first order scalar nonlinear DDE itself can exhibit not only chaotic behavior but even hyperchaotic behavior with large number of positive Lyapunov exponents, even for small delays and suitable parameter values and nonlinearity, as we will demonstrate clearly in the later Chaps. 3 and 4. Of course, higher order coupled nonlinear DDEs can also exhibit similar chaotic/hyperchaotic behavior; however, we will concentrate in the first few chapters mostly on scalar nonlinear DDEs for simplicity as even these systems typically exhibit most of the delay-related dynamical behaviors.

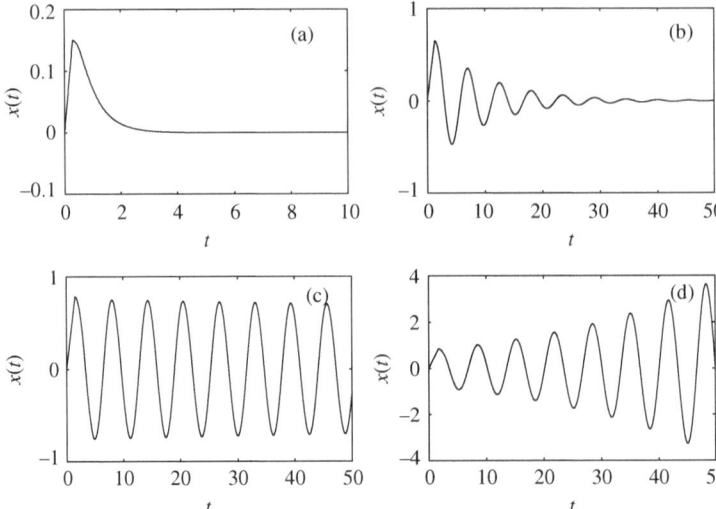

Fig. 1.2 Behavior of the solution $x(t)$ of the LDDE (1.9) for $k = 1$ and various values of delay τ, along with the initial condition $x(t) = 0.1$ in the range $t \in (-\tau, 0)$. (**a**) Monotonic decay to $x(t) = 0$ for $\tau = 0.3$, (**b**) Damped oscillatory decay to $x(t) = 0$ for $\tau = 1.4$, (**c**) Periodic oscillation for $\tau = 1.565$ and (**d**) Undamped growing oscillation for $\tau = 1.7$

Nonlinear DDEs have been used extensively to model population dynamics [38] with their inherent maturation and gestation time-delays to study epidemics [69], tumor growth [70], and immune systems [71], lossless electrical transmission lines [72], electrodynamics of interacting charged particles (the Lorenz force with Liénard-Weichert potentials) [73], etc. The necessity to generate fast chaos is an important feature in many applications. For example, information transmission with high power efficiency, generation of truly random numbers, and novel spread spectrum, ultrawide-bandwidth and optical communication schemes require fast systems, where the time it takes for the signals to propagate through device components is comparable to the time scale of the fluctuations. Hence many fast systems are most accurately described by DDEs. Examples of fast broadband chaotic oscillators that are modeled as time-delay systems include electronic [74], opto-electronic [75, 76] and microwave oscillators [77], lasers with delayed optical feedback [78] and nonlinear optical resonators [79]. An advantageous feature of these time-delay devices is that the complexity of the dynamics can be tuned by adjusting the delay time [67].

There exist several nonlinear delay dynamical systems which have been studied intensely in the recent literature in diverse areas of science and technology and in particular, in the context of chaotic dynamics. Some of the prototype scalar nonlinear delay dynamical systems are

1. Mackey-Glass system [19, 20],
2. Ikeda system [21–23],

3. Piecewise linear time-delay systems of various forms [18, 80–85],
4. Time-delay system with polynomial nonlinearity [86],
5. Time-delayed Chua's circuit [87–90],
6. Kaldor-Kalecki business model [24, 25],

to name a few. We will discuss the details of the dynamics of most of the important prototype nonlinear DDEs studied in the literature in the context of chaotic dynamics in the following chapters.

1.3 Salient Features of Chaotic Time-Delay Systems

In this section, we will present a list of important characteristic features of chaotic nonlinear time-delay systems in general. These are as follows:

1. A chaotic time-delay system has the intrinsic characteristic feature of increase in the dimension of the attractor with the value of the delay time in general and thereby can lead to an increase in the number of positive Lyapunov exponents [67, 22].
2. Transient effects play a prominent role in delay dynamical systems, as such systems require large enough transients to settle into a steady state behavior [91, 92].
3. Even a simple scalar time-delay system can exhibit hyperchaotic attractors with large number of positive Lyapunov exponents [22, 82, 83, 94, 95].
4. Feasibility of easy experimental realization of time-delay systems, particularly in electronic circuits, optical lasers, etc. [93, 86, 21, 82, 83, 94–97], is abundant.
5. Increase in the number of positive Lyapunov exponents and hence the dimension provides the possibility that the corresponding hyperchaotic attractors may be useful in secure communication applications, generating random numbers, etc. [95].
6. Chaotic time-delay systems provide a natural link to understand the dynamical features of space-time chaos [16–18, 98, 99].

References

1. M.K. Stephen Yeung, S.H. Strogatz, Phys. Rev. Lett. **82**, 648 (1999)
2. T.J. Walker, Science **166**, 891 (1969)
3. E. Sismondo, Science **249**, 55 (1990)
4. E. Tziperman, M.A. Cane, S.E. Zebiak, J. Atmos. Sci. **52**, 293 (1995)
5. I. Boutle, R.H.S. Taylor, R.A. Romer, Am. J. Phys. **75**, 15 (2007)
6. M.R. Roussel, Am. J. Phys. **75**, 15 (2007)
7. E.R. Kandel, J.H. Schwartz, T.M. Jessell, *Principles of Neural Science*, 3rd edn. (Elsevier, New York, 1991)
8. M. Dhamala, V.K. Jirsa, M. Ding, Phys. Rev. Lett. **92**, 074104 (2004)
9. G.M. Shepherd, *Neurobiology* (Oxford University Press, New York, 1983)
10. J.D. Murray, *Mathematical Biology* (Springer, New York, 1990)

11. F.M. Atay, J. Jost, Phys. Rev. Lett. **92**, 044101 (2004)
12. D.V. Ramana Reddy, A. Sen, G.L. Johnston, Phys. Rev. Lett. **85**, 3381 (2000)
13. D.V. Ramana Reddy, Ph. D. Thesis entitled *Collective Dynamics of Delay Coupled Oscillators* (Institute for Plasma Research, Gandhinagar, India, 2000)
14. J.K. Hale, *Theory of Functional Differential Equations* (Springer, New York, 1977)
15. Alwyn Scott (ed.), *Encyclopedia of Nonlinear Science* (Routledge, New York, 2005)
16. F.T. Arecchi, G. Giacomelli, A. Lapucci, R. Meucci, Phys. Rev. A **45**, R4225 (1992)
17. G. Giacomelli, R. Meucci, A. Politi, F.T. Arecchi, Phys. Rev. Lett. **73**, 1099 (1994)
18. I.G. Szendro, J.M. Lopez, Phys. Rev. E **71**, 055203 (2005)
19. M.C. Mackey, L. Glass, Science **197**, 287 (1977)
20. M.C. Mackey, L. Glass, *From Clocks to Chaos, The Rhythms of Life* (Princeton University Press, Princeton, NJ, 1988)
21. K. Ikeda, H. Daido, O. Akimoto, Phys. Rev. Lett. **45**, 709 (1980)
22. K. Ikeda, M. Matsumoto, J. Stat. Phys. **44**, 955 (1986)
23. K. Ikeda, K. Matsumoto, Physica D **29**, 223 (1987)
24. M. Szydlowski, A. Krawiec, J. Nonlinear Math. Phys. **8**, 266 (2001)
25. M. Szydlowski, A. Krawiec, Chaos Solit. Frac. **25**, 299 (2005)
26. J. Fort, V. Mendez, Phys. Rev. Lett. **82**, 867 (1999)
27. R. German, M.R. Ejtehadi, Eur. Phys. J. B **13**, 601 (2000)
28. E.W. Saad, D.V. Prokhorov, D.C. Wunsch, IEEE Trans. Neural Networks **9**, 1456 (1998)
29. L.P. Shayer, S.A. Campbell, SIAM J. Appl. Math. **61**, 673 (2000)
30. N. Buric, N. Vasovic, Int. J. Bifurcat. Chaos **13**, 3483 (2003)
31. R.D. Driver, *Ordinary and Delay Differential Equations* (Springer, New York, 1977)
32. I. Gyori, P. Ladas, *Oscillation Theory of Delay Differential Equations with Applications* (Clarendon Press, Oxford, 1991)
33. R. Ross, *The Prevention of Malaria* (John Murray, London, 1911)
34. D. Levi, P. Winterniz, J. Math. Phys. **34**, 3713 (1993)
35. S.B. Stojanovic, D.L. Delbeljkovic, Mech. Eng. **2**, 35 (2004)
36. P. K. Asea, P. J. Zak, J. Economic Dynamics and Control **23**, 1155 (1999)
37. M. Kalecki, Econometrica **3**, 327 (1935)
38. Y. Kuang, *Delay Differential Equations with Applications in Population Dynamics* (Academic Press, Boston, MA, 1993)
39. C.S. Zhou, A.E. Motter, J. Kurths, Phys. Rev. Lett. **96**, 164102 (2006)
40. A.E. Motter, C.S. Zhou, J. Kurths, Phys. Rev. E **71**, 016116 (2005)
41. J. Caperson, Ecology **50**, 188 (1969)
42. A.W. Bush, A.E. Cook, J. Theor. Biol. **63**, 385 (1975)
43. V. Volterra, *Lecons sur la theorie mathematique de la lutte pour la vieynamics* (Gauthiers-Villars, Paris, 1931)
44. N. MacDonald, *Time Lags in Biological Models*. Lectuer Notes in Biomathematics (Springer, Berlin, 1978)
45. R.M. May, *Stability and Complexity in Model Ecosystems* (Princeton University Press, Princeton, NJ, 1974)
46. K. Gopalsamy, *Stability and Oscillations in Delay Differential Equations of Population Dynamics* (Kluwer Academic Publishers, Dordrecht, 1992)
47. D.A. Baylor, A.L. Hodgkin, T.D. Lamb, J. Physiol. (London) **242**, 685 (1974)
48. A.D. Drozdov, V.B. Kolmanovskii, *Stability in Viscoelasiticity* (North-Holland, Amsterdam, 1994)
49. J.B.S. Haldane, Rev. Econ. Stud. **1**, 186 (1933)
50. J.M. Cushing, *Integrodifferential Equations and Delay Models in Population Dynamics*. Lectuer Notes in Biomathematics, Vol. 20 (Springer, Berlin, 1977)
51. F.M. Atay, Phys. Rev. Lett. **91**, 094101 (2003)
52. F.Y. Zhang, H.F. Huo, Discrete Dyn. Nat. Soc. **2006**, 27941 (2006)
53. C. Massoller, A.C. Marti, Phys. Rev. Lett. **94**, 134102 (2005)
54. S. Ruan, G.S.K. Wolkowicz, J. Math. Anal. Appl. **204**, 786 (1996)
55. P. Zitex, J. Halva, Controlling Eng. Pract. **9**, 501 (2001)

56. T. Insperger, B.P. Mann, G. Stepan, P.V. Bayly, Int. J. Machine Tools Manufacture **43**, 25 (2003)
57. T. Insperger, D.A.W. Barton, G. Stepan, Int. J. Non Linear Mech. **43**, 140 (2008)
58. Y. Li, Y. Kuang, Proc. Am. Math. Soc. **103**, 1345 (2001)
59. M.E.J. Newman, SIAM Rev. **45**, 167 (2003)
60. R. Albert, A.L. Barabasi, Rev. Mod. Phys. **74**, 47 (2002)
61. W.-H. Kye, M. Choi, M.-W. Kim, S.-Y. Lee, S. Rim, C.-M. Kim, Y.-J. Park, Phys. Lett. A **322**, 338 (2004)
62. D.V. Senthilkumar, M. Lakshmanan, Chaos **17**, 013112 (2007)
63. E. Winston, J.A. Yorke, Academie de la Republique Popolaire Roumaine **14**, 885 (1969)
64. R. Bellman, K.L. Cooke, *Differential-Difference Equations* (Academic Press, New York, 1963)
65. J.K. Hale, S.M. Verduyn Lunel, *Introduction to Functional Differential Equations* (Springer, New York, 1993)
66. K. Sriram, Ph. D thesis on *Modelling Nonlinear Dynamics of Chemical and Circadian Rhythms Using Delay Differential Equations* (Department of Chemistry, IIT, Madras, India, 2004)
67. J.D. Farmer, Physica D **4**, 366 (1982)
68. A. Wolf, J.B. Swift, H.L. Swinney, J.A. Vastano, Physica D **16**, 285 (1985)
69. F.R. Sharpe, A.J. Lotka, Am. J. Hygiene **3**, 96 (1923)
70. M. Villasana, A. Radunskaya, J. Math. Biol. **47**, 270 (2003)
71. P.W. Nelson, A.S. Perelson, Math. Biosci. **179**, 73 (2002)
72. R.K. Brayton, Auart. Appl. Math. **24**, 215 (1966)
73. R.D. Driver, J. Differ. Equ. **54**, 73 (1984)
74. G. Mykolaitis, A. Tamasevicius, A. Cenys, S. Bumeliene, A.N. Anagnostopoulos, N. Kalkan, Chaos Solit. Fract. **17**, 343 (2003)
75. H.D.I. Abarbanel, M.B. Kennel, L. Illing, S. Tang, H.F. Chen, J.M. Liu, IEEE J. Quantum Electron. **37**, 1301 (1993)
76. J.P. Goedgebure, L. Larger, H. Porte, Phys. Rev. Lett. **80**, 2249 (1998)
77. V. Dronov, M.R. Hendrey, T.M. Antonsen Jr., E. Ott, Chaos **14**, 30 (2004)
78. G.D. VanWiggeren, R. Roy, Science **279**, 1198 (1998)
79. K. Ikeda, K. Kondo, O. Akimoto, Phys. Rev. Lett. **49**, 1467 (1982)
80. H. Lu, Z. He, IEEE Trans. Circuits Syst. I **43**, 700 (1996)
81. P. Thangavel, K. Murali, M. Lakshmanan, Int. J. Bifurcat. Chaos **8**, 2481 (1998)
82. A. Tamasevicius, G. Mykolaitis, S. Bumeliene, Electron. Lett. **42**, 13 (2006)
83. A. Tamasevicius, T. Pyragiene, M. Meskauskas, Int. J. Bifurcat. Chaos **17**, 3455 (2007)
84. J. Losson, M.C. Mackey, A. Longtin, Chaos **3**, 167 (1993)
85. Y.C. Tian, F. Gao, Physica D **108**, 113 (1997)
86. H.U. Voss, Int. J. Bifurcat. Chaos **12**, 1619 (2002)
87. A.N. Sharkovsky, Y. Maisternko, P. Deregel, L.O. Chua, J. Circuits Syst. Comput. **3**, 645 (1993)
88. M. Biey, B.F. Gilli, I. Maio, IEEE Trans. Circuits Syst. I **44**, 486 (1997)
89. M. Gilli, G. M. Maggio, IEEE Trans. Circuits Syst. I **43**, 827 (1996)
90. P. Thangavel, K. Murali, D.V. Senthilkumar, M. Lakshmanan Int. J. Bifurcat. Chaos (Submitted)
91. D.V. Senthilkumar, M. Lakshmanan, Int. J. Bifurcat. Chaos **15**, 2985 (2005)
92. S.R. Taylor, S.A. Campbell, Phys. Rev. E **75**, 046215 (2007)
93. J.C. Sprott, Phys. Lett. A **366**, 397 (2007)
94. L. Wang, X. Yang, Electron. Lett. **42**, 1439 (2006)
95. M.E. Yalcin, S. Ozoguz, Chaos **17**, 033112 (2007)
96. A. Namajunas, K. Pyragas, A. Tamasevicius, Phys. Lett. A **201**, 42 (1995)
97. W. Horbelt, J. Timmer, H.U. Voss, Phys. Lett. A **299**, 513 (2002)
98. G. Giacomelli, A. Politi, Phys. Rev. Lett. **76**, 2686 (1996)
99. S. Boccaletti, D. Maza, H. Mancini, R. Genesio, F.T. Arecchi, Phys. Rev. Lett. **79**, 5246 (1997)

Chapter 2
Linear Stability and Bifurcation Analysis

2.1 Introduction

In our study of DDEs, we will mainly concentrate on equations with constant time delay (single or multiple). In particular considering Eq. (1.3), in this chapter we will consider scalar DDEs ($n = 1$ in Eq. (1.2)) and analyse the linear stability and bifurcation aspects of a class of such equations. We will use the usual method of infinitesimally displacing the solution around the equilibrium point, a geometric approach, and a more general approach to determine linear stability of equilibrium points and then illustrate them with specific examples. We will also point out the extension of these analyses to coupled DDEs/complex scalar equations.

2.2 Linear Stability Analysis

Even though it is rarely possible to completely solve a given DDE (even in the linear case) and obtain the solution exactly, one can identify the existence of several types of specific solutions, depending upon the nonlinearity, value of the delay, the number of dynamical variables, etc. These include

1. fixed points or equilibrium points,
2. simple periodic solutions,
3. decaying solutions,
4. quasiperiodic solutions,
5. strange non-chaotic attractors,
6. chaotic attractors,
7. hyperchaotic attractors, etc.

However, for some models of coupled nonlinear DDEs with exact solutions, we refer to Chap. 12.

Considerable insight into the nature of solutions of DDEs can be obtained by performing a linear stability analysis of the equilibrium solutions, similar to ordinary differential equations (ODEs). The main difference in the case of DDEs is that (as noted earlier) the phase space is infinite-dimensional, whereas in the case

M. Lakshmanan, D.V. Senthilkumar, *Dynamics of Nonlinear Time-Delay Systems*,
Springer Series in Synergetics, DOI 10.1007/978-3-642-14938-2_2,
© Springer-Verlag Berlin Heidelberg 2010

of ODEs it is finite dimensional. The (local or asymptotic) stability nature of an equilibrium point in the state space can be simply determined by analysing whether the nearby trajectories approach towards or diverge away from the fixed point. Naturally a stable fixed point X^* is the one for which all the nearby trajectories approach it asymptotically (as $t \to \infty$), and as a consequence all the time derivatives vanish identically. Hence, for a DDE of the form (1.3), any equilibrium point, $X(t) = X(t - \tau) = X^*$, satisfies the equation

$$f\left(X(t) = X(t - \tau) = X^*\right) = 0, \quad X^* = \left(x_1^*, x_2^*, \ldots, x_n^*\right)^T. \quad (2.1)$$

To identify the nature of the stability of an equilibrium point, specifically the linear stability, we perturb it in the usual way by infinitesimally displacing the solution around the equilibrium point X^* by a time dependent function $\delta X(t)$, persisting over an interval of at least the value of the longest delay, τ_{max}, in the case of multiple delays. Denoting $X = X(t)$ and $X_\tau = X(t - \tau)$, we have

$$X = X^* + \delta X, \qquad X_\tau = X^* + \delta X_\tau. \quad (2.2)$$

Then

$$\dot{X} = \delta \dot{X} = F(X^* + \delta X, X^* + \delta X_\tau), \quad (2.3)$$

where δX's are the infinitesimal displacements from the equilibrium point over the interval $(t_0 - \tau, t_0)$. Using the Taylor series expansion, the above Eq. (2.3) can be linearized about the equilibrium point as

$$\delta \dot{X} = J_0 \delta X + J_\tau \delta X_\tau, \quad (J_0)_{i,j} = \left(\frac{\partial F_i}{\partial x_j}\right)\Bigg|_{\substack{x_j = x_j^* \\ i,j=1,2,\cdots,n}}, \quad (J_\tau)_{i,j} = \left(\frac{\partial F_i}{\partial x_{\tau j}}\right)\Bigg|_{\substack{x_{\tau j} = x_j^* \\ i,j=1,2,\cdots,n}}, \quad (2.4)$$

where J_0 is the Jacobian with respect to X evaluated at the equilibrium point, while J_τ is the Jacobian with respect to X_τ again evaluated at $X = X_\tau = X^*$. As in the case of ODEs, let the solution $\delta X(t)$ to (2.3) be assumed as exponential functions of time along with the exponents given by the eigenvalue of the corresponding Jacobian matrix,

$$\delta X(t) = A e^{\lambda t}, \quad (2.5)$$

where A is a constant column matrix. Substituting (2.5) in the above Eq. (2.4) and collecting the coefficients of $e^{\lambda t}$, one obtains the matrix equation

$$\lambda A = \left(J_0 + e^{-\lambda \tau} J_\tau\right) A. \quad (2.6)$$

This equation obviously can be satisfied with nonzero displacement amplitudes A if

$$|J_0 + e^{-\lambda\tau} J_\tau - \lambda I| = 0, \tag{2.7}$$

where I is the identity matrix.

The above Eq. (2.7) is of course the characteristic equation of the equilibrium point which is now transcendental in nature. Equation (2.7) looks like the characteristic equation of an ordinary eigenvalue problem except for the appearance of the exponential term. If all of the eigenvalues of the characteristic equation have only negative real parts, then the equilibrium point is said to be stable. On the other hand, if at least one of the eigenvalues has a positive real part, then the equilibrium point is unstable. If the leading characteristic eigenvalues are zero, then the stability is undecidable to the linear order and a recourse should be taken to consider the neglected higher order terms in the Taylor expansion in Eq. (2.3). A polynomial of degree n has exactly n roots in the case of ODEs and therefore all the roots can be calculated at least in principle to determine the stability of an equilibrium point. On the other hand, transcendental equations containing quasi-polynomials usually have infinite number of roots in the complex plane, which is essentially a reflection of the infinite dimensional nature of the phase space. Hence, it is extremely difficult to find all the roots and is often impossible to do so.

Despite the existence of infinite number of roots for the Eq. (2.7), it is often possible to determine analytically whether a given equilibrium point is stable or not. Various theorems have been proposed to enable one to determine algebraically the stability of a particular equilibrium point. Before discussing these methods, the above standard stability analysis is illustrated in the following for the case of the scalar LDDE (1.9).

2.2.1 Example: Linear Delay Differential Equation

Consider the LDDE (1.9), corresponding to the linear rate equation with delay,

$$\dot{x} = -kx(t - \tau), \qquad k > 0. \tag{2.8}$$

Since the equilibrium point is the one for which

$$X(t) = X(t - \tau) = X^* \qquad \text{for all } t, \tag{2.9}$$

it can be obtained by setting $\dot{x} = 0$ and $x(t) = x(t - \tau) = X^*$ in the above LDDE. The equilibrium point is clearly

$$x = X^* = 0. \tag{2.10}$$

The Jacobians in the Eq. (2.4) for the above LDDE are

$$J_0 = 0 \text{ and } J_\tau = -k. \tag{2.11}$$

Then the characteristic equation (2.7) becomes

$$\lambda + ke^{-\lambda\tau} = 0. \tag{2.12}$$

Let $\lambda = \alpha + i\beta$, where α and β are the real and imaginary parts of λ. Then equating the real and imaginary parts of (2.12) separately to zero, we have

$$ke^{-\alpha\tau}\cos\beta\tau = -\alpha, \tag{2.13a}$$
$$ke^{-\alpha\tau}\sin\beta\tau = \beta. \tag{2.13b}$$

In order to ascertain the stability nature of the equilibrium point we have to find out whether the above Eqs. (2.13) can have solutions with positive values of α. For the moment, let $\alpha > 0$. Since $k > 0$, and $e^{-\alpha\tau} < 1$, necessarily $\cos\beta\tau < 0$ in accordance with Eq. (2.13a). Consequently, $\beta > \frac{\pi}{2\tau}$ since for any smaller positive value of β, $\cos\beta\tau$ is positive or zero. On the other hand $|\sin\beta\tau| < 1$, so that Eq. (2.13b) implies that $\beta < k$. This is in contradiction with the previous requirement of $\beta > \frac{\pi}{2\tau}$ of Eq. (2.13a). Hence β cannot simultaneously be larger than $\frac{\pi}{2\tau}$ and smaller in magnitude than k, for all values of τ and k except when $k > \frac{\pi}{2\tau}$. Therefore the real part of the eigenvalue $\lambda = \alpha + i\beta$ cannot be positive, in general, and so the corresponding equilibrium point is stable.

A detailed stability analysis using the above procedure for the case of nonlinear DDEs is given in Sect. 2.4 and later in Chaps. 3 and 4. Specifically, the linear stability analysis for the piecewise linear delay differential equation will be discussed in Chap. 3 and for the Mackey-Glass delay system it will be presented later in Chap. 4.

2.3 A Geometric Approach to Study Stability

McDonald [1] had proposed a simple geometric approach to determine the stability of a fixed point satisfying the transcendental equation (2.7), which can be rewritten in general as

$$P(\lambda) + Q(\lambda)e^{-\lambda\tau} = 0, \tag{2.14}$$

where P and Q are polynomials of degree n.

All the eigenvalues λ of the above equation should lie in the left half of the complex $\lambda-$plane for the equilibrium point X^* to be stable. That is, if for all the eigenvalues $\text{Re}(\lambda) < 0$, the corresponding solution is stable. On the other hand, even if one of the eigenvalues λ has a positive real part then the solution is unstable. Hence, a change in stability can occur only when a root of the Eq. (2.14) crosses the imaginary axis, that is $\lambda = \alpha + i\beta$ with $\alpha = 0$ is a solution of the Eq. (2.14). In other words, when the real part of a root, λ, changes from negative to positive value or vice-versa through zero (as some control parameter changes), change in stability of the steady state (equilibrium point) can occur. To check this, one can simply substitute $\lambda = i\beta$ into Eq. (2.14) so that it can be rewritten as

$$\frac{P(i\beta)}{Q(i\beta)} = -e^{-i\beta\tau}. \tag{2.15}$$

If the equilibrium point in question is stable in the absence of delay, that is for $\tau = 0$, then the change in stability can occur only if there are some real β and τ for which the Eq. (2.15) holds good. This can be determined by the simple geometric construction demonstrated by McDonald. As $\beta\tau$ is increased from 0 to 2π, the right hand side of the Eq. (2.15), namely $-e^{-i\beta\tau}$, traces out a unit circle in the complex plane. On the other hand the left hand side of Eq. (2.15), called the *ratio curve*, also defines another curve in the complex plane. If there is a change in stability then the ratio curve must intersect the unit circle. The ratio curve can also cross the unit circle more than once depending on the nature of the left hand side of the above equation. In such cases, the solution may be stable for short delays, unstable for the intermediate range of delays and then stable again for longer delays and vice-versa. Now, we will illustrate this approach for the linear DDE discussed in the previous section.

2.3.1 Example: Linear Delay Differential Equation

Considering the characteristic equation (2.12), namely $\lambda + ke^{-\lambda\tau} = 0$, the eigenvalues always have negative real parts. Correspondingly, to confirm the stable nature of the equilibrium point as discussed in Sect. 2.2, the ratio curve must intersect the unit circle at least once. On setting $\lambda = i\beta$ in $\lambda + ke^{-\lambda\tau} = 0$, it can be rewritten as

$$\frac{i\beta}{k} = -e^{-i\beta\tau}. \tag{2.16}$$

The real and imaginary parts of both the sides of the above equation are plotted in Fig. 2.1. It can be readily seen that the ratio curve indeed intersects the unit circle confirming the stable nature of the equilibrium point.

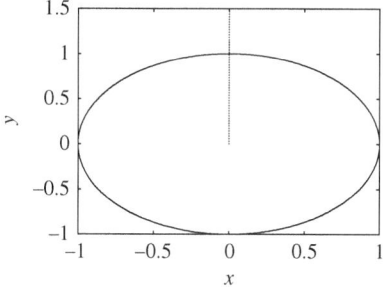

Fig. 2.1 Real (x) and imaginary (y) parts of both the sides of the Eq. (2.16)

As a second example, one can apply this approach to determine the stability nature of the equilibrium point of a nonlinear DDE, namely the Mackey-Glass delay differential equation, which we will discuss in more detail later in Chap. 4.

2.4 A General Approach to Determine Linear Stability of Equilibrium Points

The stability analysis discussed in the previous sections can be used only for a specific form of the characteristic equation, see Eq. (2.14), and hence it cannot be used for a general DDE. In this section, we will discuss a more general approach to determine the linear stability of the equilibrium point and the existence of local Hopf bifurcations in a general scalar DDE of the form

$$\dot{x} = -bx + af\left(x(t - \tau)\right), \tag{2.17}$$

where a and b are positive parameters and f is a nonlinear function. Let $x = x^*$ be an equilibrium point of Eq. (2.17). Then the analysis proceeds as follows.

2.4.1 Characteristic Equation

Considering the equilibrium point $x = x^*$ of Eq. (2.17), we perturb it linearly as $x = x^* + \rho e^{\lambda t}$, $\rho \ll 1$. Then the characteristic equation associated with the time-delay differential equation (2.17) can be written as

$$\lambda = -b + af'(x^*)e^{-\lambda \tau}. \tag{2.18}$$

2.4.2 Stability Conditions

Let $\lambda = \alpha + i\beta$ be the eigenvalue associated with the equilibrium point $x = x^*$ and the critical stability curve is the one on which $\alpha = 0$ as one can expect that there is a change in stability when the value of α crosses the imaginary axis at $\lambda = i\beta$. The stability curve can be again obtained by substituting $\lambda = i\beta$ into the characteristic equation (2.18). Then one obtains the relation

$$i\beta = -b + af'(x^*)(\cos \beta \tau - i \sin \beta \tau). \tag{2.19}$$

Separating the real and imaginary parts, we get

$$b = af'(x^*) \cos \beta \tau, \tag{2.20a}$$
$$\beta = -af'(x^*) \sin \beta \tau. \tag{2.20b}$$

Squaring and adding the above two equations, one obtains the conditions

$$\beta = \pm\sqrt{a^2 f'^2(x^*) - b^2}. \tag{2.21}$$

This is possible if and only if $|af'(x^*)| > b$. Note that $b > 0$ in Eq. (2.17). From Eq. (2.20a), it follows that

$$\beta\tau = \pm\arccos\left(\frac{b}{af'(x^*)}\right) + 2n\pi, \tag{2.22}$$

where n is any integer $(0, \pm 1, \pm 2, ...)$. Consequently for $|af'(x^*)| > b$, and for a fixed β (say $\beta > 0$), one can expect that the stability regions are confined between a set of two curves/surfaces in the (τ, a, b) parameter space if $\frac{d\alpha}{d\tau}$ on any one of these curves/surfaces is negative and on the other it is positive:

$$\tau_1(n) = \frac{2n\pi + \arccos\left(\frac{b}{af'(x^*)}\right)}{\sqrt{a^2 f'^2(x^*) - b^2}}, \qquad n = 0, 1, 2, \cdots \tag{2.23a}$$

$$\tau_2(n) = \frac{2n\pi - \arccos\left(\frac{b}{af'(x^*)}\right)}{\sqrt{a^2 f'^2(x^*) - b^2}}. \qquad n = 1, 2, \cdots \tag{2.23b}$$

In Eq. (2.23a), $n = 0, 1, 2, \cdots$ and in Eq. (2.23b), $n = 1, 2, \cdots$ for the pair of curves so that the curves have positive values of τ, which is the physically interesting case. Note that there will be a similar set of curves for negative n in (2.23) which will have identical behavior and these will correspond to the case $\beta < 0$, since τ is always positive in (2.23).

2.4.3 Stability Curves/Surfaces in the (τ, a, b) Parameter Space

In order to identify the curves $\tau_{1,2}(n)$ that encompass the stable regions for $\tau > 0$, let us evaluate $\frac{d\alpha}{d\tau}$ or $\text{Re}\left(\frac{d\lambda}{d\tau}\right)$. Differentiating the characteristic equation (2.18) with respect to τ ($\tau > 0$), it follows that

$$\frac{d\lambda}{d\tau} = af'(x^*)e^{-\lambda\tau}\left[-\lambda - \tau\frac{d\lambda}{d\tau}\right]. \tag{2.24}$$

Rewriting (2.24), using again the characteristic equation, we have

$$\frac{d\lambda}{d\tau} = -\frac{af'(x^*)\lambda e^{-\lambda\tau}}{1 + af'(x^*)\tau e^{-\lambda\tau}} \tag{2.25}$$

$$= -\frac{\lambda(\lambda + b)}{1 + \tau(\lambda + b)} \tag{2.26}$$

and hence

$$Re\left(\frac{d\lambda}{d\tau}\right) \equiv \frac{d\alpha}{d\tau} = Re\left[\frac{\beta^2 - i\beta b}{(1 + \tau b) + i\tau\beta}\right] \tag{2.27}$$

$$= \frac{\beta^2}{(1 + \tau b)^2 + \tau^2\beta^2} > 0. \tag{2.28}$$

Therefore, using (2.23) in (2.28), we conclude that

$$\frac{d\alpha}{d\tau} > 0 \text{ on both } \tau_1(n) \text{ and } \tau_2(n) \text{ for } |af'(x^*)| > b. \tag{2.29}$$

Since $\frac{d\alpha}{d\tau} > 0$ for all $\tau(n)$ in Eq. (2.23), the corresponding slopes have positive values on all the stability determining curves $\tau(n)$. This implies that there does not exist any eigenvalue with negative real part across the curves (2.23). On the other hand, we know that for $\tau = 0$ the fixed point is stable for $b - af'(x^*) > 0$, see Eq. (2.18). Therefore the condition (2.29) implies that there can be only one stable region between the $\tau = 0$ line/plane in the (a, b) parameter space and the critical curve/surface in the (τ, a, b) parameter space $\tau_1(0)$, which is closest to the line/plane in the (a, b) parameter space, $\tau = 0$.

2.4.4 Extension to Coupled DDEs/Complex Scalar DDEs

The above stability analysis can also be extended to coupled DDEs and complex scalar equations with delay/delay coupling. In fact we have applied this analysis to a system of two coupled DDEs in Sect. 3.5 of Chap. 3, where we have also brought out the existence of a single stable region between the $\tau = 0$ line/(a, b) plane and the critical curve closest to the line/plane, $\tau = 0$, in the (a, b, τ) parameter space.

We also present a similar analysis to a complex scalar equation with delay feedback exhibiting limit cycle oscillations in Sect. 5.2 of Chap. 5 and illustrate the existence of multistability regions in the corresponding parameter space which was studied by Ramana Reddy [2] and Ramana Reddy et al. [3]. For this case of complex scalar equation, indeed $\frac{d\alpha}{d\tau} < 0$ on one of the curves, $\tau_{1,2}(n)$, $n = 1, 2, \cdots$ and $\frac{d\alpha}{d\tau} > 0$ on another curve and correspondingly there exist multiple stability regions isolated by unstable regions. This analysis has been originally used by Ramana Reddy, Sen and Johnston to identify multistability regions in limit cycle oscillators, where the collective stability regions were termed as *amplitude death* regions or *death islands* [2, 3]. We refer to Sect. 5.2 for further details.

2.4.5 Bifurcation Analysis

By definition all the roots have negative real parts within the stability region bounded by the line/plane $\tau = 0$, and the curve/surface $\tau = \tau_1(0)$ in the (τ, a) plane/ (τ, a, b) space, while at least one of the roots has a positive real part on the right side of the stability curve/surface. The curve/surface on the right boundary of the stability region has purely imaginary roots with zero real parts and this curve in the (τ, a) plane (or the surface in the (τ, a, b) parameter space) corresponds to the *Hopf bifurcation curve (surface)* across which a change in stability takes place, as the real part of the eigenvalue changes from a negative to a positive value.

2.4.6 Results of Stability Analysis

Using the above line of arguments, one can arrive at the following conclusions [4].

1. If $b > |af'(x^*)|$, then all the roots of the characteristic equation (2.18) have negative real parts.
   This is because if $b > |af'(x^*)|$, then according to Eq. (2.21) β turns out to be imaginary, say $\beta = i\omega$, $\omega > 0$. Since the stability determining critical curves are the characteristic curves, where $\lambda = i\beta$, and as a consequence of β becoming imaginary the eigenvalues on the critical curves become $\lambda = i(i\omega)$. Therefore all the eigenvalues have negative real parts on the critical curves for all $\tau_{1,2}(n)$ and hence the entire parameter regime turns out to be stable.

2. If $|af'(x^*)| > b$, then there exists a sequence of values of τ, $0 < \tau(0) < \tau(1) < \cdots < \tau(k) \cdots$, such that

   i. Equation (2.18) has a pair of simple imaginary roots, $\pm i\beta$, when $\tau = \tau(n)$, $n = 0, 1, 2, \cdots$.
      Since the value of β becomes real if $|af'(x^*)| > b$, in accordance with Eq. (2.21), the critical curves are the ones on which the eigenvalue becomes $\lambda = i\beta$ and hence Eq. (2.18) has a pair of simple imaginary roots on the critical curves represented by $\tau_{1,2}(n)$ in Eqs. (2.23).

   ii. If $af'(x^*) < -b$ and $\tau \in (0, \tau(0))$, all the roots of equation (2.18) have negative real parts, if $\tau = \tau(0)$, all the roots of (2.18) except $\pm i\beta$ have negative real parts and if $\tau \in (\tau(n), \tau(n+1))$ for $n = 0, 1, 2, \cdots$ Eq. (2.18) has $2(n+1)$ roots with positive real parts.
      In accordance with Eq. (2.18), if $af'(x^*) < -b$ the eigenvalue has negative real part at $\tau = 0$. Also according to Eqs. (2.23) and (2.29), as the critical curves have positive slopes the region enclosed between the line $\tau = 0$ and the critical curve $\tau = \tau_1(0)$ closest to the line $\tau = 0$ is the only stable region (all the eigenvalues have negative real parts) as discussed above. This critical curve is the one on which there exist at least a pair of imaginary eigenvalues while the rest having negative real parts. All the other critical curves, that is $\tau_{1,2}(n)$, $n = 1, 2, \cdots$ have positive real parts in accordance with Eq. (2.29).

iii. If $af'(x^*) > b$, Eq. (2.18) has at least one root with positive real part for all
  $\tau \geq 0$.
  If $af'(x^*) > b$, then according to Eq. (2.18), the eigenvalue has only positive
  real part even for $\tau = 0$ and from Eq. (2.29) we know that there exists at least
  one eigenvalue with positive real part. Hence, if $af'(x^*) > b$ then the entire
  parameter space is unstable for any $\tau \geq 0$.

2.4.7 A Theorem on the Stability of Equilibrium Points

As a consequence of the above results the following theorem has been proposed by
Niu and Geng in [4]. For Eq. (2.17), the following hold.

1. If $|af'(x^*)| < b$, then $x = x^*$ is asymptotically stable for any $\tau \geq 0$.
2. If $af'(x^*) < -b$, then $x = x^*$ is asymptotically stable for $\tau \in (0, \tau(0))$ and
   unstable for $\tau > \tau(0)$.
3. If $af'(x^*) > b$, then $x = x^*$ is unstable for $\tau \geq 0$.
4. If $|af'(x^*)| > b$, then Eq. (2.17) undergoes a Hopf bifurcation at $x = x^*$ when
   $\tau = \tau_{1,2}(n)$ for $n = 0, 1, 2, \cdots$.

2.4.8 Example: Linear Delay Differential Equation

Now let us illustrate the above analysis for the simple functional form

$$f(x(t - \tau)) = -x(t - \tau), \tag{2.30}$$

so that the general DDE (2.17) becomes the LDDE,

$$\dot{x} = -bx - ax(t - \tau). \tag{2.31}$$

The corresponding characteristic equation turns out to be

$$\lambda + b + ae^{-\lambda\tau} = 0. \tag{2.32}$$

Substituting $\lambda = i\beta$ in the above equation, and after simple algebra, we have

$$\beta = \pm\sqrt{a^2 - b^2} \tag{2.33}$$

and

$$\beta\tau = \pm\cos^{-1}\left(\frac{b}{a}\right) + 2n\pi, \tag{2.34}$$

where n is any integer $(0, \pm 1, \pm 2, \cdots)$. Consequently one can expect that the stability regions are confined between the set of two curves

$$\tau_1(n) = \frac{2n\pi + \arccos\left(\frac{b}{a}\right)}{\sqrt{a^2 - b^2}}, \tag{2.35a}$$

$$\tau_2(n) = \frac{2n\pi - \arccos\left(\frac{b}{a}\right)}{\sqrt{a^2 - b^2}}. \tag{2.35b}$$

In Eq. (2.35a), $n = 0, 1, 2, \cdots$ and in Eq. (2.35b), $n = 1, 2, \cdots$ for the pair of curves to have positive values of τ. In order to identify those curves for $\tau > 0$ which encompass the stable regions, the critical curves should be the ones on which $\frac{d\lambda}{d\tau} > 0$. From the above characteristic equation (2.32), we have

$$\frac{d\alpha}{d\tau}\bigg|_{\alpha=0} = \beta^2 D^{-1}, \tag{2.36}$$

where

$$D = (1 + \tau b)^2 + \tau^2 \beta^2. \tag{2.37}$$

Therefore

$$\frac{d\alpha}{d\tau} > 0 \text{ on both } \tau_1 \text{ and } \tau_2. \tag{2.38}$$

The above condition implies that there can be only one stable region between the $\tau = 0$ line (where $\alpha < 0$) and the critical curve $\tau_1(0, a)$ which is closest to the line $\tau = 0$ for $|a| > b$. It is also in accordance with the result of stability analysis discussed in the previous section.

The numerical plot of the curves $\tau_1(n)$ (solid curve for $n = 0, 1, 2$) and $\tau_2(n)$ (dashed curve for $n = 1, 2$) are shown in Fig. 2.2. From the above analysis it is clear that the region between $\tau = 0$ and $\tau = \tau_1(0)$ is the only stable region (shaded region), where $\frac{d\alpha}{d\tau} > 0$ on τ_1, while passing from negative to positive values of α, whereas the other curves $\tau_2(n) < \tau < \tau_1(n)$ for $n > 0$ do not satisfy the required stability condition and hence they are all associated with unstable regions.

Now let us demonstrate the existence of a Hopf bifurcation by a numerical analysis of the LDDE (2.32). For illustration, let us choose the value of the parameters as $a = 0.5, b = 0.1$ and carry out a bifurcation analysis as a function of the delay time τ with reference to Fig. 2.2. The LDDE has been numerically integrated using the Runge-Kutta fourth order integration scheme for the aforesaid parameter values with constant initial condition in the range $(-\tau, 0)$ and the optimal step size of $\Delta h = 0.01$. It is evident from Fig. 2.2 that there exists a stable equilibrium point (the origin is the equilibrium point for the LDDE (2.32)) up to the value of delay time $\tau < 3.625$ for the above choice of parameters characterized by the eigenvalues with negative real part. This behavior is depicted in Fig. 2.3a, b for the values of

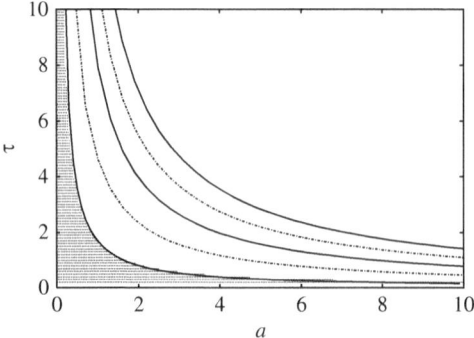

Fig. 2.2 *Curves* representing Eqs. (2.35a) and (2.35b). The *solid curves* represent τ_1 for $n = 0, +1, +2$ and *dashed curves* represent τ_2 for $n = +1, +2$. The region enclosed between the line $\tau = 0$ and the curve $\tau = \tau_1(0)$ is the only stable region (*shaded region*)

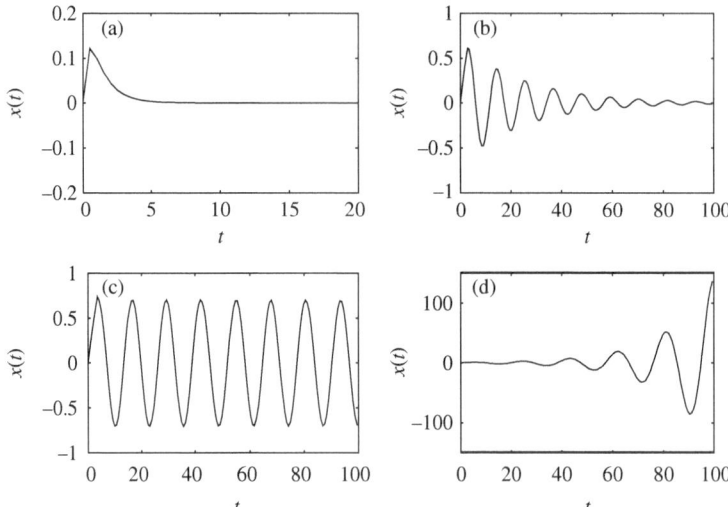

Fig. 2.3 Behavior of the solution $x(t)$ of the linear delay differential equation (2.31) for the parameter values $a = 0.5, b = 0.1$ and for various values of delay τ. (**a**) Monotonic decay to $x(t) = 0$ for $\tau = 0.5$, (**b**) Damped oscillatory decay to $x(t) = 0$ for $\tau = 3.0$, (**c**) Periodic oscillation for $\tau = 3.625$ and (**d**) Undamped growing oscillation for $\tau = 6.0$

delay times $\tau = 0.5$ and 3.0, respectively. In Fig. 2.3a, the solution of the LDDE monotonically decays to the stable equilibrium point and in Fig. 2.3b it exhibits a damped oscillatory decay to the stable equilibrium point. For the value of $\tau = 3.625$, there exist only imaginary eigenvalues corresponding to the partition curve $\tau_1(0)$ in Fig. 2.2, where the LDDE exhibits only periodic oscillations (Hopf oscillations) as shown in Fig. 2.3c, and this corresponds to the Hopf bifurcation curve. Above the value of delay time $\tau > 3.625$ for the aforesaid choices of a and b, the LDDE exhibits undamped growing oscillations as depicted in Fig. 2.3d for the value of delay time $\tau = 6.0$. Thus the change in stability across the critical curve (Hopf bifurcation curve corresponding to the imaginary eigenvalues) confirms the existence of a Hopf bifurcation. This is in accordance with the results of the stability analysis discussed in Sect. 2.4.

As noted earlier the above stability analysis has also been applied to the case of nonlinear DDEs, namely piecewise linear and Mackey-Glass time-delay systems, in the following Chaps. 3 and 4, respectively. The analysis has also been extended to coupled DDEs in Chap. 3 and to a complex scalar equation in Chap. 5, where we have demonstrated the existence of multistability regions.

References

1. N. McDonald, *Biological Delay Systems: Linear Stability Theory* (Cambridge University Press, Cambridge, 1989)
2. D.V. Ramana Reddy, Ph. D. Thesis entitled *Collective Dynamics of Delay Coupled Oscillators* (Institute for Plasma Research, Gandhinagar, India, 2000)
3. D.V. Ramana Reddy, A. Sen, G.L. Johnston, Phys. Rev. Lett. **80**, 5109 (1998)
4. H. Niu, J. Geng, Nonlinearity **20**, 2499 (2007)

Chapter 3
Bifurcation and Chaos in Time-Delayed Piecewise Linear Dynamical System

3.1 Introduction

The phenomena of bifurcations and chaos have been well studied in nonlinear dynamical systems without delay and described by nonlinear difference, differential, difference-differential, etc. equations. Routes to bifurcations, onset of chaos, nature of the chaotic attractors and their characterizations in such systems have all been analyzed extensively [1, 2]. However, such analyses have not been carried out in any greater detail in the case of nonlinear dynamical systems with delay even in the scalar systems. In this chapter we shall introduce a prototypical delay system, which is a piecewise linear one, in order to appreciate the nature of bifurcations and chaos phenomena underlying nonlinear time-delay systems, and to understand clearly the nature of transients and difficulties in numerical analysis as well as the frequent existence of hyperchaotic attractors with multiple positive Lyapunov exponents. The dynamics of other nonlinear time-delay systems will be taken up in the next chapter.

It is a widely accepted fact that chaos can occur in autonomous time continuous nonlinear systems having order greater than two and in nonautonomous time continuous nonlinear systems with order greater than one. In time discrete systems chaos can occur even in first order invertible maps and in non-invertible maps with order greater than one. Recently, Lu and He [3] have shown that chaos can occur even in a simple scalar first order delayed nonlinear (piecewise linear) dynamical system with large enough delay. Thangavel, Murali and Lakshmanan [4] have made a preliminary study of the bifurcation scenario and controlling of chaos in first order and coupled scalar piecewise linear time-delay systems including the model of Lu and He.

In this chapter, we will discuss several novel aspects of the underlying dynamics of the above scalar piecewise linear time-delay system [5]. These include the existence of a stable island in a two parameter space for the equilibrium solutions in the presence of time-delay, the significant role of transients in attaining steady state solution and thereby identifying the existence of familiar routes to chaos and the emergence of hyperchaos even for small values of time-delay for a wide range of parameters, which is confirmed by the existence of multiple positive Lyapunov exponents. The study has also been extended to the case in which the first order

M. Lakshmanan, D.V. Senthilkumar, *Dynamics of Nonlinear Time-Delay Systems*, Springer Series in Synergetics, DOI 10.1007/978-3-642-14938-2_3,

scalar time-delay system is coupled to a second scalar system without delay. Again the existence of a stable island for equilibrium points is established. Existence of different bifurcation routes, including type III intermittency route, is pointed out, where we have also discussed the transient effects. We have also plotted the two parameter bifurcation diagrams for both the cases to bring out the nature of the underlying dynamics.

3.2 Simple Scalar First Order Piecewise Linear DDE

We consider the following first order delay differential equation introduced by Lu and He [3], but with the addition of a constant external force as studied by Thangavel et al. [4],

$$\frac{dx(t)}{dt} = -ax(t) + bf\left(x(t-\tau)\right) + c, \tag{3.1}$$

where a and b are parameters, τ is the time-delay, c is a positive constant external force and f is an odd piecewise linear function defined as

$$f\left(x(t)\right) = \begin{cases} 0, & x \le -p_2 \\ -1.5x - 2, & -p_2 < x \le -p_1 \\ x, & -p_1 < x \le p_1 \\ -1.5x + 2, & p_1 < x \le p_2 \\ 0, & x > p_2. \end{cases} \tag{3.2}$$

Here p_1 and p_2 are parameters. The form of the function $f\left(x(t)\right)$ is sketched in Fig. 3.1. In our study we have fixed the parameter p_1 at $p_1 = 0.8$ and explored the dynamical behavior in the range of the external force $c \in [-0.15, 0.15]$ and delay time $\tau \in [0.0, 30.0]$ for different values of the parameter p_2 characterizing

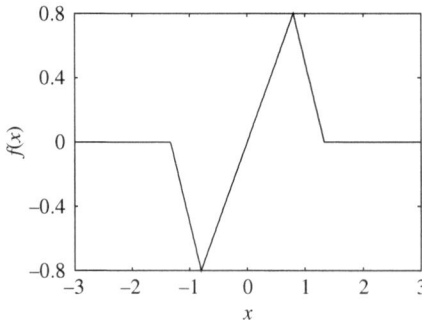

Fig. 3.1 The form of the piecewise linear function $f\left(x(t)\right)$ of Eq. (3.2) for the values $p_1 = 0.8$, $p_2 = 1.33$

the piecewise linear function $f(x(t))$. To start with let us consider the nature of the fixed (equilibrium) points of the system (3.1) in some detail.

3.2.1 Fixed Points and Linear Stability

Equation (3.1) with the piecewise linear form (3.2) for the function $f(x(t))$ can admit four distinct fixed points, $x_i(t) = x_0 = x^*, i = 1, 2, 3, 4$, depending upon the parameter values. We now carry out a detailed linear stability analysis [4, 5] of these fixed points under the linear perturbation $x = x^* + \alpha \exp(\lambda t), \alpha \ll 1$. Particularly, in this section, we will bring out the existence of a stable island in the $(\tau, 1.5b)$ plane for some of these equilibrium points, when $\tau > 0$, following the analysis discussed in Sect. 2.4. For this purpose we examine the stability nature of the fixed points of Eq. (3.1) both in the absence and in the presence of time delay τ by analyzing the eigenvalue λ of the characteristic equation and identifying the cases where $\text{Re}(\lambda) < 0$ for stability.

3.2.1.1 Time-Delay $\tau = 0$

In the absence of time-delay, the following fixed points can exist depending on the choice of parameters a, b, c, p_1 and p_2 in Eqs. (3.1) and (3.2):

1. For $|x| > p_2$, the fixed point is $x = x_0 = \frac{c}{a}$ and the characteristic equation in this region is $\lambda = -a$. The fixed point $x = x_0 = \frac{c}{a}$ is stable for positive values of a.
2. For $-p_2 < x \leq -p_1$, the fixed point is $x = x_0 = \frac{c-2b}{a+1.5b}$ and the characteristic equation in this region becomes $\lambda = -(a+1.5b)$. The fixed point is stable when $a > -1.5b$.
3. For $|x| \leq p_1$, the fixed point is $x = x_0 = \frac{c}{a-b}$ and the characteristic equation becomes $\lambda = -(a + b)$. The above fixed point is stable when $a > -b$.
4. For $p_1 < x \leq p_2$, the fixed point is $x = x_0 = \frac{c+2b}{a+1.5b}$ and the characteristic equation in this region becomes $\lambda = -(a+1.5b)$. The fixed point is stable when $a > -1.5b$.

Next we carry out the stability analysis in the presence of time-delay, $\tau > 0$.

3.2.1.2 Time-Delay $\tau > 0$

In the presence of time-delay, now we will examine the stability as follows.

1. Here again, for $|x| > p_2$, the fixed point and its characteristic equation remain the same as for the case $\tau = 0$ and the fixed point, namely $x = x_0 = \frac{c}{a}$, is stable for $a > 0$.
2. Next in the region, $-p_2 < x \leq -p_1$, for the fixed point $x = x_0 = \frac{c-2b}{a+1.5b}$, the characteristic equation becomes the transcendental equation,

$$\lambda + a + 1.5be^{-\lambda\tau} = 0. \tag{3.3}$$

Let $\lambda = \alpha + i\beta$, where α and β are real. Substituting this into the above equation and following the stability analysis discussed in Sect. 2.4, one can obtain the set of critical curves, determining the stability regions, as

$$\tau_1(n, 1.5b) = \frac{2n\pi + \arccos\left(\frac{-a}{1.5b}\right)}{\sqrt{2.25b^2 - a^2}}, \tag{3.4a}$$

$$\tau_2(n, 1.5b) = \frac{2n\pi - \arccos\left(\frac{-a}{1.5b}\right)}{\sqrt{2.25b^2 - a^2}}. \tag{3.4b}$$

In Eq. (3.4a), $n = 0, 1, 2, \cdots$ and in Eq. (3.4b), $n = 1, 2, \cdots$ for the pair of curves we have chosen so that the curves have positive values of τ. To determine those curves which enclose the stable regime, one has to analyze the nature of $\frac{d\lambda}{d\tau}$ on these curves. From Eq. (3.3), it is easy to check that

$$\frac{d\lambda}{d\tau} = \frac{1.5b\lambda \exp(-\lambda\tau)}{1 - \tau 1.5b \exp(-\lambda\tau)} \tag{3.5}$$

and that

$$\frac{d\alpha}{d\tau}\bigg|_{\alpha=0} = Re\frac{1.5b(i\beta)\exp(-i\beta\tau)}{1 - \tau 1.5b\exp(-i\beta\tau)} \tag{3.6}$$
$$= \frac{1.5b\beta \sin(\beta\tau)}{D}$$
$$= \beta^2 D^{-1},$$

where

$$D = [1 - 1.5b\tau \cos(\beta\tau)]^2 + [1.5b\tau \cos(\beta\tau)]^2,$$
$$\beta = 1.5b \sin(\beta\tau) \qquad \text{(from Eq. (2.20b))}$$

Therefore

$$\frac{d\alpha}{d\tau}\bigg|_{\alpha=0} > 0 \text{ on both the curves } \tau_1 \text{ and } \tau_2. \tag{3.7}$$

As we know from Eq. (3.3) that when the time-delay $\tau = 0$, $\lambda = -a - 1.5b$ and so $\alpha < 0$. The above condition implies that there can be only one stability region between the $\tau = 0$ line/plane in the (a, b) parameter space (where $\alpha < 0$) and the critical curve $\tau_1(0, 1.5b)$/surface in the (τ, a, b) parameter space, which is the closest to the line/plane $\tau = 0$ in the (a, b) parameter space. We note that the condition (3.7) prohibits the existence of any other stable region (that is multistability regions) because for a second stable region to exist one requires

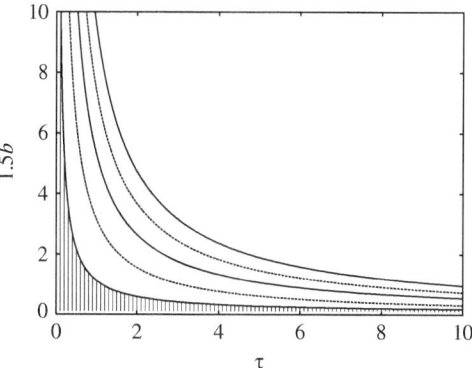

Fig. 3.2 *Curves* of Eqs. (3.4a) and (3.4b) in the range, $-p_1 < x \leq p_1$. The *solid curves* represent τ_1 for $n = 0, +1, +2$ and *broken curves* represent τ_2 for $n = +1, +2$. The region enclosed between the *line* $\tau = 0$ and the *curve* $\tau = \tau_1(0, 1.5b)$ is the only stable island (*shaded region*)

$\frac{d\alpha}{d\tau} < 0$ on any one of the other curves ($n > 0$). But this never occurs in this case. The numerical plot of the curves $\tau_1(n, 1.5b)$ (solid curve for $n = 0, 1, 2$) and $\tau_2(n, 1.5b)$ (broken curve for $n = 1, 2$) in Fig. 3.2 reveals that the region between $\tau = 0$ and $\tau = \tau_1(0, 1.5b)$ is the only stable region (shaded region), where $\frac{d\alpha}{d\tau} > 0$ on τ_1, while passing from negative to positive values of α, whereas the other curves $\tau_2(n, 1.5b) < \tau < \tau_1(n, 1.5b)$ for $n > 0$ do not satisfy the required stability condition and hence they are all associated with unstable regions. As there is a change in stability across the curve $\tau = \tau_1(0, 1.5b)$, this curve corresponds to a *Hopf bifurcation curve* on which there exists limit cycle oscillations (as this curve corresponds to purely imaginary eigenvalues).

3. As in the previous region, for the case $-p_1 < x \leq p_1$ and for the fixed point $x_0 = \frac{c}{a-b}$, the characteristic equation in this region has the form

$$\lambda + a + be^{-\lambda\tau} = 0. \tag{3.8}$$

As before, we obtain a set of critical curves as

$$\tau_1(n, b) = \frac{2n\pi + \arccos\left(\frac{a}{b}\right)}{\sqrt{b^2 - a^2}}, \tag{3.9a}$$

$$\tau_2(n, b) = \frac{2n\pi - \arccos\left(\frac{a}{b}\right)}{\sqrt{b^2 - a^2}}, \tag{3.9b}$$

where $n = 0, +1, +2, \cdots$ in Eq. (3.9a) and $n = +1, +2, \cdots$ in Eq. (3.9b). Also we get

$$\left.\frac{d\alpha}{d\tau}\right|_{\alpha=0} = \beta^2 D^{-1} > 0 \tag{3.9c}$$

where

$$D = [1 - b\tau \cos(\beta\tau)]^2 + [b\tau \cos(\beta\tau)]^2, \tag{3.9d}$$
$$\beta = b \sin(\beta\tau).$$

Using similar arguments as in the previous case, we find that there is only one stable island between the line $\tau = 0$ and the curve $\tau_1(0, b)$, which corresponds to a Hopf bifurcation curve.

4. For the case $p_1 < x \le p_2$ and for the fixed point $x = x_0 = \frac{c+2b}{a+1.5b}$, the characteristic equation turns out to be identical to Eq. (3.3) and the associated critical curves also have same form as that of Eq. (3.4). Hence, the same discussion in connection with the stability nature of the fixed point holds good in this case also.

3.3 Numerical Study of the Single Scalar Piecewise Linear Time-Delay System

In this section, we will present a discussion on the dynamics of the scalar piecewise linear time-delay system, (Eq. 3.1), in the pseudospace $(x(t), x(t - \tau))$. Further we will also discuss the significant role played by transients in attaining the steady state behavior, the computational efficiency required for achieving such steady state solution and the nature of bifurcation diagrams for both low and high transients with the external forcing c as the control parameter. We will also point out the existence of a *stable* island for equilibrium points in the two parameter (now in the delay time Vs the external forcing (τ, c) plane) bifurcation diagrams for various ranges of control parameters and nonlinearity characterized by the function $f(x(t))$ for three different values of p_2 in Eq. (3.2). In addition, we have also calculated the Lyapunov exponents associated with the system and show that there exists a large parameter range over which hyperchaos can be observed (corresponding to multiple positive Lyapunov exponents). From the two parameter bifurcation diagrams, it becomes evident that the sizes of the hyperchaotic regime and the stable fixed point regime increase with time-delay.

3.3.1 Dynamics in the Pseudospace

The dynamics of the piecewise linear time-delay system, Eq. (3.1), can be studied in a suitable phase space by plotting the numerical solution of Eq. (3.1) appropriately. Now, to calculate $x(t)$ from Eq. (3.1) for times greater than t, the function $x(t)$ over the interval $(t - \tau, t)$ must be specified. Hence, for a prescribed continuous function $x(t)$ on $(-\tau, 0)$ one can integrate Eq. (3.1) using numerical methods as in the case of ordinary differential equations, as discussed earlier in Sect. 1.2.2. We have numerically integrated Eq. (3.1) using Runge-Kutta fourth order integration routine with the parameters fixed as $a = 1.0$, $b = 1.2$ and $p_1 = 0.8$ for three different values of $p_2 = 1.0, 1.33, 1.66$ and with the initial condition (initial function) $x(t) = 0.9$ in the interval $(-\tau, 0)$. As the system is of first order in nature, the dynamics can be viewed in a pseudospace by plotting $x(t)$ against $x(t - \tau)$. (Note that the choice of $x(t - \tau)$ as the second phase variable is arbitrary; $x(t - t')$ can be equivalently used, where t' is an arbitrary delay time). One encounters typical scenario of bifurcations leading to chaos, but with an important difference: *Transients play a crucial role* and it takes a very long time before transients die down to attain steady state solutions. In particular, period doubling and inverse period doubling phenomena, besides other bifurcations, are often encountered. In many cases the system exhibits *hyperchaotic behavior* characterized by multiple positive Lyapunov exponents, which we will discuss below in Sect. 3.3.4. Figure 3.3 shows a typical chaotic attractor for $\tau = 5.0, c = 0.001$ and $p_2 = 1.33$ with the initial condition as given above in the phase space $(x(t), x(t + 5))$. The positive maximal Lyapunov exponent of this attractor for the above mentioned parameter values is $\lambda_{max} = 0.05461$. It may be noted that previously it was reported by Thangavel et al. [4] and Lu and He [3] that the system for a different set of parameters exhibits only chaos (and not hyperchaos) for large values of time-delay, that is for $\tau > 20$. On the other hand, we have identified the parameter regimes where the system exhibits chaos and even hyperchaos for small values of time-delay around $\tau = 5.0$. In the following we present the details.

3.3.2 Transients

Usually dynamical systems described by ordinary differential equations (ODEs) can attain the steady state within a few thousand transients for optimal value of time step Δt, provided due care is given for numerical accuracy and a suitable numerical

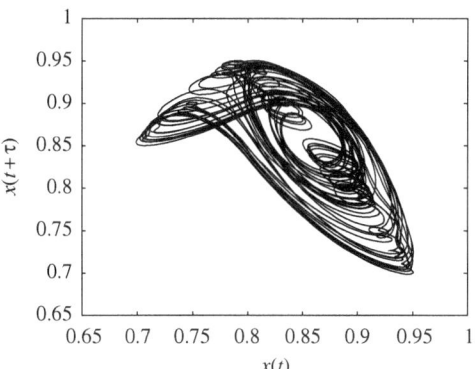

Fig. 3.3 Chaotic attractor for $\tau = 5.0, a = 1.0, b = 1.2, c = 0.001, p_1 = 0.8, p_2 = 1.33$ and $x(t) = 0.9, t \in (-5, 0)$

algorithm is chosen. However for the delay differential equation (DDE) of the form
(3.1), such low transients (of the order 10^4 and more) do not seem to lead the system to its steady state solution. In such cases one has to leave out more number
of transients so that one can have the possibility of obtaining a clear picture of the
bifurcation diagram of the known routes. In fact, the effect of transients has been
pointed out by Becker and Dorfler [6] even for the logistic map. As far as nonlinear
maps and ODEs are concerned, with some effort it is possible to realize whether the
system has reached the steady state or not by constructing the appropriate bifurcation diagrams and identifying the various bifurcation routes by leaving out sufficient
number of transients before starting the analysis for steady state solutions. The same
considerations hold good for dynamical systems modeled by finite number of coupled nonlinear maps and ODEs without time-delay.

On the other hand, when one starts studying nonlinear dynamical systems with
time-delay, effectively one considers infinite number of coupled ODEs which in
a numerical sense corresponds to several hundred coupled nonlinear ODEs. For
example, for Eq. (3.1), when the time-delay $\tau = 25.0$, in the actual numerical
analysis, we typically take the optimal time step as $\Delta t = 0.05$, so that we are
actually solving 500 coupled nonlinear ODEs. However, the optimal time step Δt
to be fixed depends explicitly on the nature of the system and on the time-delay
introduced in the system. In contrast to the nonlinear ODEs, where either a very
small or a relatively large time step Δt results in an exponentially increasing numerical error, for the time-delay system (3.1), we have found that the optimal time
step Δt varies between 0.05 to 0.0001, depending on which the actual number of
coupled nonlinear ODEs increase, and the number of transients to be left out is
found to lie anywhere between $2.0 - 3.0 \times 10^6$. For small values of time-delay
τ, the optimal time step Δt should be as small as possible to identify a typical
bifurcation scenario. However, for large values of time delay τ, a very small value
of time step Δt can again be problematic as one has to deal with a large number of
coupled ODEs and the time required for solving such huge number of coupled ODEs
becomes large. Also a large value of time step Δt does not seem to lead to the steady
state solution, as such time steps lead to propagation of numerical errors. However,
an appropriate optimal choice of time step Δt can lead to a typical bifurcation
scenario.

Thus the numerical analysis of time-delay nonlinear systems requires an appropriate time step Δt to be chosen, depending on which the actual number of coupled nonlinear ODEs increases, which in turn necessarily increases the number of
transients to be left out in order to obtain the steady state solution. In our numerical analysis we have chosen the optimal time step as $\Delta t = 0.05$ for the value of
time-delay $\tau = 25.0$. The effect of transients in reaching the steady state behavior
can be realized from the bifurcation diagrams we have obtained for the parameter
values $a = 0.16, b = 0.2$ and for two different values of p_2. Figure 3.4a shows
the one parameter bifurcation diagram for the above mentioned parameter values
with $p_2 = 1.33$, when we leave transients of the order $1.0 - 2.0 \times 10^4$, which is
sufficiently large in the case of nonlinear maps and ODEs, leaving the impression

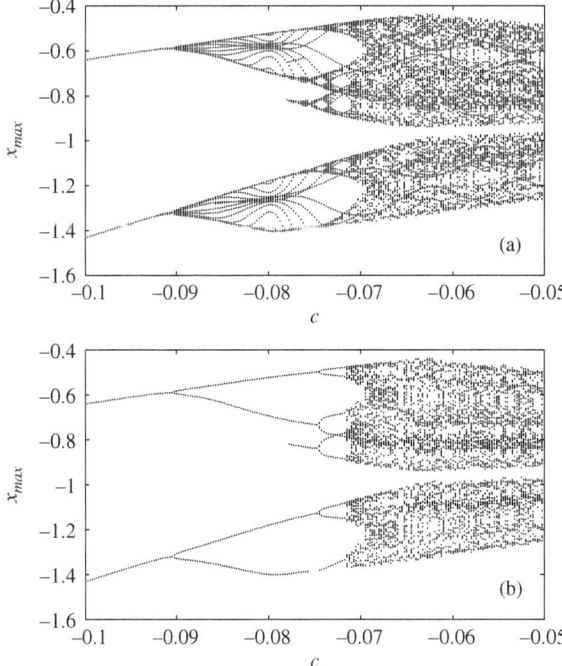

Fig. 3.4 Bifurcation diagrams of the single scalar piecewise linear time-delay system (3.1) for the parameters values $a = 0.16$, $b = 0.2$ and $\tau = 25.0$ when $p_2 = 1.33$ (**a**) for transients of the order $1.0 - 2.0 \times 10^4$ and (**b**) for transients now of the order 1.0×10^5

that the steady state has been reached with a bifurcation scenario that is quite complicated and atypical. In contrast, Fig. 3.4b shows the bifurcation diagram for the same value of the parameters except that now the transients left out are of the order of 1.0×10^5, which shows a typical bifurcation scenario. Similarly Fig. 3.5a shows the bifurcation diagram for the same values of parameters as above for τ, a and b with $p_2 = 1.66$ for the transients of the order 1.0×10^5, which shows a very complex structure, whereas Fig. 3.5b shows the bifurcation diagram for the transients of the order 1.4×10^6, in which case still there exists some stray points near the bifurcation regions, which indicates that the system requires still more transients to settle to its steady state. The complexity increases with time-delay and hence the number of transients also increases, which in turn increases the computing time enormously. This is also evident from Fig. 3.6, where the maximum value of $x(t)$, x_{max}, is plotted against the time-delay τ, for the parameter values $a = 1.0$, $b = 1.2$, $c = 0.001$ and $p_2 = 1.33$. Fig. 3.6a shows that even for transients of the order 1.0×10^5, the bifurcation diagram is still in an unsettled form for larger values of time-delay τ. On the other hand in Fig. 3.6b, the number of transients is of the order 2.5×10^5, wherein a clear bifurcation scenario has emerged. We have verified the role of transients to attain the steady state solutions for various values of time step Δt and time-delay

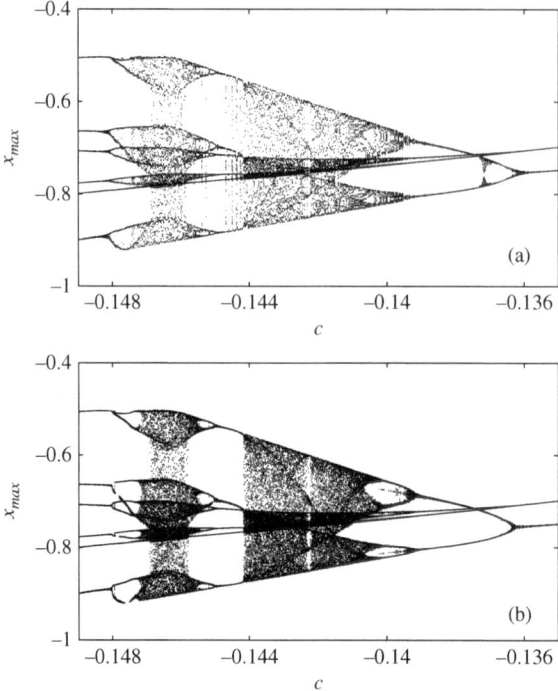

Fig. 3.5 Bifurcation diagrams of the single scalar piecewise linear time-delay system (3.1) for the parameter values $a = 0.16$, $b = 0.2$ and $\tau = 25.0$ when $p_2 = 1.66$ (**a**) for transients of the order 1.0×10^5 and (**b**) for transients of order 1.4×10^6

τ for few other delay differential equations also, including those discussed in the work of Ramana Reddy et al. [7]. The results of our detailed numerical investigation for Eq. (3.1) are tabulated in Table 3.1, where we have indicated the number of transients to be left out for given values of Δt in order to attain the steady state solutions.

As the delay time increases, the size of the delay loop (that is to be updated on every iteration) required to maintain the delay variable in numerical simulation also increases, which in turn increases the computational effort. Another important point to note in connection with the effect of transients is the following. In the above we have discussed the effect with reference to a scalar delay differential equation only. On the other hand, in recent times, there has been considerable interest to study the dynamics of multiply coupled delayed neural/nonlinear networks in biological systems [8, 9]. In order to examine the actual steady state behavior of such networks, which effectively correspond to several thousands of coupled nonlinear ODEs, the time required to obtain steady state solution increases enormously as each one of them have separate time delays (systems with multiple delays) and intrinsic transient behaviors and it becomes crucial to study the effect of the latter.

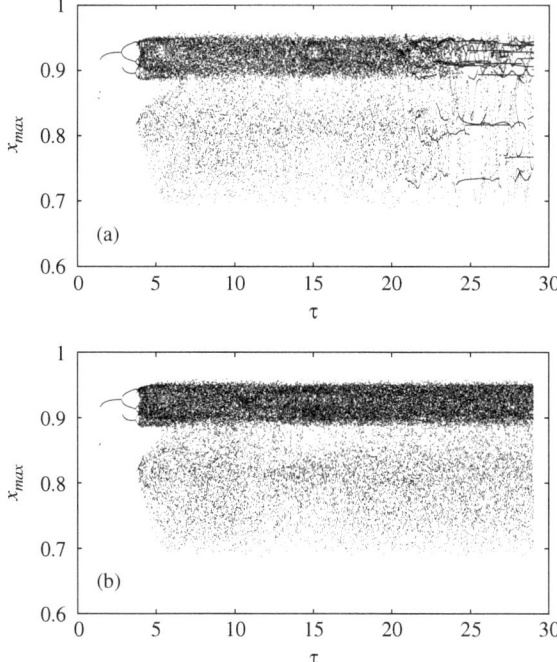

Fig. 3.6 Bifurcation diagrams of the single scalar piecewise linear time-delay system (3.1) for the parameter values $a = 1.0$, $b = 1.2$, $c = 0.001$, $p_2 = 1.33$ and $\tau \in (2, 29)$ (**a**) for transients of the order 1.0×10^5 and (**b**) for transients of the order 2.5×10^5

Table 3.1 Effect of time step Δt on transients

Value of τ	Optimal value of Δt	Number of coupled differential equations	Number of transients to be left out
5	0.002	2,500	$3.0 - 5.0 \times 10^5$
	0.001	5,000	$5.0 - 6.0 \times 10^5$
10	0.002	5,000	$1.0 - 2.0 \times 10^6$
	0.001	10,000	$> 2.0 \times 10^6$
25	0.05	500	$1.0 \times 10^5 - 2.0 \times 10^6$
	0.005	5,000	$1.0 - 2.0 \times 10^6$
	0.0025	10,000	$> 2.0 \times 10^6$

3.3.3 One and Two Parameter Bifurcation Diagrams

All the bifurcation diagrams shown in this chapter (and later chapters) have been plotted after leaving out a very large number of transients of the order $1.2 - 3.2 \times 10^6$. Figure 3.4b shows the one parameter bifurcation diagram for the values of the parameters $p_1 = 0.8$, $p_2 = 4/3$, $\tau = 25.0$ in Eq. (3.2) for $c \in [-0.1, -0.05]$. It clearly exhibits period-doubling scenario; however there exists a sudden distortion around the value of $c = -0.0785$, the cause of which

remains unexplained by any standard type of bifurcation. Similarly Fig. 3.5b shows
the bifurcation diagram with the values $p_1 = 0.8, p_2 = 5/3, \tau = 25.0$ for
$c \in [-0.15, -0.136]$, exhibiting reverse period-doubling route to chaos in the
range, $c \in [-0.136, -0.144]$, which includes a clear band merging crises and anti-
monotonicity for $c \in [-0.15, -0.144]$.

The two parameter bifurcation diagram for $\tau \in [0, 30]$ and $c \in [-0.16, 0.16]$
when $p_2 = 1.0$ is shown in Fig. 3.7a, which shows the behavior of the scalar
piecewise linear time-delay system in the combined phase space of parameters τ
and c. The following color codes are used to represent various regions: period-1
region -red, period-2 region - green, 3-blue, 4-yellow, 5-magenta, 6-cyan, 7-gray,
8-copper, chaos-black and the fixed points-white. The white region in the two
parameter bifurcation diagram corresponds to stable regions. Figure 3.7b shows
the two parameter bifurcation diagram for the same parameters as in Fig. 3.7a,
except that now $p_2 = 4/3$. We use the same colour codes as in Fig. 3.7a for all
the two parameter bifurcation diagrams in this chapter. Similarly Fig. 3.7c shows
the two parameter bifurcation diagram for the same range of the control parameters and the same values of the parameters a and b in Eq. (3.1), except that now

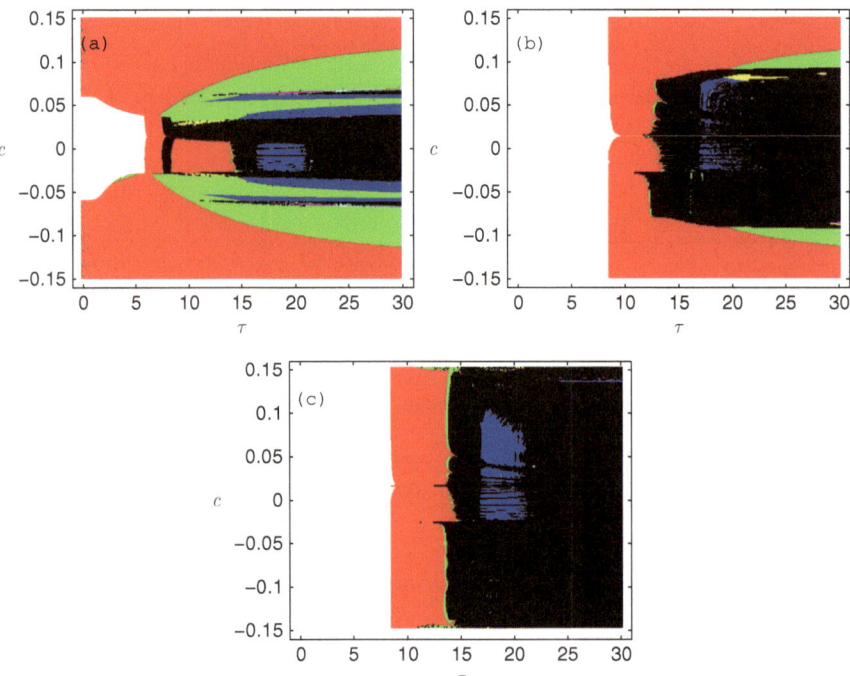

Fig. 3.7 Global bifurcation diagrams of the single scalar piecewise linear time-delay system (3.1)
for $\tau \in (0.5, 30)$ and $c \in (-0.16, 0.16)$. The following *color codes* are used to represent various
regions, period-1 region -*red*, period-2 region -*green*, 3-*blue*, 4-*yellow*, 5-*magenta*, 6-*cyan*, 7-*gray*,
8-*copper*, chaos-*black* and the fixed points-*white*. (**a**) $p_2 = 1.0$, (**b**) $p_2 = 1.33$ and (**c**) $p_2 = 1.66$

$p_2 = 5/3$. By comparing the Fig. 3.7a, b and c, we can see that the stable fixed point regions and the chaotic regions increase for small changes in the nonlinear parameters p_1 and p_2. Further one can infer from these figures that as the delay increases the chaotic nature of the system also increases and thereby contributing to the hyperchaotic nature of the system characterized by multiple positive Lyapunov exponents.

3.3.4 Lyapunov Exponents and Hyperchaotic Regimes

One of the interesting aspects of the dynamics associated with Eq. (3.1) is the existence of hyperchaos in a single first order scalar equation with time delay even for small values of delay times and for suitable values of other system parameters. As Eq. (3.1) is a first order delay differential equation, the usual procedure for calculating Lyapunov exponents is not applicable. However, the simple idea of approximating a single scalar differential equation with delay, which is essentially infinite dimensional as pointed out earlier in Sect. 1.2.2 of Chap. 1, by an N-dimensional discrete mapping [10] facilitates one to calculate the Lyapunov exponents. To simulate the behavior of such systems on a computer it is necessary to approximate the continuous evolution of an infinite dimensional system by a finite number of elements whose values change at discrete time steps (see Sect. 1.2.2). In this manner a continuous infinite dimensional dynamical system is replaced by a finite dimensional iterated mapping. This method was proposed originally by Farmer [10] to calculate the Lyapunov exponents for delay systems. As the delay parameter is increased, for most parameter values the dimension increases and the attractor generally becomes more complicated, thereby contributing to the hyperchaotic nature of the system, which gets confirmed by the increasing number of positive Lyapunov exponents. The first ten maximal Lyapunov exponents, for the parameter values $a = 1.0, b = 1.2, c = 0.001, p_1 = 0.8, p_2 = 1.33$ and $\tau \in (2, 29)$, are shown in Fig. 3.8a, where it is evident that the number of positive Lyapunov exponents increases with time-delay τ. The Kaplan-Yorke dimension obtained by using the formula

$$D_L = j + \frac{\sum_{i=1}^{j} \lambda_i}{|\lambda_{j+1}|}, \tag{3.10}$$

where j is the largest integer for which $\lambda_1 + \ldots + \lambda_j \geq 0$, for the single scalar piecewise linear time-delay system is shown as a function of the delay time τ in Fig. 3.8b. Almost the entire black regime in the two parameter bifurcation diagrams Fig. 3.7 for large values of τ is characterized by multiple positive Lyapunov exponents corresponding to the hyperchaotic nature of the system.

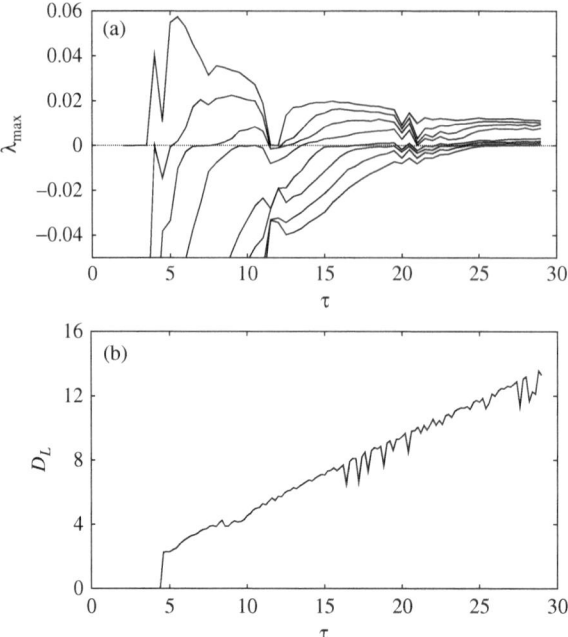

Fig. 3.8 (**a**) The first ten maximal Lyapunov exponents of the single scalar piecewise linear time-delay system (3.1) for the values $a = 1.0$, $b = 1.2$, $c = 0.001$, $p_2 = 1.33$ and $\tau \in (2, 29)$, and (**b**) Kaplan-Yorke dimension of the scalar piecewise linear time-delay system as a function of the time-delay τ

3.4 Experimental Realization using PSPICE Simulation

The simple scalar odd piecewise linear time-delay system (Eqs. 3.1 and 3.2) can also be realized experimentally using PSPICE simulation. The block diagram of the analog electronic circuit consists of a tunable delay unit, a nonlinear device and a fixed RC filter as shown in Fig. 3.9. The delay unit is of T-type LCL filter with matching resistors R at the input and output of the delay unit. Delay time can be tuned by connecting the output amplifier to a suitable output terminal i of the delay unit shown in Fig. 3.10. Delay time can be approximated by $T_d(i) = i\sqrt{2LC}$, where i is the ith output terminal of the delay unit. Only this T-type LCL filters have been used as a delay unit in most of the studies available in the literature. However, an inbuilt delay line, namely bucket brigade line MN 3011 with 3,328 stages, triggered by MN 3101 (National Panasonic) has also been used in the literature [11, 12] as a delay unit.

The circuit diagram of the piecewise linear nonlinearity denoted as **ND** in Fig. 3.9 is shown in Fig. 3.11. The first part of the circuit is the relay circuits denoted by K, which cuts off the voltage to zero on both sides of the current-voltage characteristic curve. $D1$ and $D2$ are 1N4148 diodes. $OA1$ and $OA2$ are standard operational

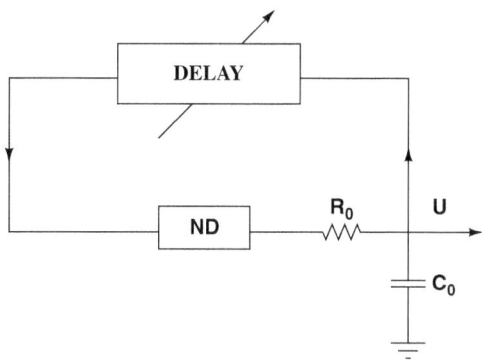

Fig. 3.9 Block diagram of the analog electronic circuit

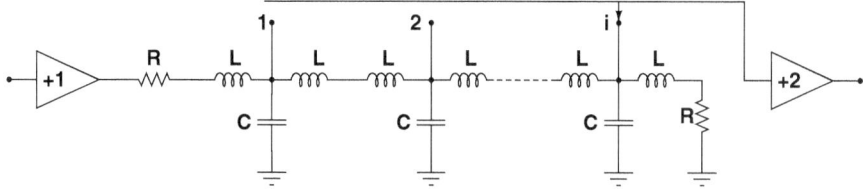

Fig. 3.10 Tunable delay unit. Different combinations of the inductance L, C and R are used in the literature. However, we have used the combinations $L = 4.7\,\text{mH}$, $C = 10\,\text{nF}$ and $R = 190\,\text{Ohms}$ in our PSPICE simulation

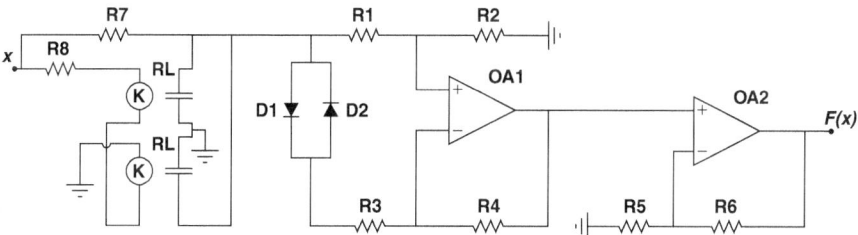

Fig. 3.11 The circuit diagram of the piecewise linear nonlinearity denoted as **ND** in Fig. 3.9

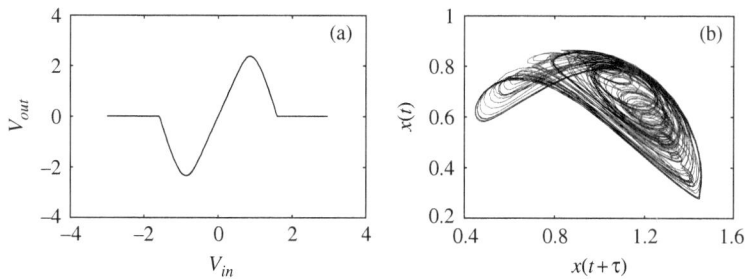

Fig. 3.12 (**a**) Input/output voltage characteristic of the piecewise linear nonlinearity (Fig. 3.11), and (**b**) the chaotic attractor exhibited by the circuit, Fig. 3.9

amplifiers. The input/output voltage characteristic of the piecewise linear nonlinearity is depicted in Fig. 3.12a and the chaotic attractor exhibited by this circuit is presented in Fig. 3.12b.

3.5 Stability Analysis and Chaotic Dynamics of Coupled DDEs

We now add a second dynamical variable evolving without delay to Eq. (3.1) to obtain the following set of coupled DDEs,

$$\frac{dx(t)}{dt} = -ax(t) + bf(x(t-\tau)) + dy(t),   \tag{3.11a}$$

$$\frac{dy(t)}{dt} = -cy(t) + ex(t),   \tag{3.11b}$$

where c and e are additional parameters and the function $f(x)$ and other parameters are the same as already defined in Eq. (3.1). As in the case of the single scalar piecewise linear time-delay system, we have fixed the parameter p_1 at $p_1 = 0.8$ and studied the system behavior in a range of variable parameter $c \in [0.0, 1.4]$ and time-delay $\tau \in [0.0, 30.0]$ for different values of the parameter p_2 characterizing the piecewise linear function $f(x)$. Now let us consider the nature of the fixed points of system (3.11).

3.5.1 Fixed Points and Linear Stability

We will now bring out the existence of stable island in the two parameter bifurcation diagram corresponding to the system of two coupled delay differential equations by linear stability analysis in the presence of time-delay τ, as in the case of single scalar piecewise linear time-delay system. Considering the fixed point $X^* = (x^*, y^*)$ of the system (3.11), we assume the linearly perturbed form as $x = x^* + \beta_1 e^{\lambda t}$ and $y = y^* + \beta_2 e^{\lambda t}$, $\beta_1, \beta_2 \ll 1$. Then, we examine the stability nature of the fixed points of Eq. (3.11) both in the absence and in the presence of time-delay τ.

3.5.1.1 Time-Delay $\tau = 0$

In the absence of time-delay, the following fixed points can exist depending on the choice of parameters a, b, c, d, e, p_1 and p_2 in Eqs. (3.11) and (3.2).

1. For $|x| > p_2$, the fixed point is $(x, y) = (x_0, y_0) = (0, 0)$ and the characteristic equation in this region is $\lambda^2 + (a+c)\lambda + ac - de = 0$. The fixed point $(x, y) = (0, 0)$ is stable when $a + c > 0$ and $ac > de$.

2. For $-p_2 < x \le -p_1$, the fixed point is $(x, y) = \left(\frac{2bc}{de - (a+1.5b)c}, \frac{2be}{de - (a+1.5b)c} \right)$ and the characteristic equation in this region becomes $\lambda^2 + (c + a + 1.5b)\lambda +$

$(ca + 1.5bc - de) = 0$. The fixed point is stable when $c + a + 1.5b > 0$ and $ac + 1.5bc > de$.

3. For $|x| \leq p_1$, the fixed point is $(x, y) = (0, 0)$ and the characteristic equation becomes $\lambda^2 + (c + a - b)\lambda + (ca - bc - de) = 0$. The above fixed point is stable when $c + a - b > 0$ and $ac > bc + de$.

4. For $p_1 < x \leq p_2$, the fixed point is $(x, y) = \left(\frac{-2bc}{de-(a+1.5b)c}, \frac{-2be}{de-(a+1.5b)c} \right)$ and the characteristic equation in this region becomes $\lambda^2 + (c + a + 1.5b)\lambda + (ca + 1.5bc - de) = 0$. The fixed point is stable when $c + a + 1.5b > 0$ and $ac + 1.5bc > de$.

Next we consider the case when time-delay is present, $\tau > 0$.

3.5.1.2 Time-Delay $\tau > 0$

In the presence of time-delay, we will examine the stability as follows.

1. For $|x| > p_2$, the fixed point and its stability remains the same as for the case $\tau = 0.0$.
2. Next in the region $-p_2 < x \leq -p_1$, for the fixed point

$$(x, y) = \left(\frac{2bc}{de - (a + 1.5b)c}, \frac{2be}{de - (a + 1.5b)c} \right),$$

the characteristic equation can be expressed as

$$(c + \lambda)(\lambda + a + 1.5b \exp(-\lambda\tau)) - de = 0, \tag{3.12}$$

which is a transcendental equation with infinite number of solutions. Let $\lambda = \alpha + i\beta$, where α and β are real. Substituting this in the above equation and equating the real and imaginary parts, we obtain two equations as

$$(c + \alpha)\left[\alpha + a + 1.5b \exp(-\alpha\tau) \cos(\beta\tau)\right] - de - \beta^2 + 1.5\beta b \exp(-\alpha\tau) \sin(\beta\tau) = 0, \tag{3.13a}$$

$$(c + \alpha)\left[\beta - 1.5b \exp(-\alpha\tau) \sin(\beta\tau)\right] + \beta\left[\alpha + a + 1.5b \exp(-\alpha\tau) \cos(\beta\tau)\right] = 0. \tag{3.13b}$$

In order to find the critical stability curves, we choose $\alpha = 0$. Then we have

$$1.5bc \cos(\beta\tau) + ca - de - \beta^2 + 1.5\beta b \sin(\beta\tau) = 0, \tag{3.14a}$$

$$1.5b\beta \cos(\beta\tau) + c\beta + \beta a - 1.5bc \sin(\beta\tau) = 0. \tag{3.14b}$$

Multiplying the above two Eqs. (3.14a) and (3.14b) with c and β, respectively, and adding we obtain

$$1.5b \cos(\beta\tau)(c^2 + \beta^2) + a(c^2 + \beta^2) - cde = 0. \tag{3.15}$$

Now multiplying the Eqs. (3.14a) and (3.14b) with β and c, respectively, and subtracting the resulting equations, we obtain

$$1.5b \sin(\beta\tau)(c^2 + \beta^2) - \beta(c^2 + \beta^2) - \beta de = 0. \tag{3.16}$$

Squaring and adding Eqs. (3.15) and (3.16), and rearranging them, we obtain the following cubic equation for β^2,

$$X^3 + uX^2 + vX + w = 0, \qquad\qquad X = \beta^2, \tag{3.17}$$

where the constants are given as

$$u = a^2 + 2(de + c^2) - 2.25b^2; \tag{3.18a}$$
$$v = (de + c^2)^2 - 2ac(de - ac) - 4.5b^2c^2; \tag{3.18b}$$
$$w = c^2(de - ac)^2 - 2.25b^2c^4. \tag{3.18c}$$

From Eq. (3.15), we obtain

$$\beta\tau = \pm \arccos\left(\frac{cde - a(c^2 + \beta^2)}{1.5b(c^2 + \beta^2)}\right) + 2n\pi, \tag{3.19}$$

where n is any integer $(0, \pm 1, \pm 2, ...)$ and the value of β can be obtained by solving the Eq. (3.17) for β^2. Consequently the stability regions are confined between the set of curves,

$$\tau_1(n, 1.5b) = \frac{2n\pi + \arccos\left(\frac{cde - a(c^2 + \beta^2)}{1.5b(c^2 + \beta^2)}\right)}{\beta} \tag{3.20a}$$

$$\tau_2(n, 1.5b) = \frac{2n\pi - \arccos\left(\frac{cde - a(c^2 + \beta^2)}{1.5b(c^2 + \beta^2)}\right)}{\beta}, \tag{3.20b}$$

where $n = 0, +1, +2, ...$ in Eq. (3.20a) and $n = +1, +2, ...$ in Eq. (3.20b). In order to check whether the region enclosed by the curves τ_1 and τ_2 forms stable islands in the (τ, β) plane, one has to again examine the sign of $\frac{d\alpha}{d\tau}$ on τ_1 and τ_2 as we have done for the case of the single scalar piecewise linear time-delay system in Sect. 3.2.1.2. We have found that there exists only one stable island between the line $\tau = 0$ and $\tau = \tau_1(0, \beta)$ curve (Hopf bifurcation curve) (shaded region) as shown in Fig. 3.13 following the similar discussion in Sect. 3.2.1.2.

3. In the region $|x| \le p_1$, for the fixed point $(x, y) = (0, 0)$, the characteristic equation is

$$(c + \lambda)(\lambda + a + b\exp(-\lambda\tau)) - de = 0,$$

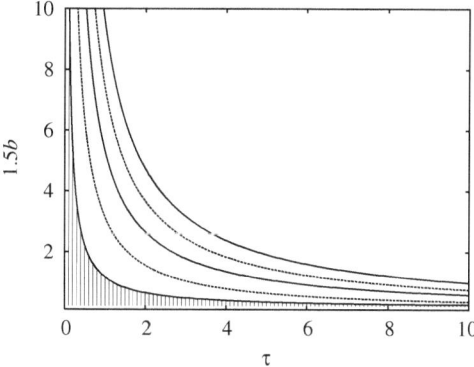

Fig. 3.13 *Curves* of Eqs. (3.20a) and (3.20b) in the range of $-p_2 < x \leq p_1$. The *solid curves* represent τ_1 for $n = 0, +1, +2$ and *broken curves* represent τ_2 for $n = +1, +2$. The *region* enclosed between the *line* $\tau = 0$ and the *curve* $\tau = \tau_1(0, 1.5b)$ (Hopf bifurcation curve) is the only stable island (*shaded region*)

which is again a transcendental equation with infinite number of solutions. Proceeding in the same way as for the case $-p_2 < x \leq -p_1$, we obtain a set of critical curves as

$$\tau_1(n, b) = \frac{2n\pi + \arccos\left(\frac{cde-a(c^2+\beta^2)}{b(c^2+\beta^2)}\right)}{\beta} \qquad (3.21a)$$

$$\tau_2(n, b) = \frac{2n\pi - \arccos\left(\frac{cde-a(c^2+\beta^2)}{b(c^2+\beta^2)}\right)}{\beta}, \qquad (3.21b)$$

where $n = 0, +1, +2, \ldots$ in Eq. (3.21a) and $n = +1, +2, \ldots$ in Eq. (3.21b). As in the previous case we found that there is only one stable region between the line $\tau = 0$ and the Hopf bifurcation curve $\tau = \tau_1(0, \beta)$.

4. For the case $p_1 < x \leq p_2$ and for the fixed point

$$(x, y) = \left(\frac{-2bc}{de - (a + 1.5b)c}, \frac{-2be}{de - (a + 1.5b)c}\right),$$

the characteristic equation turns out to be identical to Eq. (3.12) and the associated critical curves also have similar form as that of Eqs. (3.20a) and (3.20b).

3.6   Numerical Analysis of the Coupled DDE

In this section, we will discuss the dynamics of the set of two coupled time-delay systems defined by Eq. (3.11) through numerical analysis and the transient effects involved in it by studying the bifurcation diagrams for both low and high transients with c as the control parameter. The one parameter bifurcation diagram in one of the parameter regimes has the signature of type III intermittent behavior. We have also found that the dynamics at this transition possesses type III intermittent characteristic scaling behavior. We have also plotted the two parameter bifurcation diagrams in the (τ, c) plane for three different values of p_2, from which also we have pointed out the existence of stable islands of equilibrium points for suitable choice of other parameter values.

3.6.1 Transients

As we have already discussed much about the role of transients in the case of single scalar piecewise linear time-delay system in Sect. 3.3.2, we will not discuss it again in detail for the present case. We point out here only the number of transients required to obtain the bifurcation diagrams which we have shown in Figs. 3.14 and 3.15 for the system (3.11), which enables one to realize the effect of transients.

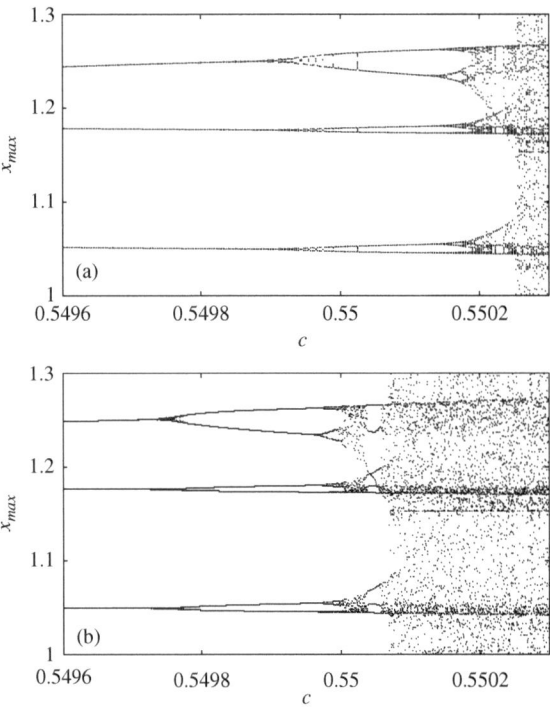

Fig. 3.14 Bifurcation diagrams of the two coupled time-delay system defined by Eq. (3.11). (**a**) For the parameters values $a = 0.16, b = 0.2, d = 0.2, e = 0.2$ and $\tau = 25.0$ for $p_2 = 1.33$ and for transients of the order 2.0×10^5 and (**b**) For the same parameter values as in (**a**) except for the transients of the order 1.75×10^6

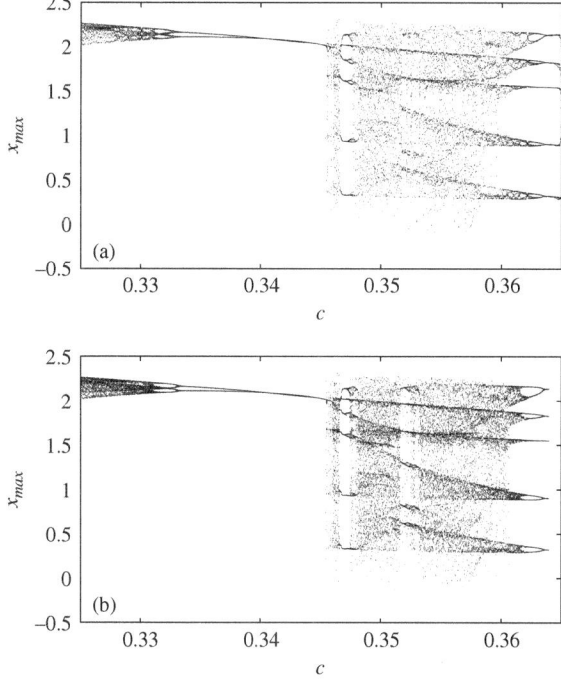

Fig. 3.15 Bifurcation diagrams of the two coupled time-delay system (3.11) (**a**) For the parameters values $a = 0.16, b = 0.2, d = 0.2, e = 0.2$ and $\tau = 25.0$ when $p_2 = 1.66$ for transients of the order 2.0×10^5 and (**b**) For the same parameter values as in (**a**) except for transients of the order 2.0×10^6

As discussed in the case of single scalar piecewise linear time-delay system, the transients predominate the evolution of the system (3.11) to its steady state solution. Figure 3.14a shows the one parameter bifurcation diagram for transients of the order 2.0×10^5 for the parameter value $p_2 = 1.33$, whereas Figure 3.14b shows the one parameter bifurcation diagram for transients of order 1.75×10^6 for the same parameter values. The stray points in the neighbourhood of bifurcation points are due to the transient effects, which suggests that it requires still more iterations need to be left out. Figure 3.15a shows the one parameter bifurcation diagram for transients of the order 2.0×10^5 for the parameter value $p_2 = 1.66$, whereas Fig. 3.15b shows the bifurcation diagram for transients of the order 2.0×10^6.

3.6.2 One and Two Parameter Bifurcation Diagrams

We have integrated Eq. (3.11) with the parameters $a = 0.16, b = 0.2, d = 0.2$, $e = 0.2$ and c as variable parameter with the initial conditions $x(t) = 0.9$ for $t \in (-25, 0)$ and $y = 0.8$. Figure 3.14b shows the one parameter bifurcation diagram in the nonlinear parameter regime characterized by the function $f(x(t))$

with $p_1 = 0.8$, $p_2 = 4/3$ and $\tau = 25.0$, which shows the period-3 doubling bifurcation route to chaos. Figure 3.15b shows reverse period-doubling in the range $c \in [0.32, 0.345]$, and at the critical value of $c_{crit} = 0.345969$ the system exhibits intermittent transition to chaos followed by reverse period-5 doubling in the range $c \in [0.346, 0.365]$ interspersed by periodic windows for the same parameter values as above except that now $p_2 = 5/3$. At the intermittent transition, the amplitude variation loses its regularity and a burst appears in the regular phase as shown in Fig. 3.16. This behavior repeats as time increases as observed in the usual type-III intermittent scenario. The duration of laminar phases is random during the transition and finally results in chaotic oscillations which is obtained by increasing the value of the control parameter c. The plot of the mean laminar length $< l >$ as a function of the parameter $f = (c_{crit} - c)$ is shown in Fig. 3.17, where c_{crit} is the critical value of the parameter for the occurrence of the intermittent transition. The phase space trajectories reveal a power law relationship of the form $< l > = f^{-\alpha}$ with the estimated value of $\alpha = 0.871$. This analysis confirms that the trajectories at the critical value of c is associated with standard intermittent dynamics of type-III

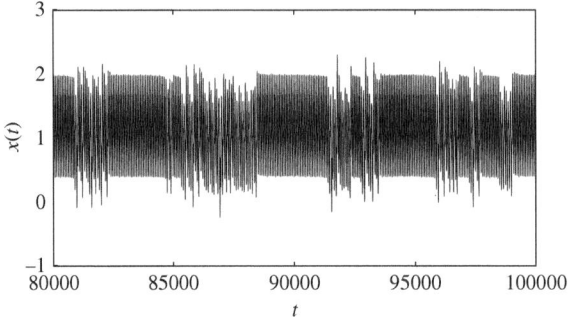

Fig. 3.16 Intermittent behavior at the parameter values $a = 0.16$, $b = 0.2$, $d = 0.2$, $e = 0.2$ and $\tau = 25.0$ when $p_2 = 1.66$ for a critical value of the parameter $c_{crit} = 0.345969$

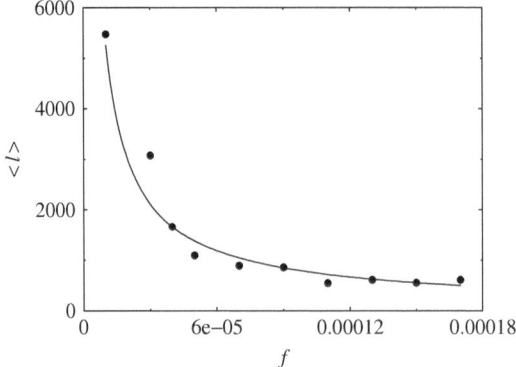

Fig. 3.17 Mean laminar length $< l >$ versus $f = c_{crit} - c$

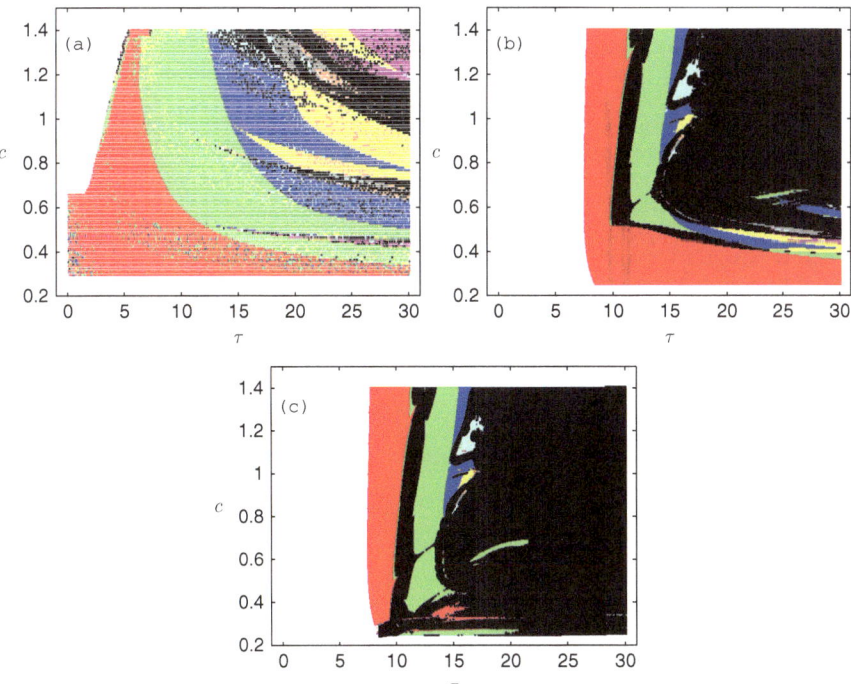

Fig. 3.18 Global bifurcation diagrams of coupled time-delay system (3.11) for $\tau \in (0.1, 30)$ and $c \in (0.3, 1.4)$. The *colour codes* are the same as in Fig. 3.7. (**a**) $p_2 = 1.0$, (**b**) $p_2 = 1.33$ and (**c**) $p_2 = 1.66$

described in the work of Pomeau and Manneville [13] and Schuster [14]. Figure 3.18 shows the global bifurcation diagrams when $p_2 = 1.0, 4/3$ and $5/3$, respectively. The colour codes used here are the same as in the previous case of single scalar piecewise linear time-delay system. By comparing the global bifurcation diagrams, we can realize that the stable island (fixed point) and the chaotic region increase in size for small changes (increase) in the parameter p_2. In addition the chaotic nature of the system increases with delay.

More detailed analysis of coupled time-delay systems, including piecewise linear systems, for their chaotic dynamics and synchronization behavior are investigated in the following chapters.

References

1. M. Lakshmanan, S. Rajasekar, *Nonlinear Dynamics: Integrability, Chaos and Patterns* (Springer, Berlin, 2003)
2. J. Guckenheimer, P. Holmes, *Nonlinear Oscillations, Dynamical System and Bifurcation of Vector Fields* (Springer, Berlin, 1983)
3. H. Lu, Z. He, IEEE Trans. Circuits Syst. I **43**, 700 (1996)

4. P. Thangavel, K. Murali, M. Lakshmanan, Int. J. Bifurcat. Chaos **8**, 2481 (1998)
5. D.V. Senthilkumar, M. Lakshmanan, Int. J. Bifurcat. Chaos **15**, 2895 (2005)
6. K.H. Becker, M. Dorfler, *Dynamical Systems and Fractals* (Cambridge University Press, Cambridge, 1989)
7. D.V. Ramana Reddy, A. Sen, G.L. Johnston, Phys. Rev. Lett. **85**, 3381 (2000)
8. M.G. Eral, S.H. Strogatz, Phys. Rev. E **67**, 036204 (2003)
9. S.A. Campbell, S. Ruan, J. Wei, Int. J. Bifurcat. Chaos **9**, 1585 (1999)
10. J.D. Farmer, Physica D **4**, 366 (1982)
11. W. Horbelt, J. Timmer, H.U. Voss, Phys. Lett. A **299**, 513 (2002)
12. H.U. Voss, Int. J. Bifurcat. Chaos **12**, 1619 (2002)
13. Y. Pomeau, P. Manneville, J. Wei, Physica D **1**, 219 (1980)
14. H.G. Schuster, *Deterministic Chaos* (Springer, Weinheim, 1988)

Chapter 4
A Few Other Interesting Chaotic Delay Differential Equations

4.1 Introduction

One of the well known properties of DDEs is that their effective dimensions increase with the delay time τ [1, 2], see Sect. 1.2.2. This allows one to select different values (sufficiently large) for the delay time τ to generate high-dimensional chaotic signals. Hence, in recent times DDEs have received increased attention in the nonlinear dynamics literature due to the possibility of generating more complex and high-dimensional chaotic attractors and also because of the feasibility of their experimental realization. Therefore, several chaotic time-delay systems and their variants have been proposed during the past few years for generating and enhancing complexity of chaotic behavior in various technological and engineering applications. In this chapter, we will briefly review the dynamical properties of some of the most important first order scalar nonlinear time-delay systems, that have been widely used in the literature, exhibiting chaotic/hyperchaotic behaviors. In addition, we will also present some of the interesting coupled (higher order) delay differential equations in different areas of science and technology.

4.2 The Mackey-Glass System: A Typical Nonlinear DDE

In order to gain further insight and clear understanding of the general features of the dynamics of nonlinear DDEs, in this section, we will consider another scalar first order nonlinear DDE, namely the Mackey-Glass equation, which has been well studied in the literature in connection with chaotic dynamics and applications. As noted in the previous chapter, it is quite intricate to obtain even the stability criteria for the fixed point solution of a scalar nonlinear DDE itself. Now we will consider the Mackey-Glass system as another example to discuss the dynamics of nonlinear DDEs. Other examples are discussed in the following sections.

4.2.1 Mackey-Glass Time-Delay System

The Mackey-Glass time-delay system [3, 4] has been a well studied nonlinear DDE equation in the literature for its chaotic dynamics [1, 5–10]. It has received a central

M. Lakshmanan, D.V. Senthilkumar, *Dynamics of Nonlinear Time-Delay Systems*,
Springer Series in Synergetics, DOI 10.1007/978-3-642-14938-2_4,
© Springer-Verlag Berlin Heidelberg 2010

importance in recent studies on synchronization in view of its hyperchaotic behavior [11–16]. Analog version of the Mackey-Glass system has also been realized experimentally using electronic circuits [14–17]. The Mackey-Glass system, which was originally deduced as a model for blood production in patients with leukemia, can be represented by the first order nonlinear DDE

$$\dot{x} = -bx(t) + \frac{ax(t-\tau)}{(1.0 + x(t-\tau)^c)}, \qquad (4.1)$$

where a, b and c are positive constants. Here, $x(t)$ represents the concentration of blood at time t (density of mature cells in bloodstreams), when it is produced, $x(t-\tau)$ is the concentration when the "request" for more blood is made and τ is the time-delay between the production of immature cells in the bone marrow and their maturation for release in circulating bloodstreams. In patients with leukemia, the time τ may become excessively large, and the concentration of blood will oscillate, or if τ is even larger, the concentration can vary chaotically, as demonstrated by Mackey and Glass [1, 3]. This model is often used as a prototype model in the literature for nonlinear delay systems exhibiting chaotic attractors and even hyperchaotic attractors for large values of delay time.

4.2.2 Fixed Points and Linear Stability Analysis

A great deal of information can be obtained about the dynamical behavior of the scalar first order Mackey-Glass DDE (4.1) by performing a linear stability analysis [18] of equilibrium states. As pointed out in Chap. 3, the difficulty lies in obtaining the conditions under which the real part of the eigenvalues of the linearized equation for the given fixed point is less than zero from the resulting transcendental characteristic equation.

The steady state or equilibrium state (fixed point) of the system (4.1) is one for which $x(t) = x(t-\tau) = x(0) = x^* \; \forall \, t$ and as a consequence all the time derivatives vanish identically. Hence substituting $x(t) = x(t - \tau)$ and $\dot{x} = 0$ in (4.1), it is easy to see that Eq. (4.1) has two fixed points,

$$x = x^* = 0, \qquad (4.2a)$$

$$x = x^* = \left(\frac{a-b}{b}\right)^{\frac{1}{c}}. \qquad (4.2b)$$

Note that the second solution exists only for $a > b$ (x is real). The stability of the fixed points is determined by examining how a small perturbation about the fixed point behaves in time for all $t > 0$. In particular, as pointed out in Chap. 3, Sect. 3.2.1, to study the linear stability we assume that the perturbations about the fixed point grows as $\rho \exp \lambda t$, where λ is in general a complex number to be determined and $\rho \ll 1$ is the amplitude of the perturbation, Then the characteristic equation turns out to be

$$\lambda = -b + \left(\frac{a(1 + x^*) - acx^*}{(1 + x^*)^2}\right) \exp(-\lambda\tau). \tag{4.3}$$

Solving the above equation for λ for either of the fixed points $x^* = 0$ or $x^* = \left(\frac{a-b}{b}\right)^{\frac{1}{c}}$ gives the criterion for the stability of the corresponding fixed points. In the following we analyze the stability of the above two fixed points given by Eqs. (4.2) both in the absence and presence of time-delay τ.

4.2.3 Time-Delay $\tau = 0$

Now, we will discuss the stability of the fixed points (4.2) in the absence of time-delay, that is $\tau = 0$.

(a) Fixed point $x^* = 0$:
From Eq. (4.3), one finds the characteristic equation for the fixed point $x = x^* = 0$ to be

$$\lambda + b - a = 0, \tag{4.4}$$

and the fixed point is stable for $a < b$.

(b) Fixed point $x^* = \left(\frac{a-b}{b}\right)^{\frac{1}{c}}$:
The characteristic equation for the second fixed point $x = x^* = \left(\frac{a-b}{b}\right)^{\frac{1}{c}}$ can be written as

$$\lambda + b - b(1 - ac + bc) = 0, \tag{4.5}$$

and the fixed point is stable if $a > b$.

Thus for $a < b$ we have $x = x^* = 0$ as the only locally stable fixed point. For $a > b$ the second fixed point, $x^* = \left(\frac{a-b}{b}\right)^{\frac{1}{c}}$, is created while the first one, $x^* = 0$, becomes unstable. Obviously $a = b$ is a bifurcation point.

4.2.4 Time-Delay $\tau > 0$

In this section, we will make use of the approach advocated in Chaps. 2 and 3 in order to (i) determine the parametric conditions due to which change in the stability of a given fixed point occurs as specified by the transcendental equation, (ii) to demarcate the asymptotically stable regions of the fixed points in the parameter space of the system and (iii) to determine the conditions on the parameters under which the given fixed point of the system is stable.

4.2.4.1 Geometric Approach

Now let us apply the geometric approach proposed by McDonald [19] to determine the stability from the general equation (4.3) as discussed in Sect. 2.3 of Chap. 2. As pointed out earlier, a change in stability of a fixed point can occur only when a root $\lambda = \alpha + i\beta$ of Eq. (4.3) crosses the imaginary axis, that is, $\lambda = i\beta$, where β is real, is a solution of Eq. (4.3). In particular let us consider the transcendental equation

$$\lambda + b - ae^{-\lambda\tau} = 0, \tag{4.6}$$

corresponding to the fixed $x^* = 0$ for $\tau > 0$. Now, if a substitution $\lambda = i\beta$ is made in the above equation, then it takes the following form

$$\frac{b + i\beta}{a} = \exp(-i\beta\tau). \tag{4.7}$$

If the steady state in question is stable in the absence of delay, that is for $\tau = 0$, an instability can occur for $\tau \neq 0$ only if there are some real β and τ for which the Eq. (4.7) holds good. This can be determined by the simple geometric construction discussed in Sect. 2.3 of Chap. 2.

Considering the right hand side of Eq. (4.7), we can easily see that it traces a unit circle as shown in Fig. 4.1 and change in stability of the fixed point is indicated by the intersection of the ratio curve, namely the left hand side of Eq. (4.7). This is examined by changing the value of a in the Eq. (4.6), whose value determines the stability of the fixed point $x^* = 0$. As a typical example, for the values of the parameters $\tau = 25.0$, $b = 0.1$ and $a = 0.09$, it turns out that $a < b$ and the ratio curve (dotted line) does not intersect the unit circle as shown in the Fig. 4.1a. Thus the fixed point $x^* = 0$ remains stable. As the value of a is increased to $a = 0.1$ so that $a = b$, one finds the ratio curve (dotted line) just touches the unit circle (Fig. 4.1b) indicating the boundary layer between the stability of the two fixed points. For $a = 0.11$, when $a > b$, the ratio curve intersects the unit circle as shown in Fig. 4.1c indicating the change in stability of the first fixed point $x^* = 0$, which becomes unstable for the latter value of a.

Similarly, using the above geometric construction, one can also determine the stability of the second fixed point $x^* = \left(\frac{a-b}{b}\right)^{\frac{1}{c}}$ from its characteristic (transcendental) equation

$$\lambda + b - b(1 - ac + bc)e^{-\lambda\tau} = 0, \tau > 0. \tag{4.8}$$

Again substituting $\lambda = i\beta$ and rewriting the above transcendental equation in the form of Eq. (4.7), one can obtain equations corresponding to unit circle and ratio curve as discussed above. As pointed out in the Sect. 4.2.3, $a > b$ (which is the condition for the instability of the first fixed point when $\tau = 0$) is also the condition for the second fixed point to become stable. Correspondingly, the results from the geometric construction reveals that for $\tau > 0$ also the second fixed point attains

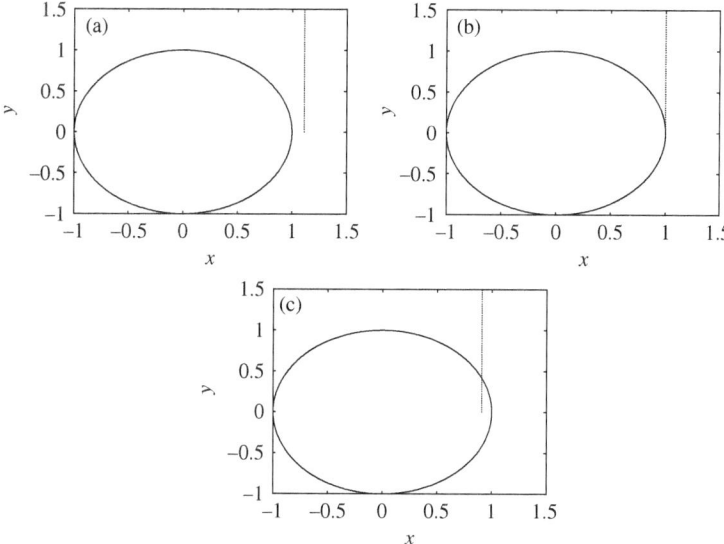

Fig. 4.1 Real and imaginary parts of Eq. (4.7) depicting the change in stability for $\tau = 25.0$ (**a**) For $b = 0.1$, and $a = 0.09$, (**b**) For $b = 0.1$ and $a = 0.1$, and (**c**) For $b = 0.1$ and $a = 0.11$

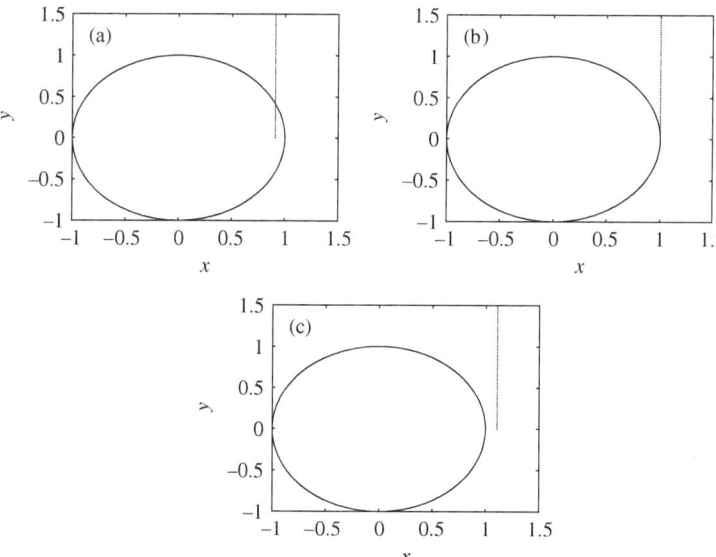

Fig. 4.2 Real and imaginary parts of the characteristic equation corresponding to the second fixed point Eq. (4.2b) at $\lambda = i\beta$ depicting the change in stability for $\tau = 25.0$ (**a**) For $b = 0.1$, and $a = 0.09$, (**b**) For $b = 0.1$ and $a = 0.1$, and (**c**) For $b = 0.1$ and $a = 0.11$

stability when $a > b$ and this is indicated by the intersection of the ratio curve with the unit circle as shown in Fig. 4.2. Here the values of the parameters a, b are kept fixed as in Fig. 4.1 and the value of $c = 10$.

4.2.4.2 General Approach to Determine Stability

Using the general approach discussed in Sect. 2.4 of Chap. 2, we will demonstrate the identification of stable regimes in the (a, τ) parameter space of the fixed point $x^* = 0$ of the Mackey-Glass system (4.1).

Let $\lambda = \alpha + i\beta$ be a root of the transcendental equation (4.6), where α and β are real. Substituting this into the Eq. (4.6) and equating real and imaginary parts, we obtain equations for α and β as

$$\alpha + b - ae^{-\alpha\tau} \cos(\beta\tau) = 0, \tag{4.9}$$

$$\beta + ae^{-\alpha\tau} \sin(\beta\tau) = 0. \tag{4.10}$$

Squaring and adding Eqs. (4.9) and (4.10), we get

$$\beta = \beta_\pm = \pm\sqrt{a^2 \exp(-2\alpha\tau) - (b + \alpha)^2}, \tag{4.11}$$

and

$$\alpha = -b - \frac{\beta}{\tan(\beta\tau)}. \tag{4.12}$$

Without loss of generality we choose the + sign in the above equation, as the eigenvalues occur in complex conjugate pairs. The change in stability occurs only when the value of α crosses the imaginary axis, $\lambda = i\beta, \beta > 0$, and hence the critical stability curve is the one at which α takes the value $\alpha = 0$. Now, we have

$$\beta|_{\alpha=0} = \pm\sqrt{a^2 - b^2}. \tag{4.13}$$

From Eq. (4.9), it follows that

$$\beta\tau = \pm \arccos\left(\frac{b}{a}\right) + 2n\pi, \tag{4.14}$$

where n is any integer $(0, \pm 1, \pm 2, \cdots)$. Consequently one can expect that the stability regions are confined between the set of two curves (for $\tau > 0$),

$$\tau_1(n) = \frac{2n\pi + \arccos\left(\frac{b}{a}\right)}{\sqrt{a^2 - b^2}}, \qquad n = 0, 1, 2, \cdots \tag{4.15a}$$

$$\tau_2(n) = \frac{2n\pi - \arccos\left(\frac{b}{a}\right)}{\sqrt{a^2 - b^2}}. \qquad n = 1, 2, 3, \cdots \tag{4.15b}$$

It is to be noted that from Eq. (4.6) when the time-delay $\tau = 0$, $\lambda = -b + a$ and so the real part of λ, $\alpha < 0$ for $b > a$. In order to identify those curves for $\tau > 0$ which encompass the stable regions, the critical curves should be the ones on which $\frac{d\lambda}{d\tau} > 0$. From Eq. (4.6), we have

$$\frac{d\lambda}{d\tau} = \frac{-a\lambda \exp(-\lambda\tau)}{1 + a\tau \exp(-\lambda\tau),} \tag{4.16}$$

and hence

$$\left.\frac{d\alpha}{d\tau}\right|_{\alpha=0} = \frac{\beta^2}{[1 + a\tau \cos(\beta\tau)]^2 + [a\tau \sin(\beta\tau)]^2}. \tag{4.17}$$

Therefore

$$\frac{d\alpha}{d\tau} > 0 \text{ on both } \tau_1 \text{ and } \tau_2. \tag{4.18}$$

The above condition implies that there can be only one stable region (where $\alpha < 0$) between the $\tau = 0$ line and the critical curve $\tau_1(0)$ which is the closest to the line $\tau = 0$ as discussed in Sect. 2.4 and also in Sect. 3.2.1 for the piecewise linear DDE. We note that the condition (4.18) prohibits the existence of any other stable region (that is multistability region) because for a second stable region to exist one requires $\frac{d\alpha}{d\tau} < 0$ on any one of the other curves ($n > 0$). But this never occurs in this case. The numerical plot in Fig. 4.3 of the curves $\tau_1(n)$ (solid curve for $n = 0, 1, 2$) and $\tau_2(n)$ (broken curve for $n = 1, 2$) reveals that the region between $\tau = 0$ and $\tau = \tau_1(0)$ is the only stable region (shaded region), where $\frac{d\alpha}{d\tau} > 0$ on τ_1, which passes from negative to positive values of α. Obviously the other curves $\tau_2(n) < \tau < \tau_1(n)$ for $n > 0$ do not satisfy the required stability condition and hence they are all associated with unstable regions.

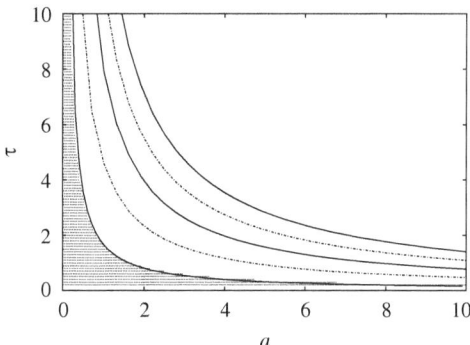

Fig. 4.3 *Curves* of Eqs. (4.15a) and (4.15b). The *solid curves* represent τ_1 for $n = 0, +1, +2$ and *broken curves* represent τ_2 for $n = +1, +2$. The *region* enclosed between the *line* $\tau = 0$ and the *curve* $\tau = \tau_1(0)$ (*shaded region*) is the only stable island of the first fixed point $x^* = 0$ where $\frac{d\alpha}{d\tau} > 0$ on $\tau_1(0)$

One can also perform similar stability analysis using the above approach to identify the stable regimes of the second fixed point $x^* = \left(\frac{a-b}{b}\right)^{\frac{1}{c}}$. Following the procedure used for the fixed point $x^* = 0$, we obtain the set of critical curves corresponding to the characteristic equation (Eq. 4.8) of the second fixed point as

$$\tau_1(n) = \frac{2n\pi + \arccos\left(\frac{b}{k}\right)}{\sqrt{k^2 - b^2}}, \qquad n = 0, 1, 2, \cdots \qquad (4.19a)$$

$$\tau_2(n) = \frac{2n\pi - \arccos\left(\frac{b}{k}\right)}{\sqrt{k^2 - b^2}}, \qquad n = 1, 2, 3, \cdots \qquad (4.19b)$$

where $k = b(1 - ac + bc)$, along with

$$\left.\frac{d\alpha}{d\tau}\right|_{\alpha=0} = \frac{\beta^2}{[1 + k\tau\cos(\beta\tau)]^2 + [k\tau\sin(\beta\tau)]^2} > 0 \text{ on both } \tau_1 \text{ and } \tau_2. \quad (4.19c)$$

With the same argument as in the previous case of first fixed point, we find that there is only one stable island between the curves $\tau = 0$ and $\tau_1(0)$ as shown in Fig. 4.4.

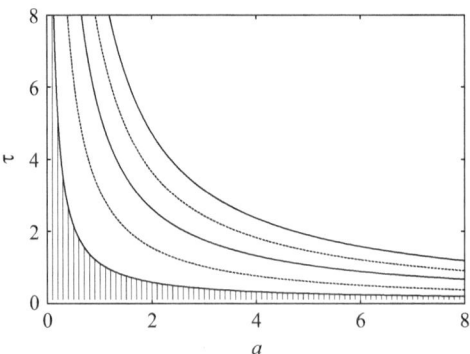

Fig. 4.4 *Curves* of Eqs. (4.19a) and (4.19b). The *solid curves* represent τ_1 for $n = 0, +1, +2$ and *broken curves* represent τ_2 for $n = +1, +2$. The *region* enclosed between the *line* $\tau = 0$ and the *curve* $\tau = \tau_1(0)$ (*shaded region*) is the only stable island of the second fixed point $x^* = \left(\frac{a-b}{b}\right)^{\frac{1}{c}}$ where $\frac{d\alpha}{d\tau} > 0$ on $\tau_1(0)$

4.2.5 Numerical Simulation: Bifurcations and Chaos

In this section, we will briefly discuss the dynamics of the Mackey-Glass time-delay system (4.1) as a function of the delay time τ using numerical simulation. For this purpose, Eq. (4.1) is integrated using Runge-Kutta fourth order integration scheme for the aforesaid parameter values with constant initial condition in

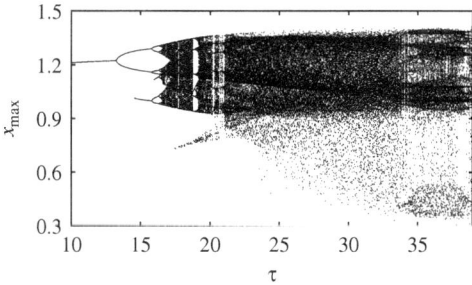

Fig. 4.5 Bifurcation diagram of the Mackey-Glass system (Eq. 4.1) for the parameter values $a = 0.2, b = 0.1, c = 10$ as a function of time-delay $\tau \in (10, 39)$

the range $(-\tau, 0)$ leaving out sufficiently large transients (of the order 1×10^6). For details on the effect of transients, see Chap. 3. Fixing the other parameters as $a = 0.2, b = 0.1$ and $c = 10$, one finds that there is a stable fixed point for $\tau < 4.53$ as we have discussed in the previous section. For $4.53 < \tau < 13.3$, numerical simulations show that there is a stable limit cycle attractor. At $\tau = 13.3$, the period of this limit cycle doubles, initiating a period doubling bifurcation sequence (as shown in Fig. 4.5) that reaches its accumulation point at $\tau = 16.8$. For $\tau > 16.8$ numerical simulations show chaotic attractors at most parameter values, with some limit cycles interspersed in between.

To depict the qualitative nature of the attractors of the Mackey-Glass system, we display a representative portion of the time series in Fig. 4.6 and the corresponding (pseudo) phase plots in Fig. 4.7 by plotting $x(t)$ against $x(t - \tau)$. It is to be noted

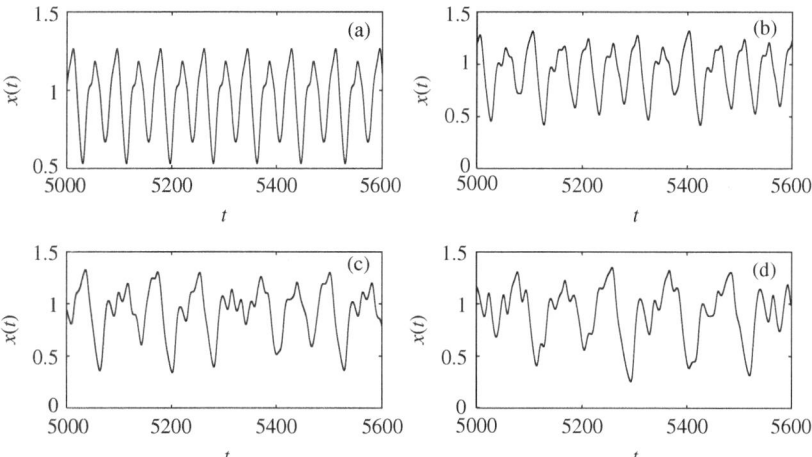

Fig. 4.6 Time series plots of the Mackey-Glass system for the parameter values $a = 0.2$, $b = 0.1, c = 10$ for different values of the delay time τ. (**a**) periodic time trajectory for $\tau = 14$, (**b**) chaotic trajectory at the "onset" for $\tau = 17$, (**c**) chaotic time series for $\tau = 23$ and (**d**) hyperchaotic trajectory for the time-delay $\tau = 32$

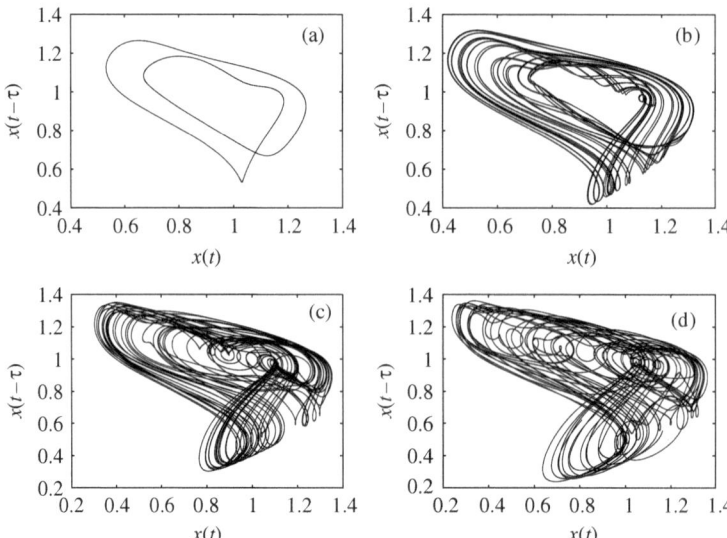

Fig. 4.7 Phase plots obtained by plotting $x(t)$ Vs $x(t - \tau)$ corresponding to the Fig. 4.6

that the choice of $x(t - \tau)$ as the second phase variable is arbitrary; $x(t - t')$ can be equivalently used, where t' is an arbitrary time-delay. A periodic time series is shown in Fig. 4.6a for the delay time $\tau = 14$ and its corresponding phase plot is shown in Fig. 4.7a. Chaotic trajectory at the "onset" of chaos is shown in Fig. 4.6b for the value of delay time $\tau = 17$ along with its phase plot in Fig. 4.7b. Chaotic time series for the value of delay time $\tau = 23$ is shown in Fig. 4.6c with phase plot for the same value of delay in Fig. 4.7c, whereas hyperchaotic trajectory and phase plots are shown in Figs. 4.6d and 4.7d, respectively, for $\tau = 32$.

It may be noted that the above plots (time series and phase plots) are adequate to distinguish periodic behavior from chaotic behavior, but are inadequate to make a sharp distinction between the properties of qualitatively different chaotic behavior and this requires a computation of the corresponding Lyapunov exponents. The first four maximal Lyapunov exponents for the same parameter values as above in the range of delay time $\tau \in (14, 39)$ is shown in Fig. 4.8a. The Lyapunov exponents are evaluated using the procedure suggested by J. D. Farmer [1], described briefly in Sect. 1.2.2. It is clear from the Lyapunov exponents that while the attractors in Fig. 4.7b, c are chaotic, the attractor of Fig. 4.7d is hyperchaotic. The Kaplan-Yorke [1, 20] dimension of the Mackey-Glass system as a function of delay time τ in the same range, obtained by using the formula given in Eq. (3.10), is shown in Fig. 4.8b.

4.2.6 Experimental Realization Using Electronic Circuit

Analog and intrinsic electronic circuits have been employed to mimic the dynamics of Mackey-Glass system [14–17]. Losson et al. [21] have made an attempt in

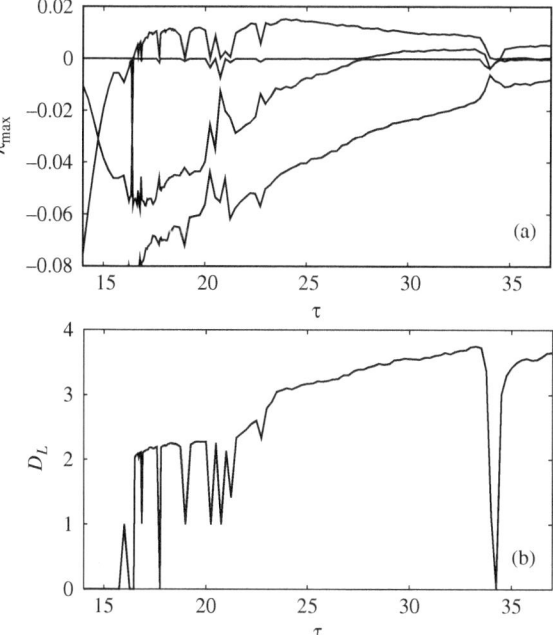

Fig. 4.8 (**a**) Four largest Lyapunov exponents for the same value of parameters as in Fig. 4.6 for the Mackey-Glass system in the range of delay time $\tau \in (14, 39)$ and (**b**) The corresponding Kaplan-Yorke dimension

1993 using analog devices to simulate a delay differential equation; however the nonlinearity used is a piecewise constant model in contrast to the nonlinearity in Mackey-Glass equation, which is a smooth hump shaped nonlinearity as shown in the numerical simulation of the Mackey-Glass nonlinearity in Fig. 4.9a. An electronic analog of the Mackey-Glass system was designed by Namajunas et al. [17] in 1995 and since then this model has been used widely to explore the dynamical and application aspects of the Mackey-Glass system [14–16, 22].

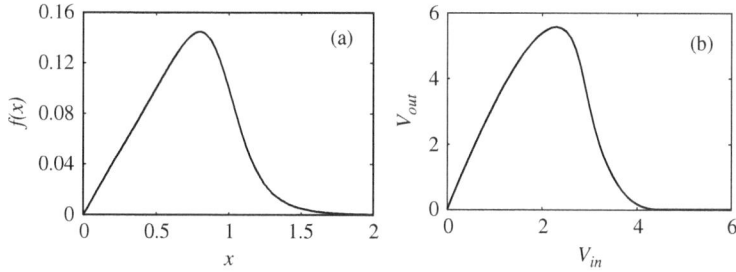

Fig. 4.9 Nonlinearity of the Mackey-Glass system (4.1). (**a**) Plot of the function $f(x) = 0.2x/(1 + x^{10})$ obtained by numerical simulation and (**b**) PSPICE simulation of V_{in} Vs V_{out} characteristics of the nonlinear device (Fig. 4.10)

4.2.6.1 Circuit Realization

The block diagram of electronic analog of the Mackey-Glass time-delay system demonstrated in [17] is the same as in Fig. 3.9 shown in Chap. 3. A description of this block diagram is already provided in Sect. 3.4 except for the nonlinearity ND corresponding to the Mackey-Glass system.

A smooth single hump nonlinearity is produced by coupling two complementary junction field-effect transistors (JFETs) [17]. The circuit diagram of nonlinearity is shown in Fig. 4.10. The output voltage from the resistor r is amplified by an operational amplifier (OA) to obtain a sufficient output signal. The output characteristic obtained by PSPICE simulation is shown in Fig. 4.9b. The parameters of the RC filter in Fig. 3.9 are chosen as $R_0 = 3$ KOhm, $C_0 = 100$ nF, and inductance, capacitance and resistance of the delay unit (Fig. 3.10) as $L = 4.7$ mH, $C = 10$ nF (while another possible combination of L and C used widely in the literature is $L = 9.5$ mH, $C = 525$ nF) and $R = 190$ Ohm, respectively. Junction field-effect transistors used in the nonlinear device (Fig. 4.10) are $Q_1 : 2N5457$, $Q_2 : 2N5460$ and $r = 470$ Ohm. The PSPICE simulation of the electronic circuit, Fig. 3.9, along with the nonlinearity shown in Fig. 4.10 of the Mackey-Glass system, is shown in Fig. 4.11, which is a quite typical attractor of the Mackey-Glass system (4.1).

Similar electronic circuits have also been considered by Dmitriyev et al. [23]. However they have used a standard delay unit with fixed delay of 64 μs, while RC value is used as a control parameter.

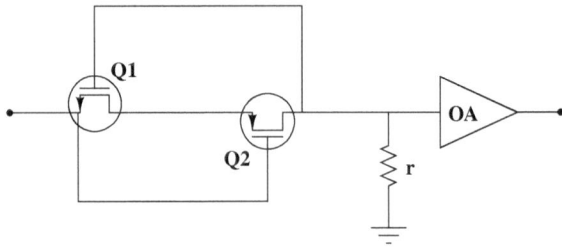

Fig. 4.10 Circuit diagram of the nonlinear device (ND) in Fig. 3.9

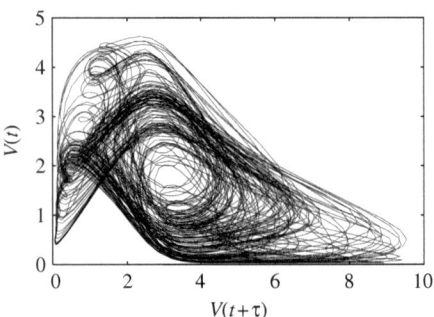

Fig. 4.11 Phase portrait of the analog Mackey-Glass electronic circuit, Fig. 3.9, obtained by PSPICE simulation of the circuit

4.3 Other Interesting Scalar Chaotic Time-Delay Systems

Several simple scalar nonlinear delay systems described by autonomous delay differential equations with suitable nonlinearity have been proposed in the literature in recent times to represent physical and biological systems in view of their potential applications in true random bit generators, global optimization of networks and secure communications, etc. In addition to the piecewise linear and the Mackey-Glass time-delay systems discussed earlier, there are a number of other important scalar chaotic time-delay systems which are being discussed in the literature in the context of chaotic dynamics and chaos synchronization. In this section, we briefly present details of some of them.

4.3.1 A Simple Chaotic Delay Differential Equation

In recent times several attempts have been made to identify simple models that are capable of generating highly complex dynamics for various technological applications. As a consequence, a number of models of delay differential equations and their variants have been proposed and demonstrated in the literature both experimentally and theoretically [24–28] for generating multiple scroll chaotic/hyperchaotic attractors .

In this connection, very recently a simple chaotic delay differential equation has been proposed by J.C. Sprott [26], which can exhibit multiple scroll chaotic/hyperchaotic attractors even for low value of delay time, say $\tau \approx 5$. The delay differential equation is of the form

$$\dot{x} = \sin x(t - \tau). \tag{4.20}$$

Details of stability and bifurcation analysis, and various dynamical behaviors, have been reported in [26]. A 6-scroll hyperchaotic attractor for the value of delay time $\tau = 5.1$ is depicted in Fig. 4.12, while the twelve-largest Lyapunov exponents and the Kaplan-Yorke dimension defined as in Eq. (3.10) in the range of delay time $\tau \in (0, 20)$ are shown in Fig. 4.13a, b, respectively.

4.3.2 Ikeda Time-Delay System

The Ikeda system was introduced to describe the dynamics of an optical bistable resonator and it was shown that the transmitted light from a ring cavity containing a nonlinear dielectric medium undergoes transition from a stationary state to periodic and nonperiodic states, when the intensity of the incident light is increased. It has also been shown that the nonperiodic state is characterized by a chaotic variation of the light intensity and associated broadband noise in the power spectrum [29]. The Ikeda system is well known for delay induced chaotic behavior [30–32] and it is

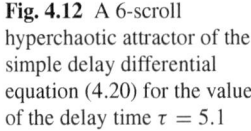

Fig. 4.12 A 6-scroll
hyperchaotic attractor of the
simple delay differential
equation (4.20) for the value
of the delay time $\tau = 5.1$

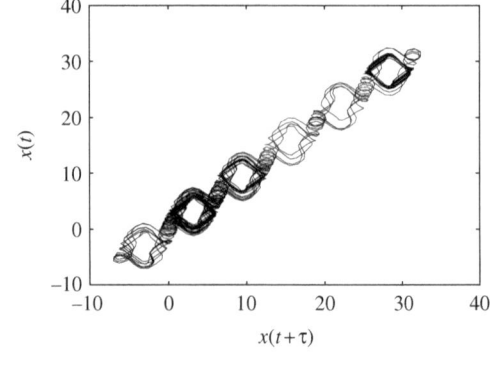

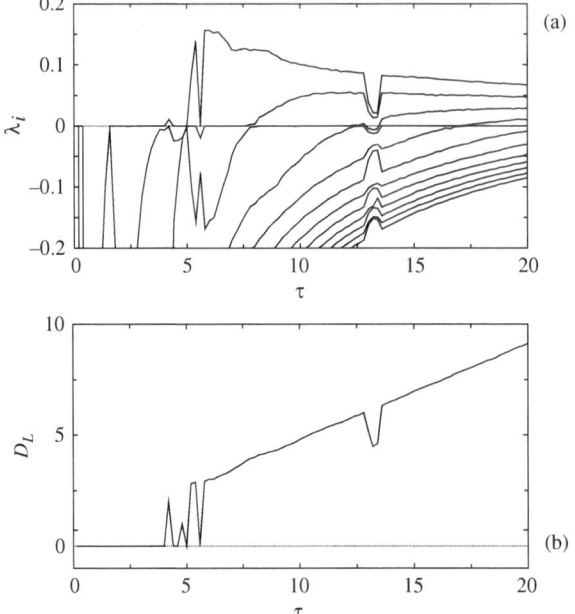

Fig. 4.13 (**a**) Largest twelve Lyapunov exponents and (**b**) Kaplan-Yorke dimension of the simple
delay differential equation (4.20) in the range of delay time $\tau \in (0, 20)$

also receiving focus on synchronization studies in recent times [33–36]. The model
is specified by the state equation

$$\dot{x} = -\alpha x(t) - \beta \sin x(t - \tau), \tag{4.21}$$

where $\alpha > 0$ and $\beta > 0$ are the parameters and τ is the delay time. Physically
$x(t)$ is the phase lag of the electric field across the resonator and thus may clearly
assume both positive and negative values, α is the relaxation coefficient, β is the
laser intensity injected into the system and τ is the round-trip time of the light in the

Fig. 4.14 Chaotic attractor of
the Ikeda system (4.21) for
the values of the parameters
$\alpha = 1.0, \beta = 20$ and $\tau = 2$

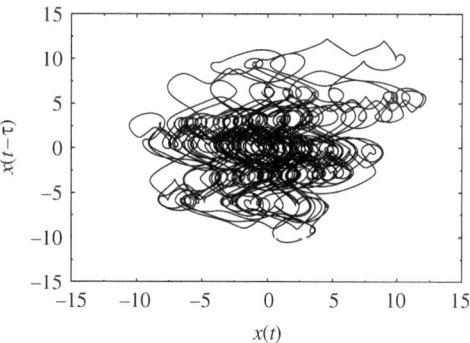

resonator. A typical hyperchaotic attractor of the Ikeda system is shown in Fig. 4.14
for the parameter values $\alpha = 1.0, \beta = 20$ and $\tau = 2$. This system also plays an
important role in electronics and physiological studies [33, 35, 37]. Dynamics and
associated complexity of the Ikeda model have been discussed in detail in [37]. In
particular the dynamics of Ikeda ring cavity laser is explained using an experimental
set up and its block diagram along with the emergence of chaos through period-
doubling cascade [30–32, 37]. The first eleven maximal Lyapunov exponents of
the Ikeda system for the parameters $\alpha = 1.0, \beta = 5$ in the range of delay time
$\tau \in (2, 25)$ are shown in Fig. 4.15a and the corresponding Kaplan-Yorke Lyapunov
dimension (4.15b) is shown in Fig. 4.15b.

4.3.3 Scalar Time-Delay System with Polynomial Nonlinearity

Recently, Voss [38] introduced a scalar time-delay system with polynomial nonlinearity experimentally in an electronic circuit with the state equation

$$\dot{x} = -\alpha x(t) - f(x(t - \tau)), \tag{4.22}$$

where

$$f(x(t - \tau)) = -10.44x_\tau^3 - 13.95x_\tau^2 - 3.63x_\tau - 0.85, \ \ x_\tau = x(t - \tau). \tag{4.23}$$

The model has been studied both experimentally and numerically for its chaotic
dynamics [38] and real time anticipation of chaotic states has been demonstrated
using this circuit. The parameters are chosen such as to closely fit the parameters of
the electronic circuit described in [38]. It is also noted that the nonlinearity (4.23)
shows a single smooth hump and resembles that of the Mackey-Glass system. The
parameter values are fixed as $\alpha = 3.24$ m/s and the delay time $\tau = 13.28$ ms
for numerical simulation and for these values the system (4.22) exhibits chaotic
dynamics as shown in Fig. 4.16.

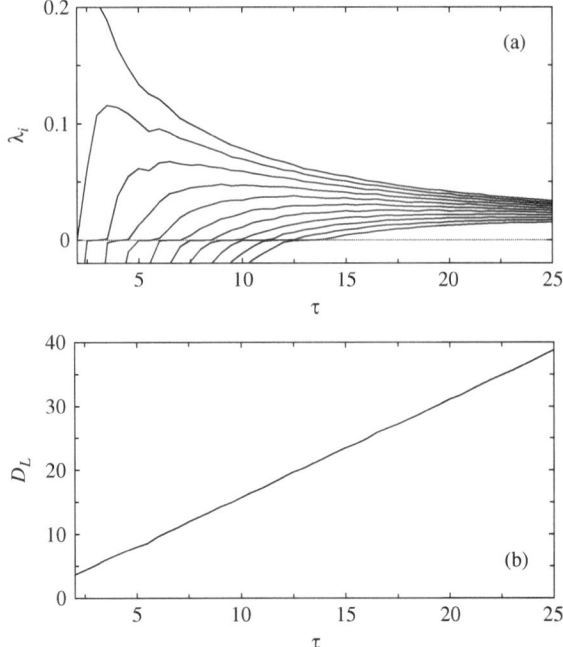

Fig. 4.15 (**a**) Largest eleven Lyapunov exponents of Ikeda system (4.21) for the values of parameters $\alpha = 1.0$, $\beta = 5$ in the range of delay time $\tau \in (2, 25)$ and (**b**) The corresponding Kaplan-Yorke dimension of the Ikeda system

Fig. 4.16 Chaotic attractor of the system (4.22) for the values of the parameters $\alpha = 3.24\,\mathrm{m/s}$ and $\tau = 13.28\,\mathrm{ms}$

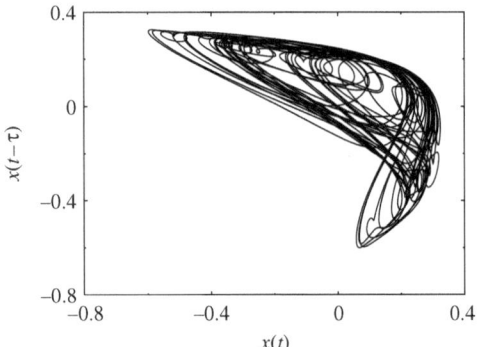

4.3.4 Scalar Time-Delay System with Other Piecewise Linear Nonlinearities

In addition to the odd piecewise linear delay differential equation discussed in the previous chapter (Chap. 3), there are several other types of piecewise linear delay differential equations which have been studied in the literature. A delay differential equation with odd piecewise linear function with only three segments can be defined as

$$\dot{x} = -ax(t) + bf(x(t - \tau)),$$ (4.24)

where $f(x)$ is a three segment odd piecewise function represented as

$$f(x) = \begin{cases} d(x + 1) - c, & x < -1 \\ cx, & -1 < x < 1 \\ d(x - 1) + c, & x > 1, \end{cases}$$ (4.25)

System (4.24) and (4.25) has been analyzed in [24, 25] both experimentally and theoretically for the parameter values $c = 2.0$ and $d = -4.0$. The only difference between the three segment odd piecewise linear function and the five segment odd piecewise linear function (2.19) is that nonlinearity does not saturate to zero at larger values of $|x|$. It has been shown that this nonlinearity can generate not only mono-scroll hyperchaotic oscillations but also more complex double-scroll hyperchaotic oscillations for suitable values of delay times. Numerically simulated three segment piecewise linear functional form is depicted in Fig. 4.17a. Mono- and double-scroll

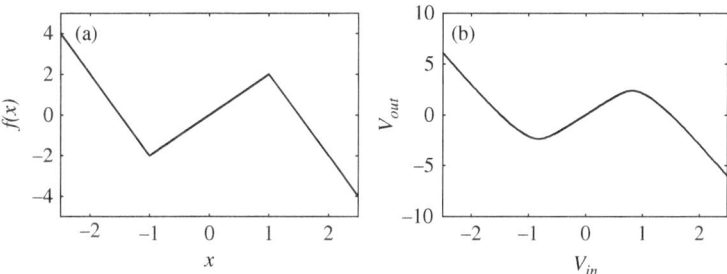

Fig. 4.17 Odd-symmetry nonvanishing three-segment nonlinear function $f(x)$ given by Eq. (4.25) (**a**) Numerical plot and (**b**) PSPICE simulation

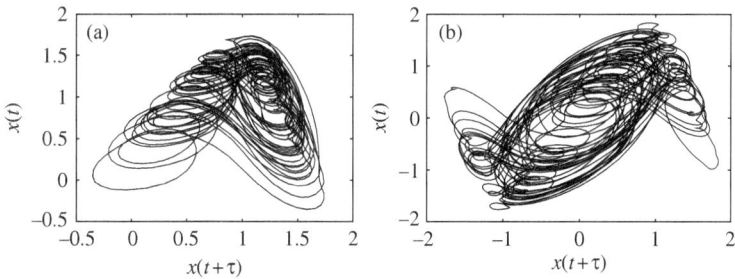

Fig. 4.18 Numerical phase portraits of the systems (4.24) for the values of the parameters $a = 1.0, b = 1.0, c = 2.0$ and $d = -4.0$ (**a**) Mono-scroll attractor for the value of the delay time $\tau = 2.0$ and (**b**) Double-scroll attractor for the value of the delay time $\tau = 8.0$

hyperchaotic attractors for the values of the parameters $a = 1.0, b = 1.0$ and the delay times $\tau = 2.0$ and $\tau = 8.0$ are shown in Fig. 4.18a, b, respectively.

The same block diagram and the delay unit depicted in Figs. 3.9 and 3.10, respectively, can be used to mimic the delay differential equation (4.24) along with the three segment piecewise linear function, Eq. (4.25), by using the nonlinear device (ND) [24, 25] shown in Fig. 4.19. The values of the resistances in the nonlinear device are chosen as $R_1 = R_2 = R_3 = 1$ KOhm and $R_4 = 3$ KOhm. Diodes and operational amplifiers in the nonlinear device are chosen as $1N4148$ and $LF356N$, respectively. PSPICE simulation of the output characteristics of the nonlinear device (ND) shown in Fig. 4.19 is depicted in Fig. 4.17b. Mono- and double-scroll hyperchaotic attractors of the circuit (Fig. 3.9) with the nonlinear device (Fig. 4.19) for the corresponding values of the parameters [24, 25] are shown in Fig. 4.20a, b, respectively. Ten largest Lyapunov exponents and the Kaplan-Yorke dimension of the delay differential equation (4.24) with the three segment piecewise linear function in the range of $\tau \in (0, 10)$ for the above values of other parameters are shown in Fig. 4.21.

Other piecewise linear models discussed in the literature include those models where the function $f(x)$ in Eq. (4.24) is of the form

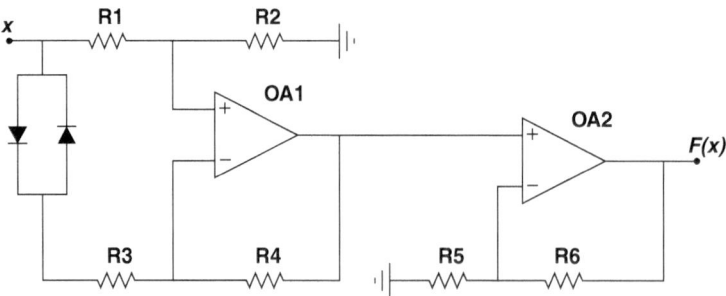

Fig. 4.19 Circuit diagram of the nonlinear device (ND) corresponding to the Eq. (4.25)

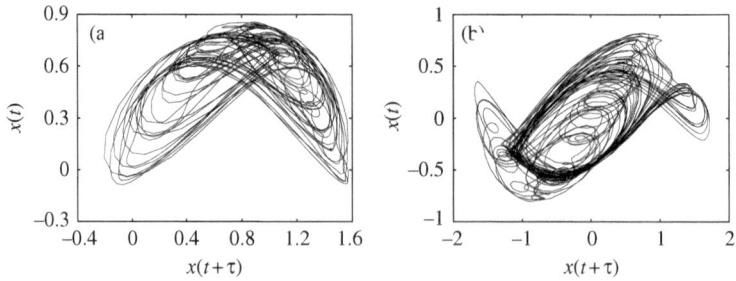

Fig. 4.20 PSPICE simulation of the nonlinear circuit Fig. 4.19 for the corresponding values of the circuit elements (**a**) Mono-scroll attractor and (**b**) Double-scroll attractor

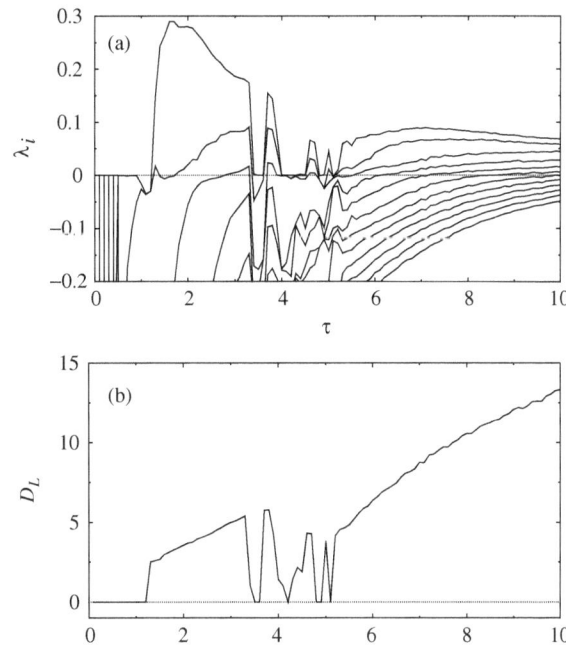

Fig. 4.21 (**a**) Largest twelve Lyapunov exponents and (**b**) Kaplan-Yorke dimension of the piecewise linear delay differential equation (4.24) in the range of delay time $\tau \in (0, 10)$

$$f(x) = \begin{cases} 2x, & x \leq 1/2 \\ 2 - 2x, & x > 1/2, \end{cases} \qquad (4.26)$$

and

$$f(x) = \begin{cases} px(1 - x), & |x| < 1 \\ 0, & |x| \geq 1. \end{cases} \qquad (4.27)$$

These piecewise linear models have been well studied in [11, 21, 39] for their chaotic dynamics and also on synchronization aspects. The chaotic attractor of the system (4.24) with the functional form (4.26) is shown in Fig. 4.22a for the parameter values $\alpha = 0.2$, $\beta = 0.2$ and $\tau = 25$, while the chaotic attractor of the system (4.24) with the piecewise linear function of the form (4.27) for the parameter values $a = 0.2$, $b = 0.2$, $p = 5$ and $\tau = 25$ is shown in Fig. 4.22b.

4.3.5 Another Form of Scalar Time-Delay System

Another form of scalar DDE which has also been used in the literature is

$$\dot{x} = -ax(t - \tau) + bf(x(t - \tau)), \qquad (4.28)$$

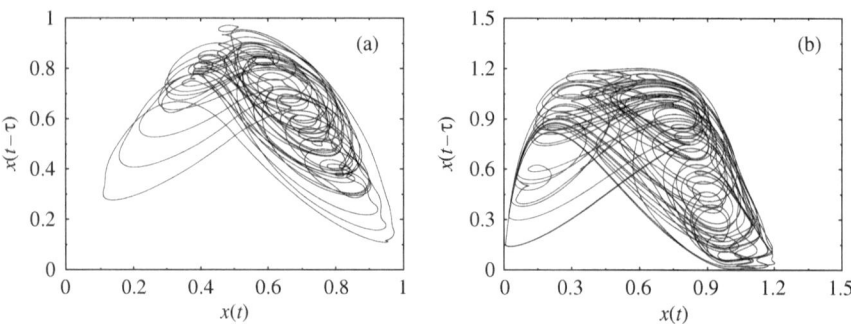

Fig. 4.22 Chaotic attractor of the system (4.24) with (**a**) functional form (4.26) for the values of the parameters $\alpha = 0.2$, $\beta = 0.2$ and $\tau = 25$ and (**b**) piecewise linear function of the form (4.27) for the parameter values $a = 0.2$, $b = 0.2$, $p = 5$ and $\tau = 25$

where a and b are parameters. As a first example to this type of scalar DDE, we will discuss the dynamical system proposed by Yalcin and Ozoguz [28] to generate an n-scroll chaotic attractor. Nonlinearity of this model is based on a hard limiter function and that the model can be easily generalized in a systematic way in order to obtain an n-scroll attractor. The nonlinear function is given by

$$f(x) = \sum_{i=1}^{M} g_{(-2i+1)/2}(x) + \sum_{i=1}^{N} g_{(2i-1)/2}(x) \qquad (4.29)$$

and

$$g_\theta(x) = \begin{cases} 1, & x \geq \theta, \theta > 0 \\ 0, & x < \theta, \theta > 0 \\ 0, & x \geq \theta, \theta < 0 \\ 1, & x < \theta, \theta < 0. \end{cases} \qquad (4.30)$$

The nonlinearity (4.29) was used for the generalization of the so called Jerk circuit [40]. The set of equilibrium points for the nonlinearity (4.29) with $a = b$ is

$$\Theta_{eq} = \{-M, -M+1, \cdots, N-1, N\}. \qquad (4.31)$$

The scrolls are located around the equilibrium points. Therefore the number of scrolls equals the number of equilibrium points. The number of scrolls generated by the nonlinearity (4.29) is equal to $M + N + 1$. Chaotic attractors with three ($M = 1, N = 1$), five ($M = 2, N = 2$), six ($M = 1, N = 4$) and nine ($M = 4, N = 4$) scrolls for the system (4.28) with the nonlinearity (4.29) for the values of the parameters $a = 0.2, b = 0.2$ and $\tau = 20$ are shown in Fig. 4.23a–d, respectively. The above dynamical system (4.28) along with the nonlinearity (4.29) has also been demonstrated using electronic circuits by the same authors.

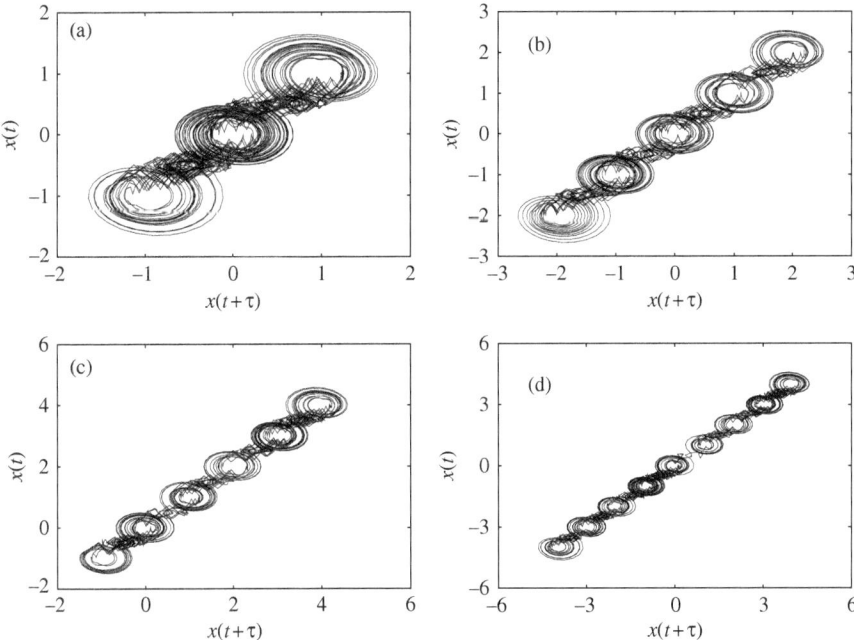

Fig. 4.23 Chaotic attractors of the system (4.28) with nonlinear functional form (4.29) for the values of the parameters $a = 0.2, b = 0.2$ and $\tau = 10$. (**a**) Three-scroll chaotic attractor for $M = N = 1$, (**b**) Five-scroll chaotic attractor for $M = N = 2$, (**c**) Six-scroll chaotic attractor for $M = 1, N = 4$ and (**d**) Nine-scroll chaotic attractor for $M = 4, N = 4$

There are also other types of nonlinear functions used in the literature for the delay differential equation (4.28) with [41]

$$f(x) = sgn(x), \tag{4.32}$$

and

$$f(x) = tanh(10x). \tag{4.33}$$

The chaotic attractors exhibited by the above nonlinearities for the values of the parameters $a = 0.2, b = 0.2$ and $\tau = 10$ are depicted in Fig. 4.24a, b, respectively.

Another interesting model with a cubic nonlinear function [26] is,

$$\dot{x} = x(t - \tau) - x^3(t - \tau), \tag{4.34}$$

for which chaos sets in from a limit cycle at $\tau \approx 1.538$. Similar dynamical behavior is also obtained when the signs of both the terms in Eq (4.34) are exchanged. Note that the right hand side of the above equation can be considered as a scaled version of the first two terms in the Taylor series for $sin(x(t - \tau))$.

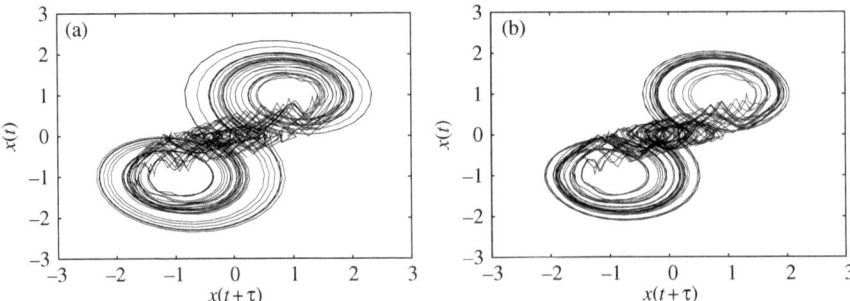

Fig. 4.24 Chaotic attractors of the system (4.28) for the values of the parameters $a = 0.2$, $b = 0.2$ and $\tau = 10$. (**a**) with the nonlinear functional form (4.32) and (**b**) with nonlinear functional form (4.33)

Fig. 4.25 Limit cycle oscillation of the delayed action oscillator (4.35) for the value of the parameters $a = 0.7$, $b = 1$ and delay time $\tau = 2$

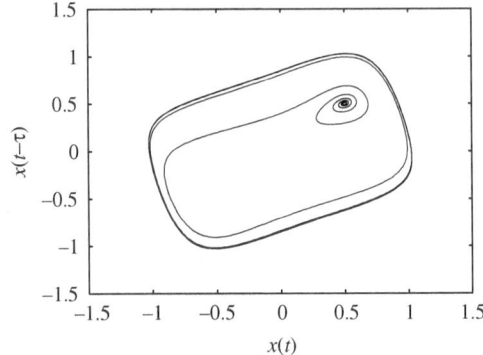

4.3.6 El Niño and the Delayed Action Oscillator

A natural atmospheric phenomenon that attracts regular public attention is the El Niño event in the equatorial Pacific. El Niño is an excellent example of the inter-action between the ocean and the atmosphere and their combined effect on climate. This phenomenon is a disruption of the ocean-atmosphere system having important consequences for weather around the globe. It results in a redistribution of rains with flooding and droughts. Along the equator, the western Pacific has some of the world's warmest ocean water, while in the eastern Pacific, cool water dwells up, carrying nutrients that support large fish populations. In every two to seven years, strong westward blowing trade wind subsides and warm water slowly moves back eastward across the Pacific, which interrupts the upwelling of cool and nutrient rich water. As a consequence fish die and climatic changes affect many parts of the world.

Historically, the term "El Niño" was used by the fisherman along the west coast of Peru and Ecuador to refer to a warm southward flowing current that moderates the low sea-surface temperature that typically appears around Christmas-time and therefore named as "El Niño" (the Little Boy or Christ Child in Spanish), which lasts several months. In certain years this current is unusually strong, bringing heavy rains and flooding inland, but also decimating fishing stocks, bird populations, and the other water based wildlife in what would normally be an abundant part of the Pacific.

Today, the term El Niño is most often used when describing a far large-scale warming that can be observed across the whole of the Pacific Ocean by certain characteristic climatic conditions. The effects of El Niño can also be seen globally. Examples include spring rainfall levels in Central Europe, flooding in East Africa, and the ferocity of the hurricane season in the Gulf of Mexico [42]. El Niño is just one phase of the El Niño Southern Oscillation (ENSO) phenomenon, that is, an irregular cycle of coupled ocean temperature and atmospheric pressure oscillations across the whole equatorial Pacific.

El Niño Southern Oscillation phenomenon is a global event arising from large-scale interaction between the ocean and the atmosphere. The Southern Oscillation refers to oscillations in the surface pressure (atmospheric mass) between the south-eastern tropical Pacific and the Australian-Indonesian regions. When the waters of the eastern Pacific are abnormally warm (an El Niño) sea level pressure drops in the eastern Pacific and rises in the west. The reduction in the pressure gradient is accompanied by a weakening of the low-latitude easterly trades.

Usually, coupled atmosphere-ocean general circulation models combined with large scale computing resources are needed to study and predict El Niño's consequences. Fortunately, a route toward qualitative and quantitative prediction based on the *delayed-action oscillator* exists. Recent works on the mechanism of the ENSO have led various simple delay oscillator models which have provided a quite satisfactory explanation for the onset, termination, and cyclic nature of ENSO events, see [43] and references therein. Very recently, a simple model described by a scalar delay differential equation has been shown to mimic most of the observed dynamics of ENSO phenomenon. It models the irregular fluctuations of the sea-surface temperature, and incorporates the full coupled Navier-Stokes dynamics of an El Niño event by a suitably chosen nonlinearity. The corresponding equation can be represented as [42]

$$\dot{x} = kx - bx^3 - ax(t - \tau),\qquad(4.35)$$

where k, b, a and τ are constants. The first term on the right-hand side of Eq. (4.35) represents a strong positive feedback within the ocean atmosphere system and the second term is an unspecified nonlinear net damping term that is present to limit the growth of unstable perturbations. The strength of the returning emerging signals relative to that of the local nondelayed feedback is denoted by a.

The delayed oscillator (4.35), as it is, exhibits only limit cycle oscillations (Fig. 4.25). The full model to include every aspect of El Niño event is represented as

$$\dot{x} = kx - bx^3 - ax(t - \tau) + y' + \beta + R(t),\qquad(4.36)$$

where $y'(t)$ is the annual cycle which plays an important role in determining the onset of El Niño, β corresponds to global warming which influences the periodicity and amplitude of El Niño events, and $R(t)$ is a stochastic term which influences the irregularity of El Niño events. The full model (4.36) represented by the delayed-action oscillator now mimics the observed richness and variance of the El Niño Southern oscillations phenomenon. Thus the oscillator modeled by delayed feedback serves as an excellent model for observed physical phenomenon. Details of stability analysis and its dynamics can be found in [42].

In addition to the above example of describing a natural phenomenon in terms of delay differential equations/models, recent investigations have also shown that the Neolithic transition (that is, transition from hunter-gatherer to agricultural economy) in Europe has also been successfully described using a time-delay model, which agrees quite well with the observations based on archaeological data. These data led to the conclusion that European farming originated in the Near East, from where it spread across Europe [44].

4.4 Coupled Chaotic Time-Delay Systems

As pointed out in the previous section, a large number of physical, biological, chemical, ecological and economic systems, as well as fluctuations in agricultural commodity prices, neural networks, etc., are successfully modeled by delay differential equations, and thereby capturing many of the inherent complex dynamics of the respective dynamical systems, cf. [4, 18, 21, 41, 45–60]. DDEs are often found not only as scalar first order equations but also as coupled higher order equations. Some of the delay models of chemical, physiological and economic dynamics, mathematical and genetic regulatory systems are represented as coupled higher order DDEs [59, 60]. Rigorous mathematical treatment of such higher order DDEs becomes much more complicated than the case of scalar DDEs and consequently details of such coupled higher order systems are available only through numerical analysis. In the following, we will discuss briefly about some of the important higher order DDEs that have been widely studied in the current literature in the context of chaotic dynamics and chaos synchronization.

4.4.1 Time-Delayed Chua's Circuit

As an example of coupled higher order DDEs in electronic circuits, we present here one of the important second order time-delay systems, namely, time-delayed Chua's circuit (TDCC) which has been realized experimentally using suitable electronic circuit. This was introduced by Sharkovsky et al. [61] and it has been studied in detail by many researchers [62–64]. In particular, the work of Sharkovsky et al. showed that the dimension of state space is increased by substituting the lumped LC resonator with an ideal (lossless) transmission line. TDCC has a constant voltage generator E in series with the Chua's diode as shown in Fig. 4.26 in order to break

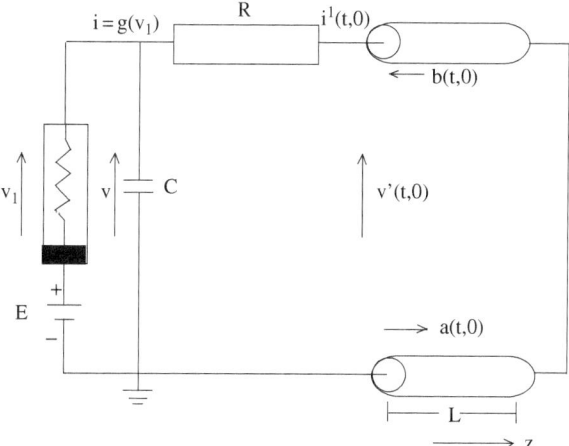

Fig. 4.26 Time-delayed Chua's circuit (TDCC)

the symmetry in the original Chua's diode $v - i$ characteristic. The other elements in the circuit are the lumped linear elements C and $G = 1/R$. The parameters of the circuit are the round trip delay T and the characteristic impedance Z of the transmission line. The line parameters relate to the line per-unit-length inductance l and capacitance c, and they are related to the line length $\mathcal{L}$ as $T = 2\mathcal{L}\sqrt{lc}$ and $Z = \sqrt{l/c}$. The circuit is represented by the coupled normalized equations of the form

$$\theta \dot{x}(t) + \lambda_q(x(t) - E) + q(\lambda_0 - \lambda_q) = (1 - \gamma)(y - 0.5E), \qquad (4.37a)$$

$$y(t + 1) = \gamma y(t) - \tfrac{\gamma+1}{2}x(t), \qquad (4.37b)$$

where $\lambda_q = \frac{(1-\gamma^2)}{2(\gamma-h_q)}$, $q = -1, 0, 1$, and θ, γ, h_0, h_1, h_{-1} are dimensionless parameters corresponding to the experimental counterparts. Chaotic attractor of the time-delayed Chua's circuit (4.37) for the parameter values $\theta = 0.1616$, $\gamma = -0.58156$, $h_0 = 0.35828$, $h_1 = h_{-1} = -4.9073$ and $E = 0.0$ for the initial conditions $y(\tau) = 0$, $\tau \in (0, 1)$ is shown in Fig. 4.27.

4.4.2 Semiconductor Lasers

A large number of systems in laser optics have been represented by coupled higher order delay differential equations that are quite complicated than the systems presented in the above sections. Also a large number of papers have appeared in recent times in studying chaos synchronization using laser systems [65–77] as models, in view of the feasibility of experimental realization of high-dimensional chaos.

Now, we will describe a well studied single mode semiconductor laser with optical feedback in the context of chaotic dynamics and its synchronization. All optical feedback in external cavity semiconductor lasers has been a subject of extensive

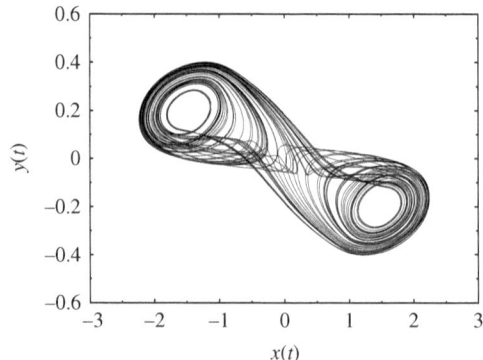

Fig. 4.27 Chaotic attractor of the time-delay Chua's circuit (TDCC) (4.37) for the parameter values $\theta = 0.1616, \gamma = -0.58156,$ $h_0 = 0.35828, h_1 = h_{-1} = -4.9073$ and $E = 0.0$ with initial conditions $y(\tau) = 0, \tau \in (0, 1)$

research during the past decade because of its importance in technical applications such as optical data storage or optical fiber communications [78]. The rate equations for the complex electric fields and the carrier densities in the lasers are the well known Lang-Kobayashi equations [65–67, 71, 74, 75, 79] represented by

$$\dot{E} = \frac{1 + i\alpha}{2}[G(t) - 1/\tau_p]E(t) + \gamma E(t - \tau)\exp[-i(\omega\tau)], \qquad (4.38)$$

$$\dot{N} = J/e - N(t)/\tau_n - G(t)|E(t)|^2, \qquad (4.39)$$

where E is the slowly varying complex field and N is the normalized carrier number. The equations are written in the reference frame where the complex optical fields of the lasers are given by $E \exp(i\omega t)$, where ω is the optical frequency of the solitary laser [66]. The other parameters are as follows: τ_p is the photon lifetime, α is the line width enhancement factor, $G(t) = G_n(N - N_0)/(1 + \varepsilon|E|^2)$ is the optical gain (where G_n is the differential gain, N_0 is the carrier number at transparency, ε is the gain saturation coefficient), $\omega\tau$ is the phase accumulation after one round trip in the external cavity, τ is the one round trip time of the external cavity, J is the injection current, e is the electric charge, and τ_n is the carrier lifetime. The model does not include multiple reflections in the external cavity, and therefore it is valid for weak feedback levels. For suitable values of the above parameters the laser is in the so-called coherence collapse regime, characterized by fast, chaotic intensity fluctuations. Semiconductor lasers subject to delayed feedback generate chaotic dynamics with intensity pulsations on subnanosecond time scales [78].

Chaos in semiconductor lasers arises in different ways: due to (periodic) modulation of the pump current [80], due to electro-optical feedback where the pump current is modulated by the emitted light intensity [81, 82], or due to an external cavity (optical feedback) [83]. Recently, optoelectronic chaos due to delayed feedback in optoelectronic circuits along with their possible applications has been discussed [84]. All these cases are described by DDEs that exhibit chaotic/hyperchaotic behaviors for suitable values of the parameters. In the above we have provided state equations for semiconductor laser with optical feedback as an example for the coupled higher order DDEs in semiconductor laser systems. It is to be noted that the semiconductor laser with polarization-rotated optical feedback is described by a 3rd order DDE containing state equations for two complex electric fields E_1 and E_2 with orthogonal polarizations, and a carrier density N [85].

4.4.3 Neural Networks

A class of delayed chaotic neural networks can be represented as the set of coupled DDEs of the form

$$\dot{x}(t) = -Cx(t) + Af\,[x(t)] + Bf\,[x(t-\tau)],\tag{4.40}$$

where $x(t) = [x_1(t), x_2(t), \cdots, x_n(t)]^T \in R^n$ is the state vector, the activation function $f\,[x(t)] = (f_1\,[x_1(t)], f_2\,[x_2(t)], \cdots, f_n\,[x_n(t)])^T$ denotes the manner in which the neurons respond to each other. C is a positive diagonal matrix, $A = a_{ij}, i, j = 1, 2, \cdots, n$ is the feedback matrix, $B = b_{ij}$ represents the delayed feedback matrix with a constant delay τ. The general class of delayed neural networks represented by the above Eq. (4.40) unifies several well known neural networks such as the Hopfield neural networks and cellular neural networks with delay.

The delayed neural networks (Eq. 4.40) corresponds to the Hopfield neural networks for the choice of activation function

$$f\,[x(t)] = \tanh\,[x(t)],\tag{4.41}$$

and for the value of the matrices

$$C = \begin{bmatrix} 1 & 0 \\ 0 & 1 \end{bmatrix},\ A = \begin{bmatrix} 2.0 & -0.1 \\ -5.0 & 3.0 \end{bmatrix},\ B = \begin{bmatrix} -1.5 & -0.1 \\ -0.2 & -2.5 \end{bmatrix}.$$

The chaotic behavior of the delayed Hopfield neural network [86, 87] for the above choice of the parameters and for the initial condition $x_0 = [0.1, 0.2]^T$ is shown in Fig. 4.28a. The above general class of delayed neural network represents delayed cellular neural network for the activation function

$$f\,[x(t)] = 0.5\,(|x_i + 1| - |x_i - 1|),\tag{4.42}$$

and for the value of the matrices

$$C = \begin{bmatrix} 1 & 0 \\ 0 & 1 \end{bmatrix},\ A = \begin{bmatrix} 1+\pi/4 & 20 \\ 0.1 & 1+\pi/4 \end{bmatrix},\ B = \begin{bmatrix} -\sqrt{2}(\pi/4)1.3 & -0.1 \\ 0.1 & -\sqrt{2}(\pi/4)1.3 \end{bmatrix}.$$

The chaotic attractor of the above delayed cellular neural network for the choice of initial condition $x_0 = [0.1, -0.1]^T$ is shown in Fig. 4.28b.

In the above two sections, we have presented a brief review of some of the important time-delay systems and, in particular, scalar first order time-delay systems with

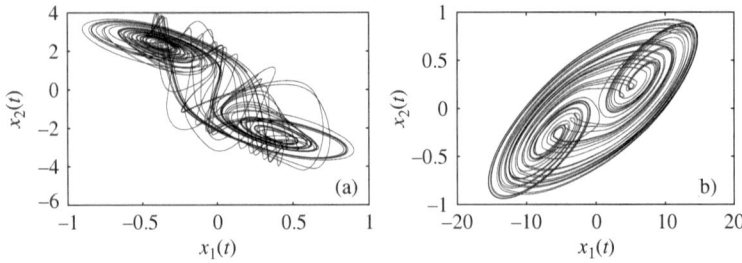

Fig. 4.28 Chaotic behavior of (**a**) Delayed Hopfield neural network and (**b**) delayed cellular neural network

different functional forms for nonlinearity that are widely studied in the literature in the context of chaotic dynamics and chaos synchronization. We have also discussed briefly some of higher order DDEs in different areas. More examples of nonlinear delay systems are given in Appendix D.

References

1. J.D. Farmer, Physica D **4**, 366 (1982)
2. K. Ikeda, M. Matsumoto, J. Stat. Phys. **44**, 955 (1986)
3. M.C. Mackey, L. Glass, Science **197**, 287 (1977)
4. M.C. Mackey, L. Glass, *From Clocks to Chaos, The Rhythms of Life* (Princeton University Press, Princeton, NJ, 1988)
5. W.-H. Kye, M. Choi, M.-W. Kim, S.-Y. Lee, S. Rim, C.-M. Kim, Y.-J. Park, Phys. Lett. A **322**, 338 (2004)
6. M.J. Bunner, M. Popp, Th. Meyer, A. Kittel, J. Parisi, Phys. Rev. E **54**, 3082(R) (1996)
7. C. Zhou, C.H. Lai, Phys. Rev. E **60**, 320 (1999)
8. V.I. Ponomarenko, M.D. Prokhorov, Phys. Rev. E **66**, 026215 (2002)
9. M.D. Prokhorov, V.I. Ponomarenko, Phys. Rev. E **72**, 016210 (2005)
10. W.-H. Kye, M. Choi, S. Rim, M.S. Kurdoglyan, C.-M. Kim, Y.-J. Park, Phys. Rev. E **69**, 055202(R) (2004)
11. I.G. Szendro, J.M. Lopez, Phys. Rev. E **71**, 055203 (2005)
12. K. Pyragas, Phys. Rev. E **58**, 3067 (1998)
13. M. Zhan, X. Wang, X. Gong, G.W. Wei, C.H. Lai, Phys. Rev. E **68**, 036208 (2003)
14. M.-Y. Kim, C. Sramek, A. Uchida, R. Roy, Phys. Rev. E **74**, 016211 (2006)
15. A. Kittle, J. Parisi, K. Pyragas, Physica D **112**, 459 (1998)
16. S. Sano, A. Uchida, S. Yoshimori, R. Roy, Phys. Rev. E **75**, 016207 (2007)
17. A. Namajunas, K. Pyragas, A. Tamasevicius, Phys. Lett. A **201**, 42 (1995)
18. R. Bellman, K.L. Cooke, *Differential-Difference Equations* (Academic Press, New York, 1963)
19. N. McDonald, *Biological Delay Systems: Linear Stability Theory* (Cambridge University Press, Cambridge, 1989)
20. J. Kaplan, J. Yorke, in *Functional Differential Equations and Approximation of Fixed Points*, ed. by H.O. Peitgen, H.O. Walther. Lecture Notes in Mathematics, vol 730 (Springer, Berlin, 1979)
21. J. Losson, M.C. Mackey, A. Longtin, Chaos **3**, 167 (1993)
22. W. Horbelt, J. Timmer, H.U. Voss, Phys. Lett. A **299**, 513 (2002)
23. A.S. Dmitriyev, Yu.N. Orlov, S.O. Starkov, Radiotekh. Elektron **34**, 1980 (1989)

24. A. Tamasevicius, G. Mykolaitis, S. Bumeliene, Electron. Lett. **42**, 13 (2006)
25. A. Tamasevicius, T. Pyragiene, M. Meskauskas, Int. J. Bifurcat. Chaos **17**, 3455 (2007)
26. J.C. Sprott, Phys. Lett. A **366**, 397 (2007)
27. L. Wang, X. Yang, Electron. Lett. **42**, 1439 (2006)
28. M.E. Yalcin, S. Ozoguz, Chaos **17**, 033112 (2007)
29. K. Ikeda, H. Daido, O. Akimoto, Phys. Rev. Lett. **45**, 709 (1980)
30. K. Ikeda, K. Matsumoto, Physica D **29**, 223 (1987)
31. C. Masoller, D.H. Zanette, Physica A **300**, 359 (2001)
32. M. Le Berre, E. Ressayre, A. Tallet, H.M. Gibbs, D.L. Kaplan, M.H. Rose, Phys. Rev. A **35**, 3020 (1987)
33. H.U. Voss, Phys. Rev. E **61**, 5115 (2000)
34. E.M. Shahverdiev, S. Sivaprakasam, K.A. Shore, Phys. Rev. E **66**, 017204 (2002)
35. E.M. Shahverdiev, S. Sivaprakasam, K.A. Shore, Phys. Lett. A **292**, 320 (2002)
36. E.M. Shahverdiev, K.A. Shore, Phys. Rev. E **71**, 016201 (2005)
37. L. Larger, J.P. Goedgebuer, V. Udaltsov, C. R. Physique **5**, 669 (2004)
38. H.U. Voss, Int. J. Bifurcat. Chaos **12**, 1619 (2002)
39. Y.C. Tian, F. Gao, Physica D **108**, 113 (1997)
40. M.E. Yalcin, S. Ozoguz, J.A.K. Suykens, J. Vandewalle, Int. J. Bifurcat. Chaos **12**, 23 (2002)
41. R.D. Driver, *Ordinary and Delay Differential Equations* (Springer, New York, 1977)
42. I. Boutle, R.H.S. Taylor, R.A. Romer, Am. J. Phys. **75**, 15 (2007)
43. E. Tziperman, M.A. Cane, S.E. Zebiak, J. Atmos. Sci. **52**, 293 (1995)
44. J. Fort, V. Mendez, Phys. Rev. Lett. **82**, 867 (1999)
45. J.K. Hale, *Theory of Functional Differential Equations* (Springer, New York, 1977)
46. D.V. Ramana Reddy, Ph. D. Thesis entitled *Collective Dynamics of Delay Coupled Oscillators* (Institute for Plasma Research, Gandhinagar, India, 2000)
47. I. Gyori, P. Ladas, *Oscillation Theory of Delay Differential Equations: With Applications* (Clarendon Press, Oxford, 1991)
48. R. Ross, *The Prevention of Malaria* (John Murray, London, 1911)
49. P.K. Asea, P.J. Zak, J. Econ. Dyn. Control **23**, 1155 (1999)
50. M. Kalecki, Econometrica **3**, 327 (1935)
51. Y. Kuang, *Delay Differential Equations with Applications in Population Dynamics* (Academic Press, Boston, MA, 1993)
52. V. Volterra, *Lecons sur la theorie mathematique de la lutte pour la vieynamics* (Gauthiers-Villars, Paris, 1931)
53. N. MacDonald, *Time Lags in Biological Models*. Lectuer Notes in Biomathematics (Springer, Berlin, 1978)
54. R.M. May, *Stability and Complexity in Model Ecosystems* (Princeton University Press, Princeton, NJ, 1974)
55. K. Gopalsamy, *Stability and Oscillations in Delay Differential Equations of Population Dynamics* (Kluwer Academic Publishers, Dordrecht, 1992)
56. A.D. Drozdov, V.B. Kolmanovskii, *Stability in Viscoelasiticity* (North-Holland, Amsterdam, 1994)
57. J.M. Cushing, *Integrodifferential Equations and Delay Models in Population Dynamics*. Lectuer Notes in Biomathematics, vol. 20 (Springer, Berlin, 1977)
58. J.K. Hale, S.M. Verduyn Lunel, *Introduction to Functional Differential Equations* (Springer, New York, 1993)
59. K. Sriram, Ph. D. thesis on *Modelling Nonlinear Dynamics of Chemical and Circadian Rhythms Using Delay Differential Equations* (Department of Chemistry, IIT, Madras, India, 2004)
60. Alwyn Scott (ed.), *Encyclopedia of Nonlinear Science* (Routledge, New York, 2005)
61. A.N. Sharkovsky, Y. Maisternko, P. Deregel, L.O. Chua, J. Circuits Syst. Comput. **3**, 645 (1993)
62. M. Biey, B.F. Gilli, I. Maio, IEEE Trans. Circuits Syst. I **44**, 486 (1997)

63. M. Gilli, G.M. Maggio, IEEE Trans. Circuits Syst. I **43**, 827 (1996)
64. P. Thangavel, K. Murali, D.V. Senthilkumar, M. Lakshmanan, Int. J. Bifurcat. Chaos (Submitted)
65. C. Masoller, Phys. Rev. Lett. **86**, 2782 (2001)
66. A. Locquet, C. Masoller, C.R. Mirasso, Phys. Rev. E **65**, 056205 (2002)
67. E.M. Shahverdiev, S. Sivaprakasam, K.A. Shore, Phys. Rev. E **66**, 037202 (2002)
68. S. Sivaprakasam, E.M. Shahverdiev, K.A. Shore, Phys. Rev. Lett. **87**, 154101 (2001)
69. S. Sivaprakasam, I. Pierce, P. Rees, P.S. Spencer, K.A. Shore, Phys. Rev. A **64**, 013805 (2001)
70. S. Tang, J.M. Liu, Phys. Rev. Lett. **90**, 194101 (2003)
71. I. Wedekind, U. Parlitz, Phys. Rev. E **66**, 026218 (2002)
72. L.B. Shaw, I.B. Schwartz, E.A. Rogers, R. Roy, Chaos **16**, 015111 (2006)
73. A. Murakami, Phys. Rev. E **65**, 056617 (2002)
74. E.M. Shahverdiev, S. Sivaprakasam, K.A. Shore, Phys. Rev. E **66**, 017206 (2002)
75. J. Ohtsubo, IEEE J. Quantum Electron. **38**, 1141 (2002)
76. L. Wu, S. Zhu, Phys. Lett. A **315**, 101 (2003)
77. A.S. Landsman, I.B. Schwartz, Phys. Rev. E **75**, 026201 (2007)
78. I. Fischer, G.H.M. van Tartwijk, A.M. Levine, W. Elsaesser, E. Goebel, D. Lenstra, Phys. Rev. Lett. **76**, 220 (1996)
79. R. Lang, K. Kobayashi, IEEE J. Quantum Electron. **QE-16**, 347 (1980)
80. J. Sacher, D. Baums, P. Panknin, W. Elsaesser, E.O. Goebel, Phys. Rev. A **45**, 1893 (1992)
81. C.-H. Lee, S.-Y. Shin, Appl. Phys. Lett. **62**, 922 (1993)
82. S.I. Turovets, J. Dellunde, K.A. Shore, J. Opt. Soc. Am. B **14**, 200 (1997)
83. J. Mork, B. Tromborg, J. Mark, IEEE J. Quantum Electron. **28**, 93 (1992)
84. L. Larger, J.M. Dudley, Nature **465**, 41 (2010)
85. D.W. Sukow, A. Gavrielides, T. McLachlan, G. Burner, J. Amonette, J. Miller, Phys. Rev. A **74**, 023812 (2006)
86. J. Meng, X. Wang, Chaos **17**, 023113 (2007)
87. H. Zhu, B. Cui, Chaos **17**, 043122 (2007)

Chapter 5
Implications of Delay Feedback: Amplitude Death and Other Effects

5.1 Introduction

For a long time, in the study of coupled nonlinear oscillators, the role of delay has often been neglected as unimportant. In many cases this approximation is physically justified and in all the cases it simplifies the mathematics. However, in recent times one has witnessed increased activities to investigate oscillator systems with delay feedback and it has been proved that delay feedback is a veritable black box which can give rise to several interesting and novel phenomena having wide applications, and these cannot be observed in the absence of delay feedback. In this chapter, we will discuss an important time-delay induced phenomenon, namely amplitude death, which has been the center of attraction in recent research on coupled oscillators with delay feedback. In addition, we will also point out some of the other important time-delay induced phenomena observed in coupled oscillators.

5.2 Time-Delay Induced Amplitude Death

The phenomenon of suppression of oscillations (*amplitude death*, see Sec. 5.2.1.1 below for more details) was realized by Bar-Eli [1] in his modeling of chemical oscillations and by Shiino and Frankowicz [2] when they considered the effects on the amplitudes in a large number of coupled limit cycles. But a rigorous and comprehensive study on the nature of amplitudes of two coupled oscillators was made by Aronson et al. [3] followed by other elegant studies extending this to a large system [4–6]. However, the concept of amplitude death has now become an active area of research due to the recent works of Ramana Reddy et al. [7–10] on the effect of time-delay feedback in limit cycle oscillators. They have shown particularly that amplitude death can occur even with zero frequency mismatch among the interacting limit cycle oscillators (identical oscillators) in the presence of time-delay feedback. This is in contrast to the earlier reports on amplitude death where such a state can occur only when there exists a broad dispersion in the natural frequencies of the coupled oscillators and that the coupling strength has to exceed a threshold value so that amplitude death

M. Lakshmanan, D.V. Senthilkumar, *Dynamics of Nonlinear Time-Delay Systems*,
Springer Series in Synergetics, DOI 10.1007/978-3-642-14938-2_5,
© Springer-Verlag Berlin Heidelberg 2010

cannot occur in a collection of identical limit cycle oscillators without delay. In this section, we will discuss the theoretical and experimental investigations of delay induced amplitude death phenomenon in single and two coupled limit cycle oscillators.

5.2.1 Theoretical Study: Single Oscillator

In this section, we will demonstrate the theoretical analysis of existence of delay induced amplitude death phenomenon in single and two coupled limit cycle oscillators and its subsequent extension to N coupled limit cycle oscillators with global and ring coupling configurations.

5.2.1.1 Single Limit Cycle Oscillator with Delay Feedback

Specifically, Ramana Reddy et al. [7–10] have considered a model equation, which represents the dynamics of a single Hopf oscillator that is driven autonomously by a time-delayed feedback term, given by

$$\dot{Z}(t) = (a + i\omega - |Z(t)|^2)Z(t) - KZ(t - \tau), \tag{5.1}$$

where $Z(t) = X + iY$ is a complex quantity, ω is the frequency of oscillation, a is a real constant and $\tau > 0$ is the time-delay of the autonomous feedback term. In the absence of the feedback term (5.1) is often called the Stuart-Landau equation, which exhibits a stable limit cycle of amplitude $\sqrt{a}$ with angular frequency ω. In the absence of time-delay, Eq. (5.1) has a time-asymptotic periodic solution given by $Z(t) = \sqrt{a - K} \exp i\omega t$ for $a > K$. If $a \leq K$, then the origin is the only stable solution; that is, no oscillatory time-asymptotic solution is possible. At $a = K$, the oscillator undergoes a supercritical Hopf bifurcation. Carrying out a linear stability analysis for the above equation around the fixed point $Z = Z^* = 0$, one can straightforwardly obtain the characteristic equation

$$\lambda = a + i\omega - Ke^{-\lambda\tau}, \tau > 0. \tag{5.2}$$

For $\tau = 0$, one can have $\lambda = a + i\omega - K$ and hence the origin is stable for $K > a$. So the critical curve is given in this case by $K = a$. When $\tau \neq 0$, Eq. (5.2) remains a transcendental equation with infinite number of roots. Now, one has to find conditions on K, ω and τ such that real parts of all the roots are negative for stability of the fixed point Z^*. Substituting $\lambda = \alpha + i\beta$, where α and β are real, in Eq. (5.2), one can obtain the real and imaginary parts as

$$\alpha = a - Ke^{-\alpha\tau} \cos(\beta\tau), \tag{5.3}$$

$$\beta = \omega + Ke^{-\alpha\tau} \sin(\beta\tau). \tag{5.4}$$

Following the analysis in Sect. 2.4 in Chap. 2, one can obtain a set of critical curves,

$$\tau_1(n, K) = \frac{2n\pi + \arccos\left(\frac{a}{K}\right)}{\omega + \sqrt{K^2 - a^2}}, \qquad n = 0, +1, +2, \cdots \qquad (5.5a)$$

$$\tau_2(n, K) = \frac{2n\pi - \arccos\left(\frac{a}{K}\right)}{\omega - \sqrt{K^2 - a^2}}, \qquad n = +1, +2, |3, \cdots \qquad (5.5b)$$

Also we get

$$\left.\frac{d\alpha}{d\tau}\right|_{\alpha=0} = \frac{\beta(\beta - \omega)}{[1 - K\tau\cos(\beta\tau)]^2 + [K\tau\sin(\beta\tau)]^2}. \qquad (5.6)$$

Hence, from (5.4) and (5.6), one can easily see that

$$\left.\frac{d\alpha}{d\tau}\right|_{\alpha=0} \begin{cases} > 0 & \text{on} & \tau_1, \\ > 0 \text{ on} & \tau_2 & \text{if} & K > f(\omega), \\ = 0 \text{ on} & \tau_2 & \text{if} & K = f(\omega), \\ < 0 \text{ on} & \tau_2 & \text{if} & K < f(\omega), \end{cases} \qquad (5.7)$$

where $f(\omega) = \sqrt{a^2 + \omega^2}$. The above condition implies that there can be only one stability region if $K > f(\omega)$. On the other hand, there is a possibility of multiple stability regions if $K < f(\omega)$. The numerical plot in Fig. 5.1a of the curves $\tau_1(n, K)$ (represented by continuous curves) and $\tau_2(n, K)$ (represented by discrete curves) reveals that the region between $\tau = 0$ and $\tau = \tau_1(0, K)$ is the only stable region (indicated as the shaded region) possible for small values of ω. However, as the value of ω is increased, so that the condition $K < \sqrt{a^2 + \omega^2}$ is satisfied, the number of stability regions also increases. These regions are specified by $0 \leq \tau < \tau_1(0, K)$ and $\tau_1(n, K) < \tau < \tau_2(n, K)$, where the integer $n > 0$. In Fig. 5.1b the critical

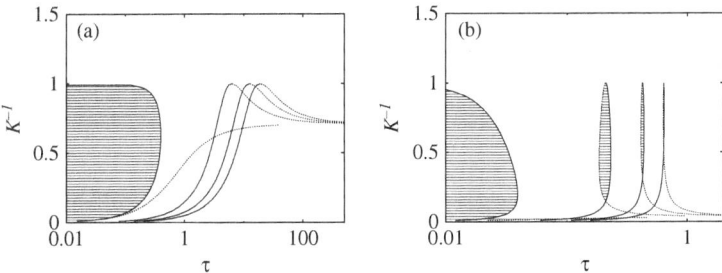

Fig. 5.1 Curves of Eq. (5.5a) and (5.5b). The *solid curves* represent τ_1 for $n = 0, +1, +2, +3$ and *broken curves* represent τ_2 for $n = +1, +2, +3$. (**a**) The *region* enclosed between the *line* $\tau = 0$ and the *curve* $\tau = \tau_1(0, K)$ is the only amplitude death region for $\omega = 1$. (**b**) The *region* enclosed between the *line* $\tau = 0$ and the *curve* $\tau = \tau_1(0, K)$, and that between $\tau_1(n, K)$ and $\tau_2(n, K)$ for $n > 0$ are the multiple amplitude death regions for $\omega = 30$

curves are plotted from Eqs. (5.5a) and (5.5b) for such a large value of ω, namely $\omega = 30$, and the multiple stability regions are represented by shaded regions enclosed between the continuous curves, $\tau_1(n, K)$, and the discrete curves, $\tau_2(n, K)$, for $n > 0$ in addition to that enclosed between the $\tau = 0$ and $\tau = \tau_1(0, K)$ curves. The above collective stability regions have been termed as amplitude death regions or death islands in the $(K - \tau)$ space.

Note that the above analysis shows that when the parameters in (5.1) are such that (i) $K < a$ in the absence of delay ($\tau = 0$), the system has a stable periodic solution $Z = \sqrt{a - K}e^{i\omega t}$, and (ii) in the presence of delay ($\tau \neq 0$), with $K < \sqrt{a^2 + \omega^2}$, this solution becomes unstable, while the fixed point $Z = 0$ becomes stable. This phenomenon is nothing but the amplitude death of the periodic oscillation.

5.2.1.2 Two Delay Coupled Limit Cycle Oscillators

The above analysis has also been extended to a system of two coupled limit cycle oscillators with delay [7–10]. Here, we will briefly describe the existence of delay induced death phenomenon in a set of two delay coupled limit cycle oscillators represented by

$$\dot{Z}_1(t) = (1 + i\omega - |Z_1(t)|^2)Z_1(t) - K[Z_2(t - \tau) - Z_1(t)], \qquad (5.8a)$$
$$\dot{Z}_2(t) = (1 + i\omega - |Z_2(t)|^2)Z_2(t) - K[Z_1(t - \tau) - Z_2(t)]. \qquad (5.8b)$$

The parameters are the same as discussed above for the case of a single limit cycle oscillator with the delay feedback. Following a linear stability analysis, one can obtain the characteristic equation associated with the equilibrium solution $Z_1 = 0 = Z_2$ for the above delay coupled limit oscillators, Eq. (5.8), as

$$\lambda = 1 - K + i\omega \pm Ke^{-\lambda\tau}. \qquad (5.9)$$

After simple algebra, one can obtain a set of critical curves,

$$\tau_1(n, K) = \frac{2n\pi + \arccos\left(1 - \frac{1}{K}\right)}{\omega - \sqrt{2K - 1}}, \qquad n = 0, +1, +2, \cdots \qquad (5.10a)$$

$$\tau_2(n, K) = \frac{2n\pi - \arccos\left(1 - \frac{1}{K}\right)}{\omega + \sqrt{2K - 1}}, \qquad n = +1, +2, +3, \cdots \qquad (5.10b)$$

Also we get

$$\frac{d\alpha}{d\tau}\bigg|_{\alpha=0} = \frac{\beta(\beta - \omega)}{[1 \pm K\tau]^2 + [K\tau\sin(\beta\tau)]^2}. \qquad (5.11)$$

Hence, as in the case of the single oscillator,

$$
\left.\frac{d\alpha}{d\tau}\right|_{\alpha=0}
\begin{cases}
< 0 \text{ on} & \tau_1 & \text{if} & K < f(\omega), \\
> 0 \text{ on} & \tau_1 & \text{if} & K > f(\omega), \\
= 0 \text{ on} & \tau_1 & \text{if} & K = f(\omega), \\
> 0 & & \text{on} & \tau_2
\end{cases}
\tag{5.12}
$$

where $f(\omega) = \sqrt{1+\omega^2}$. In analogy with the arguments as in the case of single limit cycle oscillator with the delay feedback, there exists only a single death island enclosed between the critical curves $\tau_1(0, K)$ and $\tau_2(1, K)$ for small values of ω. However, for large values of ω, one can obtain the higher order islands as demonstrated [7–10] above.

The results have also been extended to the case of a large assembly of delay coupled limit cycle oscillators with global [7, 8, 10] and ring [11] coupling configuration in order to show the general nature of amplitude death phenomenon. The same authors have also provided experimental evidence for the existence of delay induced death phenomenon in two delay coupled limit cycle oscillators using electronic circuits, which will be discussed in detail in the next section.

5.2.2 Experimental Study

Experimental observation of time-delay induced amplitude death in two coupled limit cycle oscillators has also been reported by Ramana Reddy et al. [9]. They have provided experimental evidence for time-delay induced death islands and their multiple connectedness as predicted by their theoretical studies in the parameter space defined by the coupling strength, time-delay and frequency for the case of two coupled identical limit cycle oscillators using a suitable electronic circuit.

A schematic representation of the system of two coupled electronic circuits that are individually capable of exhibiting limit cycle oscillations is shown in Fig. 5.2. Each individual oscillator circuit is a nonlinear LCR circuit, which may be considered as a variant of the well known Chua circuit. Each one of these circuits consists

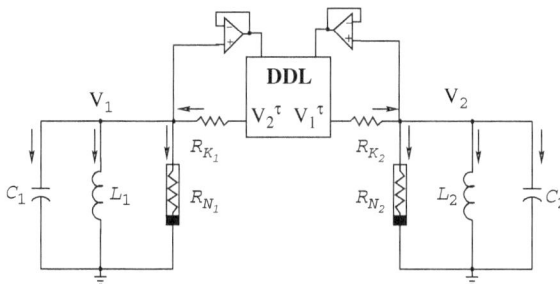

Fig. 5.2 The circuit diagram of two delay coupled nonlinear LCR circuits that are capable of exhibiting limit cycle oscillations [9]

of a capacitance C, an inductance L and a nonlinear resistive element R_N that are coupled through a digital delay line (DDL). The OP-AMPs are buffer amplifiers and the coupling strength $K = 1/(CR_K)$ is varied by changing the resistances R_K that couple the two oscillator circuits. The coupling is linear, resistive, and proportional to the difference in the signal strengths of the two oscillators with a time-delay. Each oscillator is capable of oscillating with a characteristic frequency $\omega = 1/\sqrt{LC}$. The authors have fixed the values of capacitances as $C_1 = C_2 = 0.1 \, \mu F$ and the values of the inductances $L = L_1 = L_2$ and the resistances $R = R_{K1} = R_{K2}$ are varied

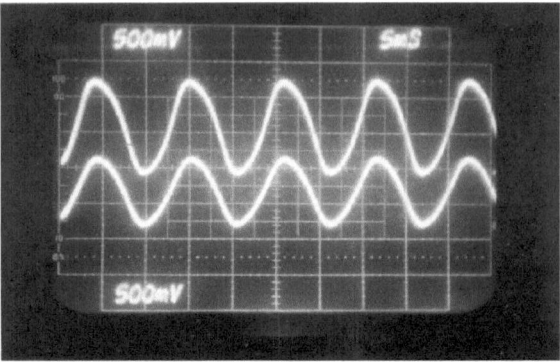

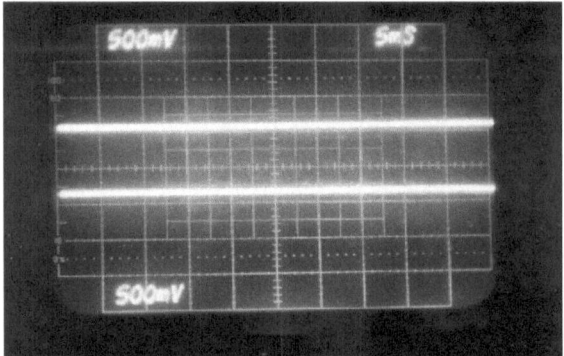

Fig. 5.3 Experimental results of two delay coupled limit cycle oscillators for different values of delay times with $K = 1,000/s$ and $\omega = 837$ s. The *top panel* shows the in-phase oscillations for $\tau = 0.514$ ms, the middle panel shows amplitude death for $\tau = 2$ ms and the *bottom panel* displays anti-phase oscillations for $\tau = 4.428$ ms. This figure is adapted from the work of Ramana Reddy et al. [9]

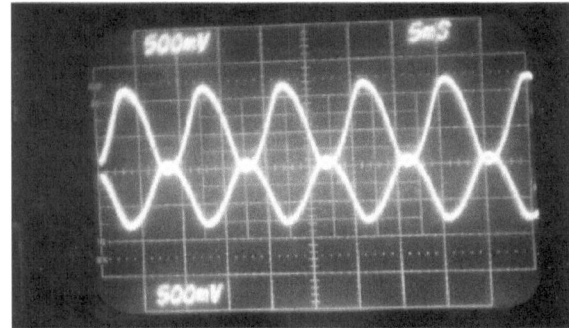

such that the frequencies $\omega = \omega_1 = \omega_2$ and the coupling strengths $K = K_1 = K_2$ vary in the ranges of 100–1,000/s and 10–10^6/s, respectively.

The experimental results of the time evolution of the oscillator voltages as a function of the delay parameter are shown in Fig. 5.3 for $K = 1,000$/s and $\omega = 837$/s. In-phase oscillations of the coupled identical limit cycle oscillators are shown in the top panel of Fig. 5.3 for $\tau = 0.514$ ms. Quenching of oscillations, that is, amplitude death is observed for $\tau = 2$ ms as shown in the middle panel of Fig. 5.3 and anti-phase oscillations are observed for $\tau = 4.428$ ms as depicted in the bottom panel of Fig. 5.3. Thus the phenomenon of time-delay induced amplitude death has been experimentally observed in coupled identical nonlinear LCR circuits with delay feedback for suitable values of the experimental parameters.

Amplitude death has also been shown to occur in a dual-wavelength class-B laser with modulated losses by Kuntsevich and Pisarchik [12], which is a nonautonomous system since the losses in a channel are modulated by an external sinusoidal force.

The phenomenon of amplitude death has also been reported experimentally [13] in a pair of opto-thermal oscillators that are coupled by heat transfer. Existence of amplitude death for stronger couplings and its relation to Hopf bifurcations of the uncoupled and coupled systems have also been experimentally verified. The authors of [13] have also complimented their experimental results with a theoretical analysis of the corresponding model equations. However, the role of delay in the death phenomenon in this experiment has not been clearly understood.

5.3 Amplitude Death with Distributed Delay in Coupled Limit Cycle Oscillators

F.M. Atay [14] has shown that distributed delay increases the parameter regime in which the phenomenon of amplitude death occurs. He has demonstrated that distributed delay enlarges and merges death islands and the death region becomes unbounded if the variance of the delay distribution is larger than a certain threshold. Since most of the studies are concerned with constant or discrete delays, propagation of information among the corresponding physical systems occurs only at fixed time intervals while the dynamical systems themselves are evolving. In contrast, physical systems in such situations can be better understood if the propagation delay from one unit to the other is also evolving dynamically or uniformly distributed over an interval. This approach is particularly significant in biological systems (in particular in neurobiology) where the system can be described by networks of coupled dynamical units evolving in time, for instance as in the case of evolving networks.

Realizing the importance of time-delay in inducing amplitude death in identical limit cycle oscillators, now we will discuss briefly the effects of distributed delay on these systems. Atay has considered a system of two coupled limit cycle oscillators, each one of the form of Eq. (5.1), with a distributed delay coupling [14]. The dynamical equation can be written as

$$\dot{Z}_1(t) = (1 + i\omega - |Z_1(t)|^2)Z_1(t) + K \left[\int_0^\infty f(\tau')Z_2(t - \tau')d\tau' - Z_1(t) \right],$$

$$\text{(5.13a)}$$

$$\dot{Z}_2(t) = (1 + i\omega - |Z_2(t)|^2)Z_2(t) + K \left[\int_0^\infty f(\tau')Z_1(t - \tau')d\tau' - Z_2(t) \right],$$

$$\text{(5.13b)}$$

where K is the coupling strength and f represents the distribution of delay time. In particular, the distributed delay is considered to be uniformly distributed over the interval $\tau \pm \delta$, that is $f(\tau)$ in the above Eq. (5.13) has been chosen as

$$f(\tau) = \begin{cases} \frac{1}{2\delta}, & |\tau - \tau'| \le \delta \\ 0, & \text{else} \end{cases} \tag{5.14}$$

Carrying out a linear stability analysis about the fixed point $Z = 0$ in Eq. (5.13), as discussed in Sect. 2.4, and in the previous section, one can obtain the pair of stability curves as

$$(1 - \gamma^2)K^2 - 2K + 1 = -(\beta - \omega)^2, \tag{5.15}$$

$$\tan(\beta\tau) = \frac{\beta - \omega}{1 - K}, \tag{5.16}$$

where

$$\gamma = \gamma(\beta, \delta) = \begin{cases} \sin(\beta\delta)/(\beta\delta), & \beta\delta \ne 0 \\ 1, & \beta\delta = 0. \end{cases} \tag{5.17}$$

The stability regime confined by the above critical curves in the (τ, K) parameter space is again determined from the knowledge of $Re(d\lambda/d\tau)$ on the critical curves as was done in the previous section. The stability regime in the (τ, K) space for the value of the parameter $\omega = 30$ is shown in Fig. 5.4 for the above mentioned uniformly distributed delay for various values of δ. For the value of $\delta = 0.0$ the distributed delay becomes a discrete delay and the corresponding stability regime for this parameter value consists of three amplitude death islands, Fig. 5.4a, which continuously deforms as the degree of distribution determined by the value of the parameter δ increases as seen in Fig. 5.4b–d. The connected stability regimes become unbounded in the direction of the delay time τ after certain threshold value of δ, Fig. 5.4c, d.

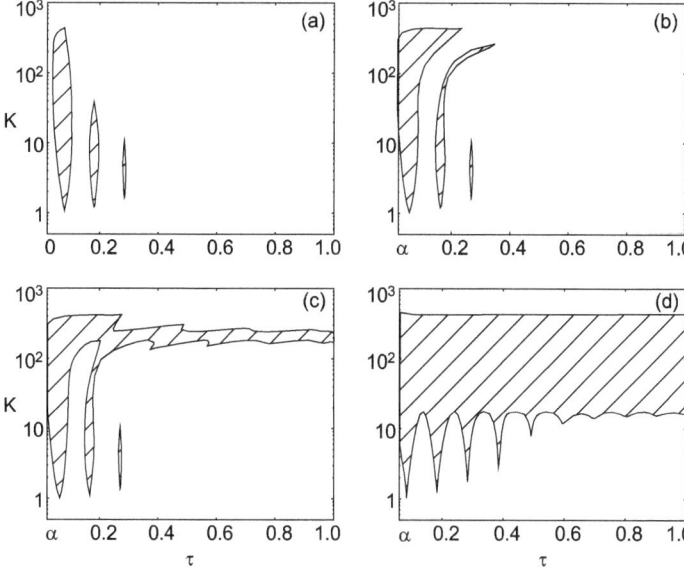

Fig. 5.4 Spread of amplitude death islands as the distribution of the delay increases. (**a**) $\delta = 0$, (**b**) $\delta = 0.007$, (**c**) $\delta = 0.008$ and (**d**) $\delta = 0.02$. This figure is adapted from the work of F. M. Atay [14]

5.4 Amplitude Death in Coupled Chaotic Oscillators

In the previous sections, we have discussed the phenomenon of amplitude death in a single/coupled limit cycle oscillator driven autonomously by a time-delay feedback and also in two coupled limit cycle oscillators with distributed delay. We also note that this phenomenon has been shown to occur in a large ensemble of coupled limit cycle oscillators with time-delay coupling [8, 10]. It has also been demonstrated that this phenomenon is generic and that it can also occur in chaotic dynamical systems with time-delay coupling, similar to the case of coupled limit cycles. As this phenomenon is quite general, it occurs for identical as well as nonidentical coupled chaotic systems. The existence of amplitude death with time-delay coupling has been discussed in coupled Lorenz and Rössler chaotic oscillators by A. Prasad [15].

It has also been shown that the transition from chaos to amplitude death via a limit cycle occurs in analogy with the same mechanism as that of the coupled limit cycle oscillators [3, 7, 8], that is, a pair of complex conjugate eigenvalues cross the imaginary axis from right to left and simultaneously the unstable fixed point becomes stable, initiating the amplitude death at a Hopf bifurcation. However there exist two distinct nature of transitions to the fixed point from the limit cycle behavior, namely in-phase and out-of-phase transitions. The shift in the nature of these distinct transitions occurs at a critical value of delay time τ_c at which the phase relationship of the two oscillators changes from in-phase to out-of-phase oscillations

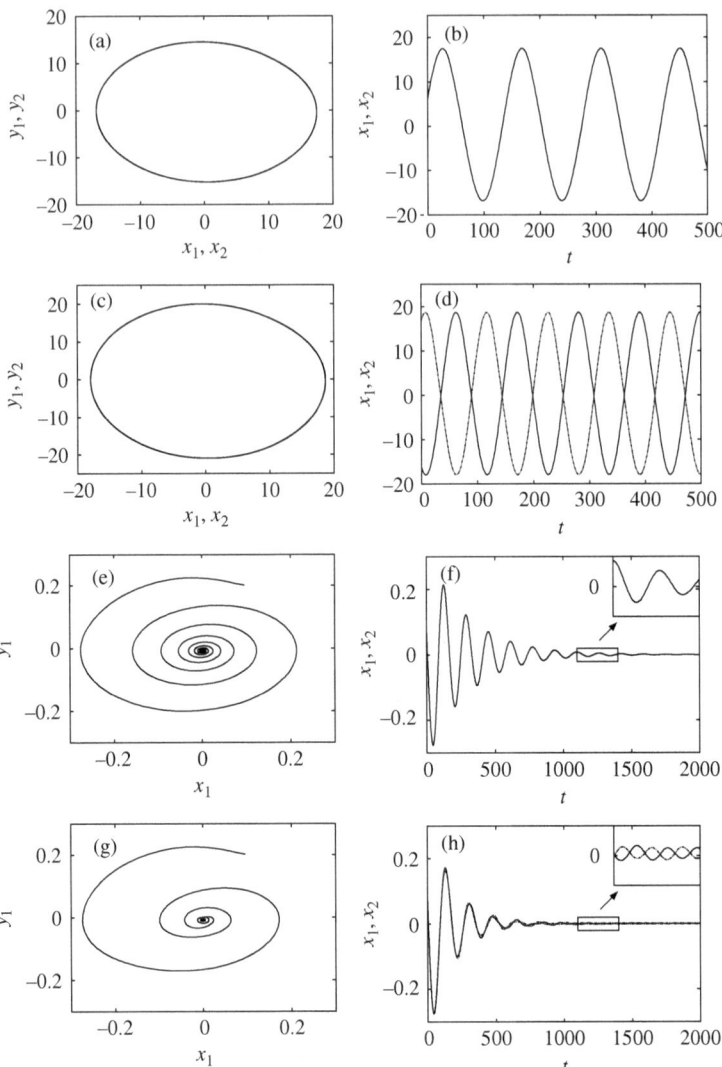

Fig. 5.5 Phase space plots (*left panel*) and time series plots (*right panel*) of the coupled identical Rössler oscillators (5.18). (**a**) and (**b**) at $\tau = 0.6$, (**c**) and (**d**) at $\tau = 2.25$, (**e**) and (**f**) at $\tau = 1.5$ and (**g**) and (**h**) at $\tau = 2$, respectively, see [15]

as a function of delay time τ. These distinct transitions were later realized as a *phase flip* bifurcation [16] which is discussed below.

To be specific, consider a system of two delay coupled identical Rössler oscillators [15] represented by

$$\dot{x}_i = -y_i - z_i, \tag{5.18a}$$
$$\dot{y}_i = x_i + ay_i + \varepsilon(y_j(t - \tau) - y_i), \tag{5.18b}$$
$$\dot{z}_i = b + z_i(x_i - c), \qquad i, j = 1, 2, i \neq j, \tag{5.18c}$$

where $a = b = 0.1, c = 14, \varepsilon = 0.5$ and the delay time τ is considered as the control parameter. In the absence of coupling $\varepsilon = 0$, the individual systems (5.18) evolve chaotically for the above values of the parameters. Dynamical behavior of the coupled identical Rössler oscillators for various values of time-delay τ are shown in Fig. 5.5. Limit cycle oscillations of both the coupled systems are shown in Fig. 5.5a for the delay time $\tau = 0.6$ and the corresponding time series plot is depicted in Fig. 5.5b, where there is no phase difference between both the oscillators x_1 and x_2. This implies that both the oscillators x_1 and x_2 are exhibiting in-phase oscillations. Similar limit cycle oscillations and their time trajectory plots of both the coupled oscillators are illustrated in Fig. 5.5c, d, respectively, for the value of delay time $\tau = 2.25$. In contrast to the previous case, it can be now noted from the time series plot that the variables x_1 and x_2 exhibit out-of-phase oscillations. For the value of delay time $\tau = 1.5$, the coupled chaotic oscillators transit to their fixed point as shown in Fig. 5.5e, f, where the oscillators are in in-phase state which can be seen clearly in the inset of Fig. 5.5f. Transition to the fixed point of both the oscillators by out-of-phase oscillations for the value of time-delay $\tau = 2$ is depicted in Fig. 5.5g, h. These distinct transitions from in-phase to out-of-phase oscillations can also be identified from the phase-difference $(\Delta\phi)$ between the two oscillators defined as $\Delta\phi = \langle|\phi_1(t) - \phi_2(t)|\rangle$, where $\langle\cdot\rangle$ denotes the time average while $\phi = \arctan(y/x)$. The phase-difference $(\Delta\phi)$ between the two oscillators as a function of the delay time $\tau \in (0, 2.5)$ is shown in Fig. 5.6.

These distinct transitions to amplitude death via limit cycle oscillations and their mechanism of transitions in coupled chaotic oscillators are generic in nature and that they can also be demonstrated in coupled identical Lorenz oscillators, coupled non-identical Rössler and Lorenz oscillators and also in mixed chaotic oscillators [15].

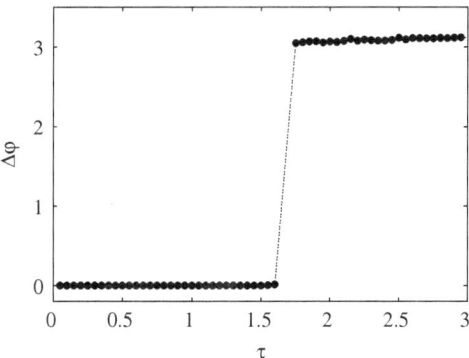

Fig. 5.6 Phase difference $\Delta\theta = \langle|\theta_1(t) - \theta_2(t)|\rangle$ (which before and after the transition are equal to 0 and π) between the oscillators, (5.18), as a function of delay time in the range $\tau \in (0, 3)$

5.5 Amplitude Death with Conjugate (Dissimilar) Coupling

Recently, it was shown that amplitude death can also occur by coupling through conjugate (dissimilar) variables [17]. The scenario for the occurrence of amplitude death in this case is quite different from our previous discussions in time-delay coupled set of identical oscillators. In [17] it has been demonstrated that amplitude death can occur in identical limit cycle oscillators, and even in identical chaotic oscillators, with conjugate coupling without time-delay. It was realized that coupling via conjugate variables provides time-delayed interaction in the sense that the other variables of a dynamical system are reconstructed from a known time series using a time-delay, a procedure in embedding. Indeed Takens' embedding theorem [18] asserts that the topological properties of the reconstructed system match those of the true system for suitable choices of embedding dimension and time-delay. The similarity between time-delayed variables and conjugate variables has been extensively employed in the process of attractor reconstruction [19]. Hence, the analysis of [17] asserts that some of the time-delay coupling effects can also be realized by means of conjugate coupling and this reduces computing efforts enormously when arrays or networks of oscillators are considered. It is also to be noted that it is easy to implement conjugate coupling in experiments when compared to time-delay feedback or coupling. However, the conjugate coupling does not lead to the infinite dimensionality of a dynamical system thereby leading to hyperchaotic attractors with large number of positive Lyapunov exponents, a hallmark property of a dynamical system with time-delay feedback or coupling.

Coupling conjugate (dissimilar) variables is natural in a variety of experimental situations where subsystems are coupled by feeding the output of one into the other. As an example for employing the conjugate coupling, recently Kim et al. [20] in their experiments on coupled semiconductor laser systems used photon intensity fluctuation from one laser to the other to modulate the injection current and vice versa.

Consider, as an example, the Landau-Stuart oscillator specified by the equation of motion

$$\dot{Z}(t) = (1 + i\omega - |Z(t)|^2)Z(t), \qquad (5.19)$$

where ω is the frequency and $Z(t) = x(t) + iy(t)$. Two such dynamical equations coupled through conjugate (dissimilar) variables can be expressed in Cartesian coordinates as

$$\dot{x}_i(t) = P_i x_i - \omega_i y_i, \qquad (5.20a)$$

$$\dot{y}_i(t) = P_i y_i + \omega_i x_i + \varepsilon x_j, \qquad (5.20b)$$

where $P_i = 1 - |Z_i|^2, i, j = 1, 2$, and $j \neq i, \varepsilon$ is the coupling strength and $\omega_i = \omega = 2.0$. The largest Lyapunov exponent of the coupled system (5.20) is shown in Fig. 5.7 in the range of coupling strength $\varepsilon \in (1, 3)$. For $\varepsilon < 2$, the largest

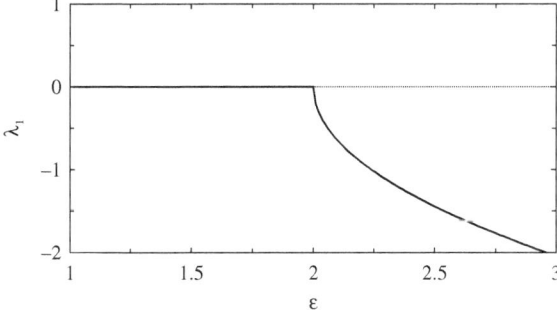

Fig. 5.7 Largest Lyapunov exponent of the coupled Landau-Stuart oscillator (5.20) as a function of the coupling strength in the range $\varepsilon \in (1, 3)$

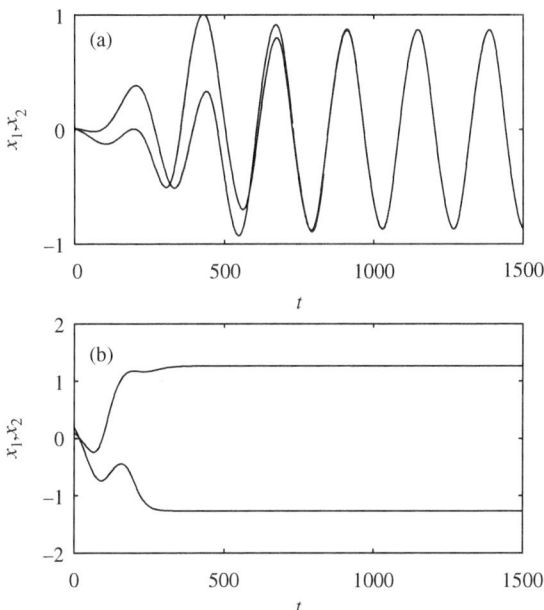

Fig. 5.8 Trajectories of the x components of the two oscillators (5.20). (**a**) Limit cycle behavior for the coupling strength $\varepsilon = 1.0$, and (**b**) Monotonic decay to the fixed points for $\varepsilon = 2.5$

Lyapunov exponent is zero and the second largest one is negative (which is not shown here), which is indicative of the limit cycle behavior as shown in Fig. 5.8a. For $\varepsilon > 2$, transient trajectories decay monotonically to the fixed points as shown in Fig. 5.8b for the value of the coupling strength $\varepsilon = 2.5$.

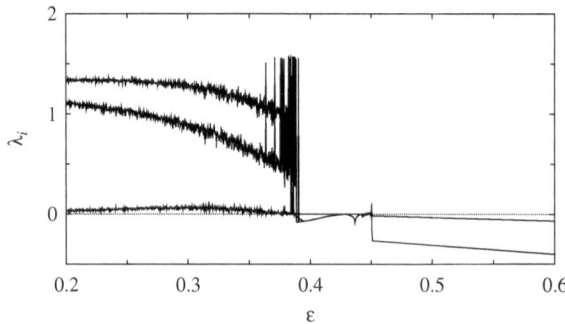

Fig. 5.9 Largest three Lyapunov exponents of the coupled Lorenz systems (5.21) as a function of the coupling strength in the range $\varepsilon \in (0.2, 0.6)$

It has also been shown that amplitude death occurs in coupled Lorenz oscillators coupled through conjugate variables [17],

$$\dot{x}_i = 10(-y_i - x_i), \tag{5.21a}$$

$$\dot{y}_i = -x_i z_i + 28x_i - y_i + \varepsilon(x_j - y_i), \tag{5.21b}$$

$$\dot{z}_i = x_i y_i - \frac{8}{3}z_i, \qquad i, j = 1, 2, i \neq j. \tag{5.21c}$$

The largest three Lyapunov exponents of the coupled Lorenz system is depicted in Fig. 5.9 as a function of the coupling strength $\varepsilon \in (0.2, 0.6)$. All the Lyapunov exponents become negative for $\varepsilon > 0.44$, indicating the occurrence of amplitude death in the coupled Lorenz system. Preceding the regime of amplitude death the largest Lyapunov exponents show wild fluctuations due to the presence of coexisting attractors, namely multistability, an impact of time-delay. Chaotic trajectories of both the coupled Lorenz systems for the value of the coupling strength $\varepsilon = 0.3$ is shown in Fig. 5.10a, while the dynamical behavior in the amplitude death regime is plotted in Fig. 5.10b for the value of the coupling strength $\varepsilon = 0.5$.

5.6 Amplitude Death with Dynamic Coupling

In the above studies on amplitude death, the coupling signal is proportional to the difference between the dissimilar or conjugate states of the two oscillators. The proportionality factor is of a constant value and hence the couplings are considered *static*. It was observed that *dynamic* coupling, which has not only the proportionality factor but also has its own dynamics, can induce amplitude death without time-delay [21]. It is to be noted that RC-ladder coupling [22], a kind of dynamic coupling, can be used as an approximation of RC wire delay connections in VLSI

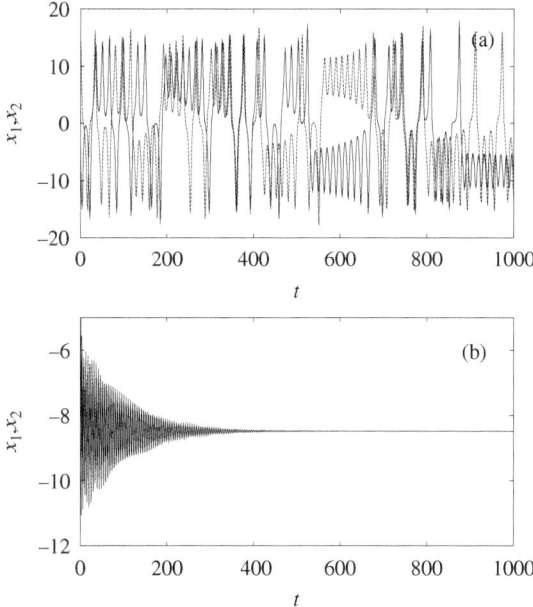

Fig. 5.10 Transient trajectories of the x components of the coupled Lorenz systems (5.21).
(**a**) Chaotic behavior for $\varepsilon = 0.3$ and (**b**) Monotonic decay to the fixed points for $\varepsilon = 0.5$

chips [23]. Consequently, dynamic coupling may also be used to produce some of
the time-delay effects in appropriate situations.

To illustrate the existence of amplitude death due to dynamic coupling, let us
consider the two identical limit cycle oscillators

$$\dot{x}_i = x_i \left(1 - y_i^2 - x_i^2 \right) - \omega y_i + u_i, \tag{5.22a}$$

$$\dot{y}_i = y_i \left(1 - y_i^2 - x_i^2 \right) - \omega x_i, \qquad i = 1, 2, \tag{5.22b}$$

where the dynamic coupling involves an additional variable with a time evolution,

$$\dot{z}_i = -z_i + x_j, \qquad u_i = \varepsilon(z_i - x_i) \qquad i, j = 1, 2, i \neq j, \tag{5.23}$$

where ε is the coupling strength. The limit cycle oscillations of the coupled system
is shown in Fig. 5.11a in the absence of coupling, that is $\varepsilon = 0$. As the value of the
coupling strength is increased from zero for a fixed value of the natural frequency
$\omega = 4$, oscillatory behavior is found to exist up to a certain threshold value of
the coupling strength, followed by a sudden quenching of oscillations above the
threshold value, leading to amplitude death of the oscillators. The full scenario is
shown as a bifurcation diagram in Fig. 5.12 as a function of the coupling strength
in the range $\varepsilon \in (0, 12)$. Both the oscillators exhibit chaotic oscillations in the
range $\varepsilon \in (0, 2.3)$ and amplitude death occurs in the range $\varepsilon \in (2.3, 8.5)$, where
all the variables converge to the origin. The stable fixed point, which differs from

Fig. 5.11 Behavior of the
system of coupled limit cycle
oscillators (5.22), (5.23). (**a**)
Limit cycle oscillations of the
two uncoupled systems in the
absence of coupling, (**b**)
Quenching of oscillations
(amplitude death) of the
coupled oscillators for the
value of the coupling strength
$\varepsilon = 4.0$ and (**c**) Both the limit
cycle oscillations and
quenching of oscillations
before and after the dynamic
coupling, respectively,
switched at $t = 1500$

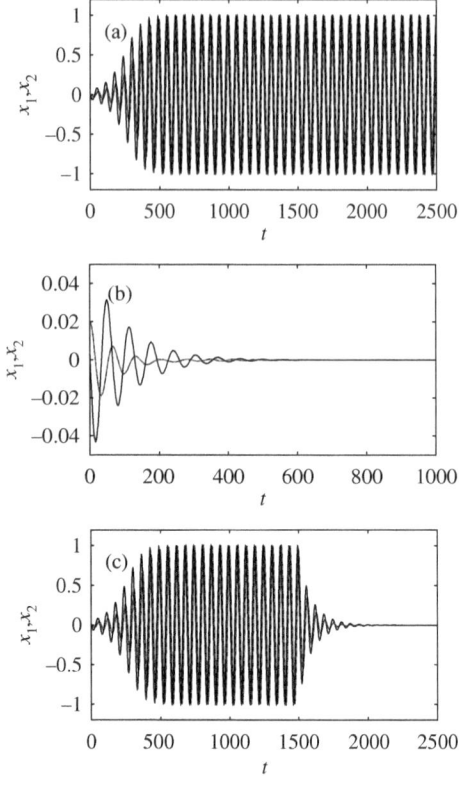

Fig. 5.12 Bifurcation
diagram of the coupled
limit-cycle oscillators (5.22),
(5.23) as a function of the
coupling strength $\varepsilon \in (0, 12)$
for a fixed value of frequency
$\omega = 4.0$

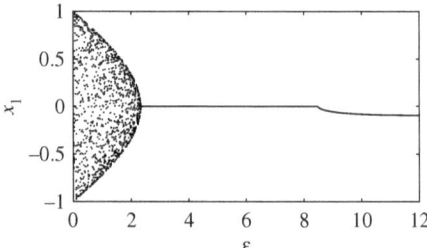

the origin, appears for $\varepsilon > 8.5$. Figure 5.11b shows the limit cycle oscillations
being damped out to reach the fixed point for $\varepsilon = 4.0$, while Fig. 5.11c shows the
existence of the limit cycle oscillations and quenching of oscillations, respectively,
before and after the dynamic coupling is switched on at $t = 1500$. Quenching of
oscillations due to dynamic coupling can also be demonstrated in coupled identical
Rössler systems [21].

Recently, the phenomenon of amplitude death has also been reported in coupled
time-delay systems [24] with both delay and dynamic couplings. Stability condition
for the stabilization of the oscillatory behavior in coupled time-delay systems has

also been derived and it is also shown that static connection never induces amplitude death. These results have also been confirmed experimentally using electronic
circuits.

5.7 Time-Delay Induced Bifurcations

It has also been shown that phase flip bifurcation occurs in a general class of nonlinear oscillators coupled through time-delay coupling [16]. Here, the relative phase
between the coupled oscillators changes abruptly from zero to π as a function of
the delay time for fixed values of the other system parameters and hence it is named
as a phase flip bifurcation, which is a general feature of the time-delay coupled
systems. This bifurcation phenomenon has a broad range of occurrence, that is,
it is observed for periodic as well as chaotic oscillators, for identical as well as
nonidentical coupled systems, and in a variety of dynamical systems [16].

For illustration, let us consider a pair of diffusively coupled Rössler oscillators
represented by Eqs. (5.18) with the same values of the parameters a, b and c as
used in the Sect. 5.4. They evolve chaotically as indicated in Sect. 5.4. The coupling
strength ε and delay time τ are chosen as control parameters. The phase difference
$\Delta\phi = \langle|\phi_1(t) - \phi_2(t)|\rangle$, where $\langle\cdot\rangle$ is the time average, is depicted in Fig. 5.13 in the
(ε, τ) parameter space. From this figure it is evident that the relative phase between
both of the coupled systems is zero for a fixed value of the coupling strength ε up
to a certain threshold value of delay time τ and above this threshold value there is a
difference of π in the relative phase. As the phase flips from zero to π as a function
of the delay time, this phenomena is termed as phase flip bifurcation. The dynamics
of this phase flip bifurcation, namely transition from in-phase oscillations to out of
phase oscillations has been discussed in Sect. 5.4 (see Fig. 5.5) for both limit cycle
oscillations and chaotic oscillations.

It was also observed that Neimark-Sacker-type bifurcation [25] is prevalent in
delay coupled networks for larger values of delay times, which results in high-period
solutions followed by more complex behavior [26]. It was shown that the synchronized solution of delay coupled logistic map exhibits such complex bifurcation

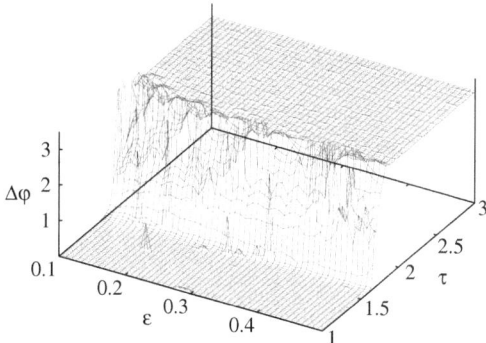

Fig. 5.13 Phase difference
$\Delta\phi$ in the (ε, τ) plane

scenario for finite values of connection delays whereas such bifurcation scenario (Neimark-Sacker-type bifurcation) cannot arise in one-dimensional maps [26], a clear implication of delay coupling. It has also been shown that a variety of rich bifurcation structures can arise in the synchronized solution as a function of the coupling strength for finite values of connection delays.

Bifurcations such as inverse, direct Hopf and fold limit cycle were observed in time-delay coupled FitzHugh-Nagumo excitable systems [27]. It was also identified that for an intermediate range of time lags, inverse sub-critical Hopf and fold limit cycle bifurcations lead to the phenomenon of oscillator death [27]. Bifurcations in two coupled excitable FitzHugh-Nagumo systems ($N = 2$) has been studied analytically, and it is also numerically confirmed that the same bifurcations are relevant for $N > 2$ in the presence of delay coupling.

5.8 Some Other Effects of Delay Feedback

In the following, we will briefly point out the main features and the emergence of other types of behaviors in delay coupled systems (delay feedback) which are not possible in the absence of time-delay feedback.

1. The first systematic investigation of time-delayed coupling was done by Schuster and Wagner [28] who studied two coupled phase oscillators and found multistability of synchronized solutions.
2. A novel form of frequency depression was observed for small delays in the limit cycle oscillators that interact via time-delayed diffusive coupling and for larger delay metastable synchronized state was observed [29].
3. Bistability between synchronized and incoherent states have been observed in Kuramoto oscillators coupled via time-delay coupling. Exact formulas for the stability boundaries of the incoherent and synchronized state as a function of the delay has been established [30].
4. Delay induced chaos has been demonstrated in catalytic surface reaction [31]. A mathematical model has been proposed which explains the origin of chaos in this reaction as being due to delays in the response of a population of reacting adsorbate islands globally coupled via the gas phase. The dynamical equations of this model yields a sequence of period-doubling bifurcations resulting in chaos [31].
5. It has also been shown that delay coupling in networks enhances the synchronizability of networks and interestingly it leads to the emergence of a wide range of new collective behavior (see [26, 32] and reference therein). On the other hand, it has also been shown that connection delays can actually be conducive to synchronization so that it is possible for delayed systems to synchronize whereas the undelayed system does not [26].
6. Enhancement of neural synchrony, that is, the existence of stable synchronized state even for a very low coupling strength for significant time-delay in the coupling has also been demonstrated [33].

7. Time-delay feedback has also been demonstrated to be used for bifurcation control of nonlinear models of chaotic cardiac activity [34].
8. It has also been demonstrated that time-delay feedback can be used for suppressing a pathological period-2 rhythm (cardiac alternans) in a atrioventricular nodal conduction model [35].
9. In delay coupled fiber lasers, it was demonstrated that a reduction in dynamical complexity occurs for short coupling delays while a logarithmic growth is observed as the coupling delay is increased [36].

References

1. K. Bar-Eli, J. Phys. Chem. **94**, 2368 (1990)
2. M. Shiino, M. Frankowicz, Phys. Lett. A **136**, 103 (1989)
3. D.G. Aronson, G.B. Ermentrout, N. Koppel, Physica D **41**, 403 (1990)
4. K. Okuda, Y. Kuramoto, Prog. Theor. Phys. **86**, 1159 (1991)
5. G.B. Ermentrout, Physica D **41**, 219 (1990)
6. R.E. Mirollo, S.H. Strogatz, J. Stat. Phys. **60**, 245 (1990)
7. D.V. Ramana Reddy, A. Sen, G.L. Johnston, Phys. Rev. Lett. **80**, 5109 (1998)
8. D.V. Ramana Reddy, A. Sen, G.L. Johnston, Physica D **129**, 15 (1999)
9. D.V. Ramana Reddy, A. Sen, G.L. Johnston, Phys. Rev. Lett. **85**, 3381 (2000)
10. D.V. Ramana Reddy, Ph. D. Thesis entitled *Collective Dynamics of Delay Coupled Oscillators* (Institute for Plasma Research, Gandhinagar, India, 2000)
11. R. Dodla, A. Sen, G.L. Johnston, Phys. Rev. E **69**, 056217 (2004)
12. B.F. Kuntsevich, A.N. Pisarchik, Phys. Rev. E **64**, 046221 (2001)
13. R. Herrero, M. Figueras, J. Rius, F. Pi, G. Orriols, Phys. Rev. Lett. **84**, 5312 (2000)
14. F.M. Atay, Phys. Rev. Lett. **91**, 094101 (2003)
15. A. Prasad, Phys. Rev. E **72**, 056204 (2005)
16. A. Prasad, J. Kurths, S.K. Dana, R. Ramaswamy, Phys. Rev. E **74**, 035204(R) (2006)
17. R. Karnatak, R. Ramaswamy, A. Prasad, Phys. Rev. E **76**, 035201(R) (2007)
18. F. Takens, in *Detecting Strange Attractors in Turbulence*, ed. by D. Rand, L. Young. Lecture Notes in Mathematics (Springer, New York, 1981)
19. N.H. Packard, J.P. Crutchfield, J.D. Farmer, R.S. Shaw, Phys. Rev. Lett. **45**, 712 (1980)
20. M.Y. Kim, R. Roy, J.L. Aron, T.W. Carr, I.B. Schwartz, Phys. Rev. Lett. **94**, 088101 (2005)
21. K. Konishi, Phys. Rev. E **68**, 067202 (2003)
22. J.P. Fishburn, C.A. Schevon, IEEE Trans. Circuits Syst. I : Fundam. Theory Appl. **42**, 1020 (1995)
23. X. Zhan, IEEE Circuits Devices Mag. **12**, 12 (1996)
24. K. Konishi, K. Senda, H. Kokame, Phys. Rev. E **78**, 056216 (2008)
25. S.N. Elaydi, *Discrete Chaos* (Chapman & Hall/CRC, Boca Raton, FL, 2000)
26. F.M. Atay, J. Jost, Phys. Rev. Lett. **92**, 044101 (2004)
27. N. Buric, D. Todorovic, Phys. Rev. E **67**, 066222 (2003)
28. H.G. Schuster, P. Wagner, Prog. Theor. Phys. **81**, 9392 (1987)
29. E. Niebur, H.G. Schuster, D.M. Kammen, Phys. Rev. Lett. **67**, 2753 (1991)
30. M.K. Stephen Yeung, H.G. Schuster, Phys. Rev. Lett. **82**, 648 (1999)
31. N. Khrustova, G. Veser, A. Mikhailov, Phys. Rev. Lett. **75**, 3564 (1995)
32. C. Massoller, A.C. Marti, Phys. Rev. Lett. **94**, 134102 (2005)
33. M. Dhamala, V.K. Jirsa, M. Ding, Phys. Rev. Lett. **92**, 074104 (2004)
34. M.E. Brandt, G. Chen, IEEE Trans. Circuits Syst. I : Fundam. Theory Appl. **44**, 1031 (1997)
35. M.E. Brandt, H.T. Shih, G. Chen, Phys. Rev. E **56**, 1334(R) (1997)
36. A.L. Franz, R. Roy, L.B. Shaw, I.B. Schwartz, Phys. Rev. Lett. **99**, 053905 (2007)

Chapter 6
Recent Developments on Delay Feedback/Coupling: Complex Networks, Chimeras, Globally Clustered Chimeras and Synchronization

6.1 Introduction

Time-delay systems are ubiquitous in nature and occur in connection with various aspects of physical, chemical, biological and economic systems as pointed out in earlier chapters. The study of time-delay induced modifications in the collective behavior of systems of coupled nonlinear oscillators is a topic of much current interest both for its fundamental significance from a dynamical systems point of view and for its practical applications. There have been several reports of interesting phenomena in time-delay systems such as multistable states [1, 2], oscillation death [3], chimera states [4] and globally clustered chimera states [5], with vast investigations on synchronization of coupled time-delay systems [6]. Current research on collective dynamics of coupled dynamical systems have been focusing on the so called *chimera states* and also on the dynamics of complex networks with delay. The discovery of chimera states came as a surprise because an array of *identical* oscillators splits into two domains: one coherent and phase locked and the other incoherent and desynchronized. This characteristic behavior of complex systems reminded the discoverers of a fire-breathing monster in Greek mythology, which has a lion's head, a goat's body, and a serpent's tail (see Fig. 6.1), and hence it was named so. In this chapter, we will discuss some of the recent developments on the existence of this kind of interesting chimera states and on the synchronization of complex networks with connection delays. We will also discuss the controlling aspects using time-delay feedback.

6.2 Complex Networks

A network is nothing but a digraph (directed graph) with weighted edges. The study of networks is directed towards examining common principles, algorithms and tools that govern the network behavior which in turn are applicable in analyzing

M. Lakshmanan, D.V. Senthilkumar, *Dynamics of Nonlinear Time-Delay Systems*,
Springer Series in Synergetics, DOI 10.1007/978-3-642-14938-2_6,
© Springer-Verlag Berlin Heidelberg 2010

Fig. 6.1 The monster
according to Greek
mythology which has a lion's
head, a goat's body, and a
serpent's tail. This figure is
obtained from
www.freewebs.com/chimeraclan/

interactions in and among diverse physical or engineered networks, information
networks, biological networks, cognitive and semantic networks, social networks,
etc. [7–9]. A complex network is a network with features that do not occur in simple
networks such as latices or random graphs. The current interest in networks is part
of a broader movement towards complex networks. In the study of complex networks, the anatomy of the concerned network is always considered important. This
is because the structure always affects function. For instance, the topology of social
networks affects the spread of information and disease and the topology of the power
grid affects the robustness and stability of power transmission. Networks of dynamical systems have been used to model everything from earthquakes to ecosystems,
neurons and neutrinos [10–17]. The nature of couplings in such complex network
models has been conventionally considered as instantaneous during earlier studies.
One of the main reasons for this assumption is that it substantially simplifies the
analysis of the system. In addition, such an approximation is more often physically
justified. However, the fact is that consideration of time-delay is vital for modeling
real life systems. Furthermore, as we will demonstrate in this chapter, certain interesting dynamical phenomena in complex systems are characteristic of time-delay
and they will not occur in systems without time-delay. Since the introduction of
time-delay increases the effective dimension of the system, one can expect certain
complex phenomena to be explained in a better way in models of real physical
systems when delay is included.

A few types of networks to mention are (see Fig. 6.2 for schematic
representation)

- Small world network – A network in which any two arbitrary nodes are connected by only six degrees of separation, i.e. the diameter of the corresponding
  graph of connections is not much larger than six.
- Scale free network – A network whose degree of distribution, i.e., the probability
  that a node selected uniformly at random has a certain number of links (degree),
  follows a particular mathematical function called a power law. The power law
  implies that the degree of distribution of these networks has no characteristic
  scale.

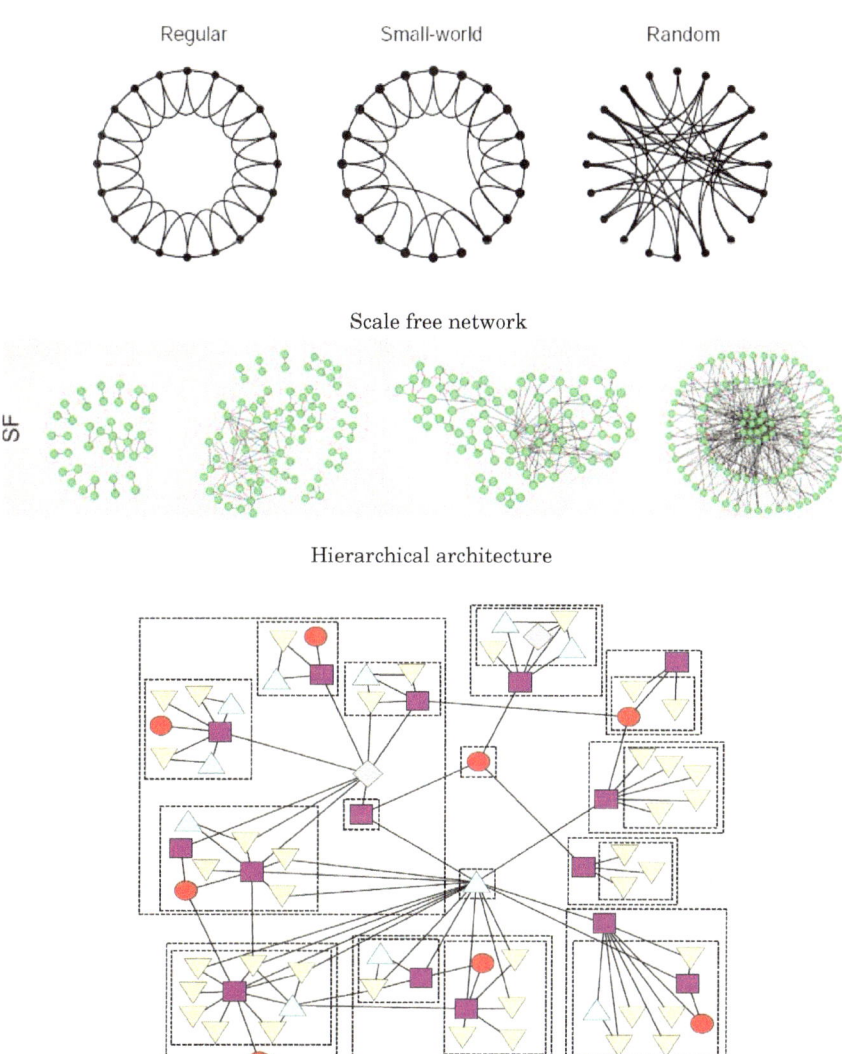

Fig. 6.2 Schematic representation of different types of networks. These figures are taken from google images

- Regular network – A regular network is one in which the nodes are normally connected to their nearest neighbours and/or next nearest neighbours.
- Random network – A random network is generated by some random process; starting with a set of n nodes and adding nodes between them at random.
- Hierarchical architectures – These are nothing but networks of networks, where each node of the larger network is itself a network (usually called as a sub-population).

In the early stages of chaos synchronization studies certain regularity in the connection was maintained whereas later on more general networks with random, small world, scale free and hierarchical architectures have been emphasized as suitable models of real world networks. However, more realistic modeling of many large networks with nonlocal interaction inevitably requires connection delays [18] to be taken into account, since they naturally arise as a consequence of finite information transmission and processing speeds. Hence, it is important to consider the individual dynamical units of networks as delay dynamical systems [19] with constant and time varying delays to mimic most of the real networks and to unravel their actual hidden dynamics. For example, in populations of spatially separated neurons, the synaptic communications between them which depend on the propagation of action potentials over appreciable distances involve distributed delays [19].

6.3 Chimera States in Delay Coupled Identical Oscillators

6.3.1 Discovery of Chimera States

Mathematically, as pointed out in the introduction, the phenomenon of the coexistence of coherent and incoherent states of nonlocally coupled identical oscillators was named chimera states by Abrams and Strogatz [20], see for example the occurrence of chimera in a system of coupled phase oscillator populations [5]. Actually Kuramoto et al. [21] were studying arrays of identical limit-cycle oscillators that are coupled nonlocally. They found that for certain choices of parameters and initial conditions, the array would split into two domains: one composed of coherent, phase-locked oscillators, coexisting with another composed of incoherent, drifting oscillators. They also found that the coexistence of locking and drifting was robust and that the phenomenon could occur in both one and two spatial dimensions and also for various kinds of oscillators including the Fitzhugh – Nagumo model, complex Ginzburg – Landau equations, phase oscillators and an idealized model of biochemical oscillators. It also occurs in a wide class of reaction-diffusion equations, under particular assumptions on the local kinetics and diffusion strength that lead to effective nonlocal coupling [22, 23]. This discovery came as a surprise because, in general identical oscillators would settle into one of a few basic patterns [10, 11], the simplest of them being synchrony, with all the oscillators moving in unison, executing identical motions at all times. Another common pattern is wave propagation, typically in the form of solitary waves in one dimension, spiral waves in two dimensions and scroll waves in three dimensions. The common feature in these cases is that all the oscillators are locked in frequency, with a fixed phase difference between them. At the opposite end of the spectrum is incoherence, where the phases of all the oscillators drift quasiperiodically with respect to each other, and the system shows no spatial structure whatsoever. However, this coexistence (unnamed then) was so odd because the locking and incoherence were present in the same system, simultaneously (see Fig. 6.3, where synchronized and desynchronized populations are

Fig. 6.3 Numerical illustration of a snapshot of a chimera state reproduced from [24]. (**a**) Synchronized population, (**b**) Desynchronized population and (**c**) the density of desynchronized oscillators

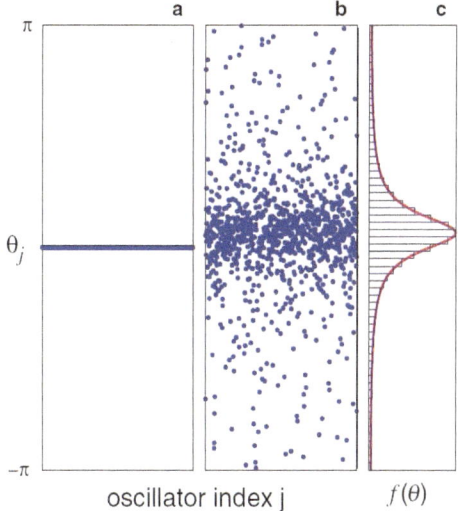

oscillator index j $f(\theta)$

obtained for same values of the parameters but for different initial conditions [24]). Furthermore, this combination could not be attributed to the simplest mechanism of pattern formation (a supercritical instability of the spatially uniform oscillation), because it can occur even if the uniform state is linearly stable, as indeed it was for the parameter values used by Kuramoto and his colleagues. It also has nothing to do with the classic partially locked/partially incoherent states that occur in populations of nonidentical oscillators with distributed natural frequencies [11, 25].

6.3.1.1 Coupled Limit Cycle Oscillators in the Continuum Limit

Abrams and Strogatz, reported the existence of chimera and named them so while studying a system of a ring of phase oscillators. This is nothing but the same system studied by Kuramoto and Battogtokh [26–31] as a simplest model for densely and uniformly distributed oscillators with nonlocal coupling. The system is governed by the Ginzburg-Landau equation in one spatial dimension as

$$\frac{\partial}{\partial t} A(x,t) = (1 + i\omega_0)A - (1 + ib)|A|^2 A$$
$$+ K(1 + ia)(Z(x,t) - A(x,t)). \tag{6.1}$$

Here $A(x,t)$ is the complex amplitude of oscillation and $Z(x,t)$ is the mean field that represents the effect of the nonlocal coupling and is given by

$$Z(x,t) = \int G(x - x')A(x',t)dx'. \tag{6.2}$$

The coupling function G changes with distance as $G(y) = \frac{k}{2}exp(-k|y|)$ and is normalized. The reductive derivation of Eq. (6.1) from a certain class of reaction-diffusion systems introduces the exponential form of G when Eq. (6.1) involves an inactive diffusive component to be eliminated adiabatically [27–30]. When the strength of the coupling K is small, which is actually the case, then Eq. (6.1) is reduced to a phase equation

$$\frac{\partial}{\partial t}\phi(x,t) = \omega - \int_{-L}^{L} G(x - x')$$
$$\times \sin[\phi(x,t) - \phi(x',t) + \alpha]dx', \tag{6.3}$$

which is much easier to analyze. Here ω is the natural frequency of the phase oscillator and α is a tuning parameter. Abrams and Strogatz presented an exact solution for the chimera state in this system coupled by a cosine kernel. Letting $\phi = \theta + \Omega(t)$, where Ω is the angular frequency of the rotating frame, Eq. (6.3) becomes

$$\frac{d\theta}{dt} = \omega - \Omega - R\sin[\theta - \Theta + \alpha]. \tag{6.4}$$

Here $Re^{i\theta} = \int_{-\pi}^{\pi} G(x - x')e^{i\theta(x',t)}dx'$ is the complex order parameter. Focusing only on stationary solutions, a self consistent equation for the complex order parameter can be given as

$$R(x)e^{i\theta(x)} = e^{i\beta}\int_{-\pi}^{\pi} G(x - x')e^{[i\Theta(x')]}\frac{\Delta - \sqrt{\Delta^2 - R^2(x')}}{R(x')}dx' \tag{6.5}$$

where, $\beta = \frac{\pi}{2} - \alpha$ and $\Delta = \omega - \Omega$. Exact solution for the stationary state can be obtained by solving Eq. (6.5) for the three unknowns $R(x)$, $\Theta(x)$ and Δ using perturbation analysis. Using this solution they showed that the stable chimera state bifurcates from a spatially modulated drift state, and died in a saddle-node bifurcation with an unstable chimera state.

Chimera states are considered to be nongeneric, as they emerged only for a particular set of initial conditions [4, 20, 21, 31]. The nature and properties of this exotic collective state as well as its potential applications have not been fully explored or understood and hence it remains unclear whether such a state exists in delay coupled oscillators. Recently, two independent studies revealed the existence of chimera states in different model systems that are coupled in a time-delayed and spatially nonlocal fashion [4, 32]. Furthermore an analytical description of spiral wave chimera has been provided by Martens et al. [33]. They have also calculated the rotation speed and the size of the incoherent core of the spiral wave using perturbation theory.

6.3.2 Chimera States in Delay Coupled Systems

Sethia et al. [4] reported that chimera states do indeed exist in delay coupled systems but acquire an additional spatial modulation such that the single spatially connected phase coherent region of the usual chimera state is now replaced by a number of spatially disconnected regions of coherence with intervening regions of incoherence. Furthermore, it was shown that the adjacent coherent regions of this clustered chimera state are found to be in antiphase relation with respect to each other. These authors have considered a model equation representing the continuum limit of a chain of identical phase oscillators arranged on a circular ring C,

$$
\frac{\partial}{\partial t}\phi(x, t) = \omega - \int_{-L}^{L} G(x - x')
$$
$$
\times \sin\left[\phi(x, t) - \phi(x', t - \tau_{x,x'}) + \alpha\right] dx', \tag{6.6}
$$

where $2L$ is the system length and a closed chain configuration is ensured by imposing periodic boundary conditions. The kernel $G(x - x')$, appropriately normalized to unity over the system length, is taken as

$$
G(x - x') = \frac{k}{2(1 - e^{-kL})} e^{-kd_{x,x'}}, \tag{6.7}
$$

which provides a non-local coupling among the oscillators over a finite spatial range of the order of k^{-1} which is taken to be less than the system size. The coupling is time-delayed through the argument of the sinusoidal interaction function, namely the phase difference between two oscillators located at x and x' is calculated by taking into account the temporal delay for the interaction signal to travel the intervening geodesic (i.e. shortest) distance determined as $d_{x,x'} = min\{|x - x'|, 2L - |x - x'|\}$. The time-delay term is therefore taken to be of the form, $\tau_{x,x'} = d_{x,x'}/v$, where v is the signal propagation speed. In the absence of time-delay the above equation reduces to the one investigated in [21, 31], Eq. (6.3) above. The constant phase shift term α in the undelayed model breaks the odd symmetry of the sinusoidal coupling function and as discussed in [20] it is needed as a tuning parameter for obtaining chimera solutions in the undelayed case. In the presence of time-delay however it is found that α no longer plays such a critical role since the time delay factor also fulfills a similar function.

The system parameters chosen for the simulations are, $2L = 1.0$, $\alpha = 0.9$, $k^{-1} = 0.25$, $\omega = 1.1$, $v = 0.09765625$ and the number of oscillators in the ring $N = 256$ corresponding to a maximum delay time (τ_{max}). As discussed in the above studies [21, 31], the choice of appropriate initial conditions is very important for numerically accessing a chimera state. Kuramoto used a random distribution with a Gaussian envelope for the initial distribution of the phases to obtain a chimera solution. For the present time-delayed system, it is found that choosing the initial phases of the oscillators from a uniform random distribution between 0 and 2π and

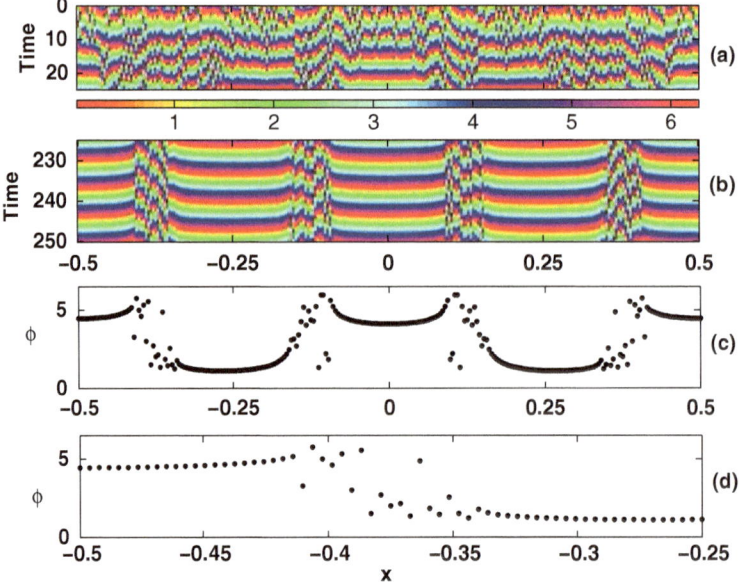

Fig. 6.4 (**a**) The space-time plot of the oscillator phases ϕ for the parameters $2L = 1.0$, $k = 4.0$, $1/v = 10.24$, $\omega = 1.1$ and $\alpha = 0.9$ in the early stages of evolution from a random set of initial phases. **Panel** (**b**) shows a later time evolution and **panel** (**c**) gives a snapshot of the final stationary state. **Panel** (**d**) is a blowup of the region between $x = -0.5$ to $x = -0.25$ giving an enlarged view of an incoherent region and portions of the adjacent coherent regions. This figure is adapted from the work of Gautam C. Sethia et al. [4]

then arranging them in a mirror symmetric distribution in space provides a rapid access to a clustered chimera state. The existence of such time-delay induced phase clustering is also further supported through solutions of a generalized functional self-consistency equation of the mean field [4].

Figure 6.4a, b show a space time plot for the parameters mentioned above in the early stages of evolution (starting from random initial phases) and in the final stages of the formation of a clustered chimera state, respectively. Figure 6.4c shows a snapshot of the spatial distribution of the phases in the final stationary state. Four coherent regions are shown interspersed by incoherence and also note that the adjacent coherent regions are in anti-phase states. Blowup of the region between $x = -0.5$ to $x = -0.25$ giving an enlarged view of an incoherent region and portions of the adjacent coherent regions is shown in Fig. 6.4d.

As pointed above, the chimera states revealed so far are nongeneric in the sense that they occur for specific set of initial conditions. However, Oleh et al. [32] demonstrated that the chimera states generically emerge already in a rather simple network of globally coupled oscillators, provided the latter is subject to spatially modulated delayed feedback. Spatial modulation here means that the strength of the delayed feedback is maximal at the site of injection and decreases with increasing distance from the injection site. This feature is typical for spatially extended systems under the influence of nonhomogeneous, local control forces [34].

Oleh et al. [32] have investigated the role played by chimera states in delay feedback systems. Two main aspects of their report included (i) the generic nature of chimera states as induced by delayed feedback and (ii) the link between coherent and incoherent states is nothing but a chimera state. They have considered an ensemble of identical, densely and uniformly distributed Landau-Stuart oscillators, representing a normal form of a supercritical Andronov-Hopf bifurcation governed by

$$\frac{\partial}{\partial t} W = (1 + i\omega - |W|^2)W + C\left(\frac{1}{2}\int_{-1}^{1} W(x,t)dx - W\right)$$

$$+\frac{K}{2}\rho(x)\int_{-1}^{1} W(x, t-\tau)dx, \tag{6.8}$$

where $W(x,t)$ stands for a complex amplitude of the oscillator at position x and at time t. Positive parameters ω and C correspond to the natural frequency and the global coupling strength, respectively. The term $\int_{-1}^{1} W(x,t)dx$ is the ensemble's mean field and the last term in Eq. (6.8) corresponds to the mean field delivered with the delay τ and the spatial profile $\rho(x)$ to each oscillator with the delayed feedback strength K. They showed that chimera states can be robustly induced by delayed feedback stimulation with a variety of exponentially or linearly decaying stimulation profiles, provided the key stimulation parameters, the delay and strength, are tuned appropriately.

6.4 Chimera States in Delay Coupled Subpopulations: Globally Clustered States

So far, we have discussed the existence of chimera states in a population of identical oscillators with nonlocal delay coupling. Recently, globally clustered chimera state (GCC) has been identified by Sheeba et al. [5] in a system of identical oscillators with two subpopulations with time-delay coupling. It is demonstrated that coupling delay can induce globally clustered chimera (GCC) states in systems having more than one coupled identical oscillator (sub) populations. By GCC state here we mean that a system, which has more than one (sub) population, splits into two different groups, one synchronized and the other desynchronized, each group comprising of oscillators from both the populations.

Sheeba et al. [5] have considered a system of two populations of identical oscillators coupled through a finite delay, represented by the following equation of motion,

$$\dot{\theta}_i^{(1,2)} = \omega - \frac{A}{N}\sum_{j=1}^{N} f\left(\theta_i^{(1,2)}(t) - \theta_j^{(1,2)}(t - \tau_1)\right)$$

$$+\frac{B}{N}\sum_{j=1}^{N} h\left(\theta_i^{(1,2)}(t) - \theta_j^{(2,1)}(t - \tau_2)\right), i = 1, 2, ..., N. \tag{6.9}$$

Population–I　　　　　　　　　Population–II

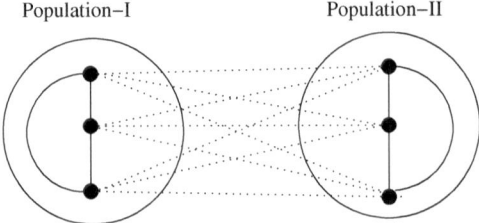

Fig. 6.5 Schematic representation of system (6.9) with $N = 3$ comprised of two populations of all – to – all coupled oscillators; the oscillators within each population are identical. Here *solid lines* represent coupling within a population (with strength A) and *dotted lines* represent coupling between populations (with strength B)

A schematic representation of the system is given in Fig. 6.5. The occurrence of various synchronization states in system (6.9) is schematically represented in Fig. 6.6 (details given below). Here ω is the natural frequency of the oscillators in the populations and it is the same for all oscillators in both the populations making all of them identical. However, the two populations are distinguished by the initial distribution of their phases; the phases are uniformly distributed between 0

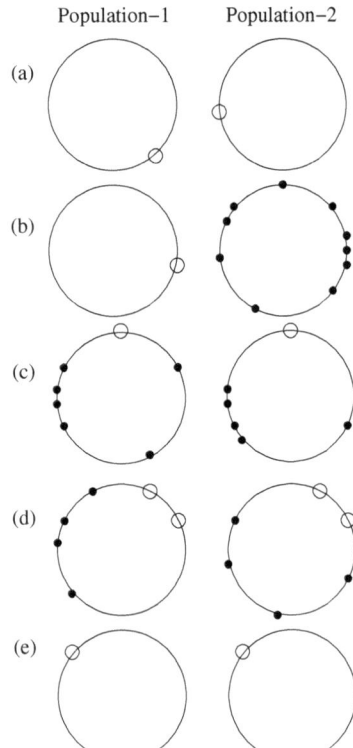

Fig. 6.6 Schematic representation of phase portraits of the states of synchronization in system (6.9). (**a**) Individual synchronization in both the populations, (**b**) chimera, (**c**) GCC, (**d**) multi-clustered GCC, and (**e**) global synchronization. *Open circles* represent synchronized group of oscillators and the *closed circles* represent the desynchronized oscillators

and π for the first population and between π and 2π for the second population. A and B refer to coupling strengths within and between populations, respectively. The functions f and h are 2π periodic that describe the coupling. N refers to the size of the populations. The complex mean field parameters, $X^{(1,2)} + iY^{(1,2)} = r^{(1,2)}e^{i\psi^{(1,2)}} = \frac{1}{N}\sum_{j=1}^{N}e^{i\theta_j^{(1,2)}}$, characterize synchronization within a population but not global clustering. Therefore, in order to quantify a GCC numerically, after allowing the transients, they have identified those oscillators whose θ_is are equal for all times and neglect them so as to end up with the desynchronized group (that comprises oscillators from both the populations, whose size is N^{DS}) and calculate its order parameter as

$$r^{\mathrm{DS}}e^{i\psi^{\mathrm{DS}}} = \frac{1}{N^{\mathrm{DS}}}\sum_{j=1}^{N^{\mathrm{DS}}}e^{i\theta_j^{\mathrm{DS}}}, \tag{6.10}$$

where $N^{\mathrm{DS}} = 2N - N^{\mathrm{s}}$. τ_1 and τ_2 quantify coupling delay within and between populations, respectively.

Their simulation results show that the coupling delay can induce splitting of identical delay coupled populations into desynchronized frequency suppressed (vanishing oscillating frequencies) clusters and synchronized clusters. This splitting can occur either within the populations (Fig. 6.7 Bottom Panel) representing chimera states or between the populations (Fig. 6.7 Top Panel) representing the globally coupled chimera states. Further, it is also shown that the GCC state need not be stable but it can either breathe or can be unstable depending upon the value of the coupling delay. A GCC state is called as a breather state when one of the groups is completely

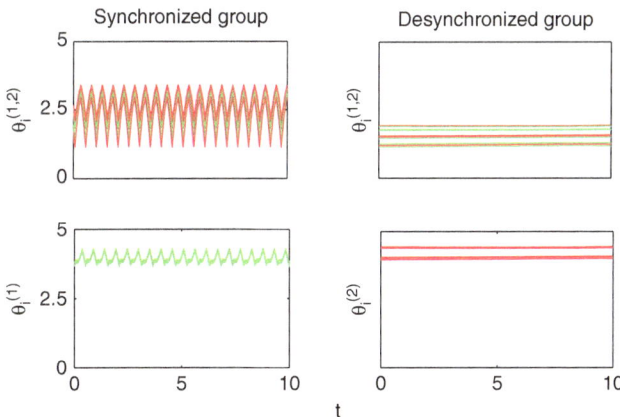

Fig. 6.7 Occurrence of (stable) GCC in system (6.9). **Top panel**: Global clustering phenomenon – synchronized and desynchronized (frequency suppressed) groups have oscillators from both the populations. **Bottom panel**: One of the populations is synchronized and the other is desynchronized (frequency suppressed). *Green* (*light gray*) and *red* (*dark gray*) *lines* represent oscillators in the first and the second populations, respectively. Here $\{f, h\} = \{\sin(\theta), \cos(\theta)\}$, $\tau_1 = n\tau_2 = n\tau$ with $n = 1$ (**top panel**) $A = 1.2$, $B = 1$ and $\tau = 2$, (**bottom panel**) $A = 1.6$, $B = 1$ and $\tau = 1$

synchronized while the other group is desynchronized and fluctuates. Breathers can be long, short (defined in a relative sense), periodic or aperiodic. A breather can also be unstable when the desynchronized group remains desynchronized for a while after which this state loses stability and all the oscillators lock to one phase. The different types of breather states are reported to be a steady dynamical states. Typical illustration of breather and unstable states are shown in Fig. 6.8. Investigations on the phenomenon of chimera states and exploring its possible applications have been

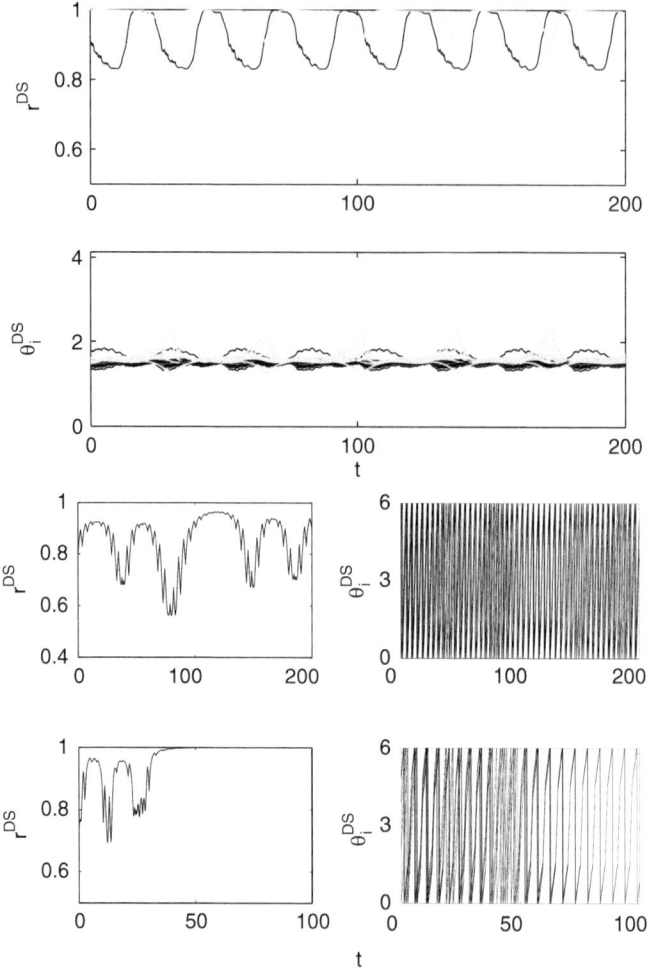

Fig. 6.8 Topmost row: Illustration of a breathing GCC state with initial condition close to the GCC state. Grey and black lines represent the long- and short-periodic breather with $\tau = 3.6$ and $\tau = 4$, respectively, where the time evolution of the order parameter r^{DS} is plotted. **Second row**: The corresponding phases θ_i^{DS} are plotted against time. **Third row**: Aperiodic breathing GCC with $\tau = 5$. **Fourth row**: Unstable breathing GCC with $\tau = 4$. For third and fourth rows, left panels correspond to the time evolution of r^{DS} and right panels represent θ_i^{DS}

only in their initial stage. There are many open problems such as mechanism of emergence of chimera states, existence in higher dimensional systems, their generic nature, etc., need to be explored in this direction. Furthermore, experimental realization of existence of chimera states remains to be explored.

6.5 Synchronization in Complex Networks with Delay

It has been shown that delay in networks enhances their synchronizability and interestingly it leads to the emergence of a wide range of new collective behavior [18, 19] (for details on synchronization, see Appendix B and later chapters). On the other hand, it is also shown that connection delays can actually be conducive to synchronization, so that it is possible for the delayed system to synchronize where the undelayed system does not [18].

To be specific, Atay and Jost [18] consider a finite connected graph Γ with nodes (vertices) i, writing $i \sim j$ when i and j are neighbors, that is, connected by an edge, and with the number of neighbors of i denoted by n_i. On Γ, one can have a dynamical system with discrete time $t \in \mathbf{Z}$, with the state x_i of i evolving according to

$$x_i(t+1) = f(x_i(t)) + \varepsilon \left(\frac{1}{n_i} \sum_{\substack{j \\ j \sim i}} f(x_j(t - \tau)) - f(x_i(t)) \right). \tag{6.11}$$

Here f is a differentiable function mapping some finite interval, say $[0, 1]$, to itself, $\varepsilon \in [0, 1]$ is the coupling strength, and $\tau \in \mathbf{Z}^+$ is the transmission delay between vertices. For simulation, the function $f(x)$ is chosen as $f(x) = \rho x(1 - x)$ with the value of $\rho = 4$, for which the individual uncoupled systems are fully chaotic. Synchronized regions in the parameter space for scale-free, random, small-world and nearest-neighbour coupling are shown in Fig. 6.9a–d, respectively. The authors had used the same size, $N = 10000$, and the same number of average connections, even though the architectures may be different.

It is evident from Fig. 6.9 that the scale-free and random networks can synchronize for a large range of parameters whereas more regular networks with nearest-neighbor and small-world type coupling do not. In the case of scale-free and random networks, for strong coupling (roughly for $\varepsilon > 0.6$) synchronization is achieved regardless of the actual value of the delay, as long as it is positive as seen in Fig. 6.9a, b. For intermediate coupling in the range $0.4 < \varepsilon < 0.6$, the value of the delay becomes decisive for synchronization. There are also smaller regions of synchronization that exist for weaker coupling ($0.15 < \varepsilon < 0.20$) and only for odd delays. Note that for zero delay synchronization can occur only for a rather limited range ($\varepsilon > 0.85$). Also, the small regions of synchronization in Fig. 6.9c, d occur for nonzero delay only. Hence, the presence of delay can indeed facilitate synchronization.

In a recent work [35], the authors have calculated the master stability function of networks of chaotic units with time-delayed couplings. They have shown that

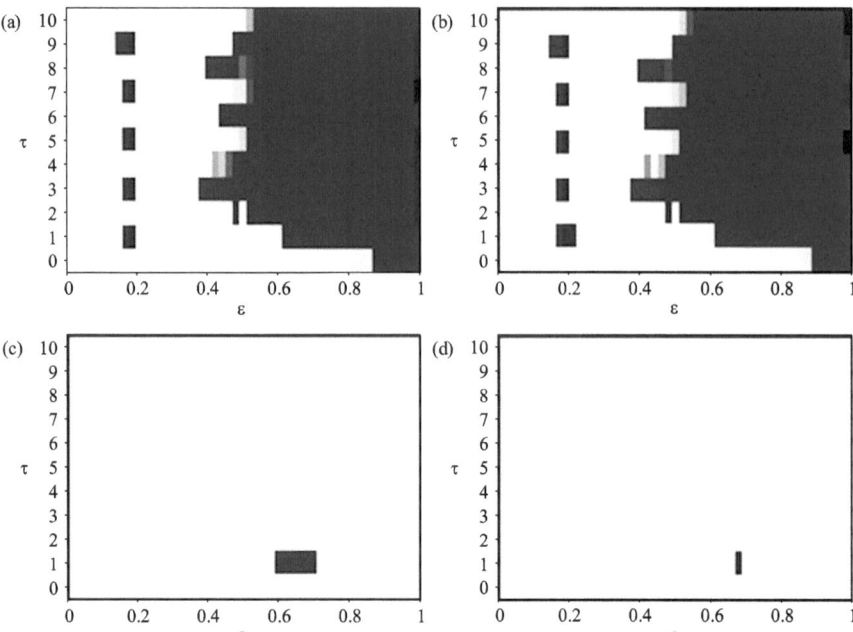

Fig. 6.9 Synchronization of coupled logistic maps for different values of coupling strength ε and connection delays τ, for the cases of (**a**) scale-free, (**b**) random, (**c**) small-world, and (**d**) nearest-neighbor coupling. The *grayscale* encodes the degree of synchronization, with *black regions* corresponding to complete synchronization. This figure is adapted from the work of Atay et al. [18]

when the delay times of transmission are much larger than any characteristic time scales of the individual units of any arbitrary finite network, the individual units are not synchronized. It is also shown that this holds for any network including the case where the individual units contain self-feedback delays. For several models including chaotic flows and maps, the authors have calculated the master stability function and determined the maximal value of delay time for which synchronization can occur.

Among these interesting reports on synchronization in complex networks with connection delays, there are also few studies available in the literature on the effect of connection delays on the synchronization dynamics of several network motifs of different architectures [36–38]. However, synchronization in complex networks with individual units as intrinsic time-delay systems with and without connection delays have not yet been considered and this remains an open problem.

6.6 Controlling Using Time-Delay Feedback

Controlling complex dynamics has been emerging as an important issue in modern nonlinear science [39–44]. Major progress has been made by extending the methods of chaos control, in particular by employing time-delay feedback, to control unstable periodic orbits or stationary states of dynamical systems. In particular, time-delay feedback control has been used to control pattern formation in neuroscience, to prevent the pathological activity in cortical tissues [45, 46], and developing applications in bio-medical engineering [47, 48], in technological application such as congestion control, load balancing control, networked control systems, etc in communication networks, to remote control and robotized operations of mechanical systems, stabilization of planar vertical takeoff and landing aircraft (PVTOL) models [49], control of gantry cranes, control of fuel injection rate in combustion engines [43], in population dynamics, economics, nonlinear optics, fluid dynamics [43, 44], etc.

In short, very often whenever one observes a state of a system to control its dynamics there exists a delayed feedback of the corresponding state to the system in its control mechanism. For instance in our daily life, steering control of a car, controlling/tuning a tap for optimal flow of hot water, etc are made by observing the occurred (past) state to control their current state and hence there appears a delayed feedback. Hence time-delay feedback control is ubiquitous and has emerged as a highly interdisciplinary subject and a large amount of material on it is now available in the literature [40, 41, 50, 42–44]. In the following, we will describe briefly the classical control scheme of Pyragas [51] and its advantage over the other methods and its further variations. We will also discuss briefly a recent work on transient behavior in systems with time-delayed feedback [52] by Hinz et al., which is closely related to the transient behaviors we have discussed in Chap. 3.

6.6.1 Pyragas Time-Delay Feedback Control

Chaotic behavior has been regarded as unwanted signal in the initial stage of its identification as it restricts the operating range of many electronic and mechanical devices [51]. Hence, several methods have been introduced to control chaotic dynamics to a desired or a targeted behavior, which includes feedback and nonfeedback methods. Nonfeedback methods require prior knowledge of the dynamical system and is less flexible. However, in nonfeedback methods control procedures can be applied at any time to control its dynamics to desired behaviour and it does not require to follow the trajectory as in feedback methods. Nevertheless, the feedback methods do not alter the system much as the feedback becomes negligibly small on reaching the targeted behavior, whereas nonfeedback methods change the systems by small parameter shift to change the behavior [40, 41, 50].

Ott-Grebogi-Yorke [53] (OGY) have proposed an efficient method of chaos control, which stabilizes the desired unstable periodic orbit (UPO) embedded in the chaotic attractor by a small time-dependent perturbation in the form of feedback to an accessible system parameter. OGY scheme has been demonstrated successfully in different experiments including Chua's oscillator, chaotic diode resonator, chaotic laser, fluid mechanical systems, etc. [51, 40]. However, OGY scheme requires continuous tracking of the trajectory and it stabilizes only those periodic orbits whose maximal Lyapunov exponent is small compared to the time interval of the parameter changes as this scheme deals with the Poincaré map. Further, this scheme is highly sensitive to noise and then control becomes less efficient.

To overcome these difficulties continuous control in the form of feedback was proposed by Pyragas [51]. Two schemes, namely a combined feedback with a periodic external force and a self controlling delayed feedback, were proposed. In the first method an external oscillator which generates a signal $y_i(t) = y(t + T_i)$ is combined with the feedback $F(t) = K\left[y_i(t) - y(t)\right]$, where T_i is the period of ith UPO, K is the adjustable weight of the perturbation and $y(t)$ is the observable scalar signal from the actual system to which the feedback signal $F(t)$ is applied. However, in the delayed feedback method no such external oscillator oscillating with the desired/target period is employed.

The delayed feedback is designed in such a way that whenever the length of the delay time, $\tau = T_i$, is equal to the period of ith UPO the corresponding UPO is stabilized and the perturbation

$$F(t) = K\left[y(t - \tau) - y(t)\right] = KD(t) \tag{6.12}$$

becomes zero. This implies that the above perturbation does not change the solution of the system corresponding to the ith UPO and stabilization is achieved for suitable value of K. In the case of OGY scheme the perturbation (feedback) is applied only when the trajectory is sufficiently close to the targeted fixed point/periodic orbit, whereas in the present case one can apply the feedback at any time to effect the control.

We present the results of the above time-delay feedback controlling scheme using the paradigmatic Rössler system represented as

$$\dot{x} = -y - z, \tag{6.13a}$$
$$\dot{y} = x + 0.2y + F(t), \tag{6.13b}$$
$$\dot{z} = 0.2 + z(x - 5.7), \tag{6.13c}$$

where $F(t)$ is as specified in Eq. (6.12). For the above chosen parameters the Rössler system exhibits chaotic oscillations in the absence of $F(t)$. As soon as the feedback delay is introduced the Rössler system exhibits periodic oscillations of period $\tau = T_i$ of the ith UPO for appropriate value of K. For the value of $K = 0.2$ and $\tau = 17.5$ the Rössler system exhibits period-three cycle as shown in Fig. 6.10 after introducing the perturbation $F(t)$ for $t > 300$. Chaotic oscillations are observed for the value of $t \leq 300$ and period-three oscillations for $t > 300$ in Fig. 6.10a. The origin of perturbation $F(t)$ in Fig. 6.10b indicates the switching of the perturbation and period-three attractor is depicted in Fig. 6.10c, plotted for $t > 700$. The amplitude of the feedback signal $D(t) = \left[y(t + \tau) - y(t)\right]$ can be considered as a criterion for stabilization of UPOs, the value of which becomes negligibly small compared to the amplitude of oscillation when the system resides on the UPOs. The dependence of the amplitude of $D(t)$ on the delay time τ for the Rössler system is illustrated in Fig. 6.11a. The dispersion of the perturbation $\left(D^2(t)\right)$ calculated for 20 different initial conditions is depicted in the range of delay time $\tau \in (4, 20)$, which indicates a sequence of resonance curves with deep minima. These minima are located at the points of delay time coinciding with the periods of the UPO,

Fig. 6.10 Results of
stabilization for
$K = 0.2, \tau = 17.5$ **(a)**
dynamics of the output signal
$y(t)$, **(b)** perturbation $F(t)$
and **(c)** the period-three cycle
of the Rössler system after
transients. The origin of the
curve F corresponds to the
moment of switching on the
perturbation

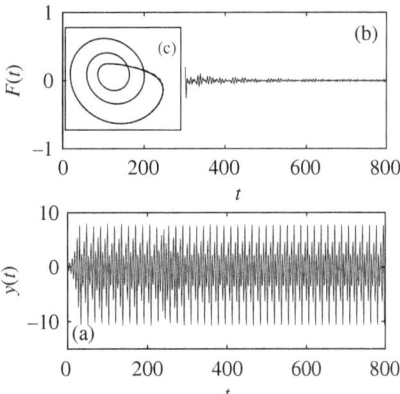

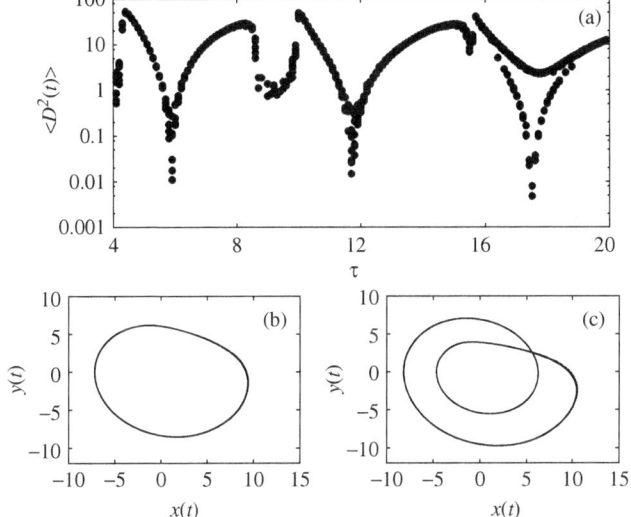

Fig. 6.11 **(a)** Dependence of the dispersion of perturbation on the delay time $\tau \in (4, 20)$ for
$K = 0.2$, **(b)** period-one cycle at $\tau = 5.9$ and **(c)** period-two cycle at $\tau = 11.7$

$\tau = T_i$. The phase portraits shown in Figs. 6.11(b,c) and 6.10c at these minima
correspond to period-one, -two and -three cycles, respectively.

   One of the major advantage of this method is that the experimental implemen-
tation of this scheme is simple when compared to the other schemes. Only two
parameters K and τ are to be tuned appropriately and a simple delay line is required
to implement this feedback. This scheme has also been shown to be robust against
noise. Further, this scheme has been successfully applied in experiments in different areas of research such as chemical systems [54], mechanical systems [55, 56],
optics [57, 58], neuroscience [59], semiconductor devices [60], etc. Pyragas control scheme is also known as time-delay autosynchronization [51, 61] (TDAS), its

further generalizations as extended time-delay autosynchronization [61, 62] (ETDAS), N time-delay autosynchronization [63] (NTDAS), multiple delay feedback control [64] (MDFC), etc., and limitations of this powerful control method [62, 65, 66] have also been investigated.

6.6.2 Transient Behavior with Time-Delay Feedback

In this section, we will discuss briefly the recent results on transient behavior in systems with time-delay feedback by Hinz et al. [52], which is interrelated to the transient effect discussed in Chap. 3. For this purpose consider a single Hopf oscillator (also called the Stuart-Landau equation in the absence of feedback term) that is driven autonomously by a time-delayed feedback term,

$$\dot{Z}(t) = \left[\lambda + i\omega - (a + ib)|Z(t)|^2\right] Z(t) - K\left[Z(t) - Z(t - \tau)\right], \qquad (6.14)$$

where, $Z(t) = x + iy$ is a complex quantity, ω is the frequency of oscillation, $\lambda, a < 0, b$ are real constants, K is the feedback gain and $\tau > 0$ is the time-delay of the autonomous feedback term. In analogy with the discussion in Sect. 5.2, the above Eq. (6.14), exhibits stable limit cycle/periodic oscillation of amplitude $\sqrt{-\lambda/a}$ for $\lambda > 0$, corresponding to supercritical Hopf bifurcation as $a < 0$, and unstable fixed point at the origin in the absence of the feedback $K = 0$.

For numerical simulation, the parameters have been fixed as $\lambda = 0.5, a = -0.1, b = 1.5, \omega = \pi, \tau = 1 = T_0/2$ and the initial conditions as $x = r_0, y = 0$ for $t \in (-\tau, 0)$ as in [52]. Dependence of the transient time τ_{tr} on the feedback gain K and initial amplitude r_0 is shown in Fig. 6.12. The unshaded regimes correspond to the parameter values for which the trajectory does not reach the fixed point. As indicated above the fixed point is unstable in the absence of the feedback gain K. However, the fixed point is stabilized by the time-delayed feedback for some finite value of the feedback gain. It is evident from the figure that the transient time τ_{tr} increases with the feedback gain K until the stabilization is lost. In order to get more clear insights on the transient time of control of time-delay feedback scheme, the trajectory in the (x, y) phase space and the amplitude $r = |Z(t)|$ as a function

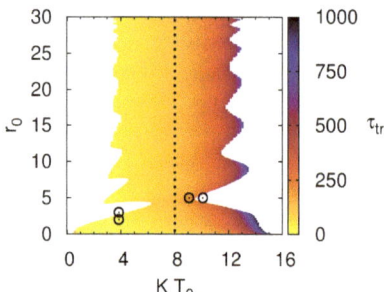

Fig. 6.12 Transient times to reach the fixed point at the origin in the (K, r_0) parameter space. The *circles* corresponds the choice of parameters used to depict Fig. 6.13. This figure is adapted from the work of Hinz et al. [52]

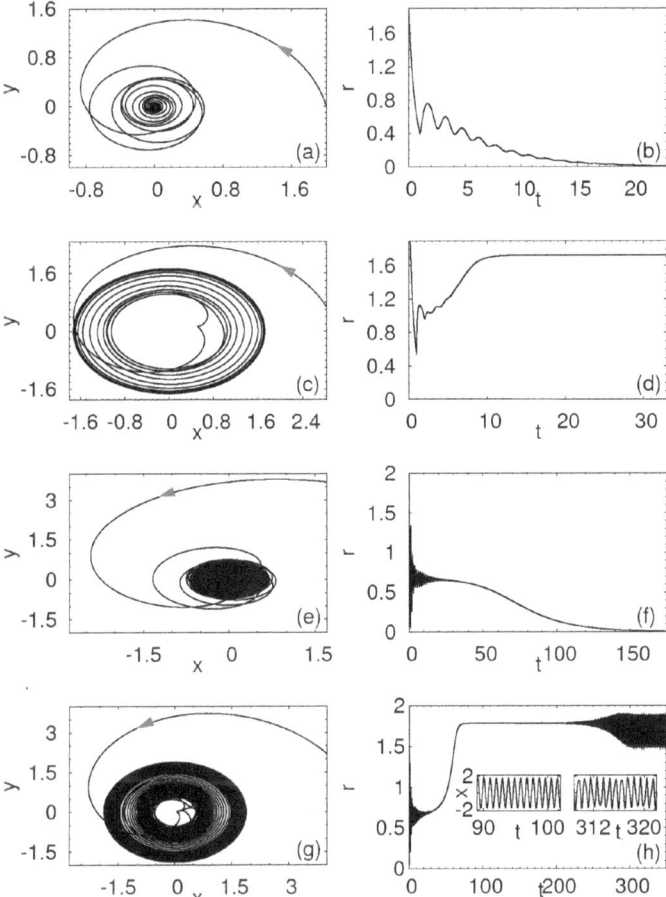

Fig. 6.13 Phase portraits and the amplitude $r = |Z(t)|$ as a function of time t for different combinations of KT_0 and r_0. (**a, b**) $KT_0 = 4, r_0 = 2$; (**c, d**) $KT_0 = 4, r_0 = 3$; (**e, f**) $KT_0 = 9, r_0 = 5$; (**g, h**) $KT_0 = 10, r_0 = 5$. This figure is adapted from the work of Hinz et al. [52]

of time t are illustrated in the left and right panels of Fig. 6.13 for four different combinations of the feedback gain K and the initial amplitude r_0, corresponding to the circles in Fig. 6.12. The fixed point in the origin is stabilized for the initial amplitude $r_0 = 2$ and the feedback gain $KT_0 = 4$ as in Fig. 6.13a, b, whereas a delay-induced stable periodic orbit is asymptotically reached on increasing the initial amplitude to $r_0 = 3$ for the same value of the feedback gain $KT_0 = 4$ (Fig. 6.13c, d).

Again the fixed point at the origin is stabilized on increasing the feedback gain and it is depicted in Fig. 6.13e, f for $KT_0 = 9$ and $r_0 = 5$. Further increase in the value of the feedback gain ($KT_0 = 10$) for the same value of the initial amplitude results in a delay-induced torus (see inset in Fig. 6.13h) as shown in Fig. 6.13g, h. It is also clear from these figures that the transients increase with the value of feedback gain until the stabilization is lost. It also explains the modulation of the stability range in Fig. 6.12 due to resonances with delay-induced periodic or quasi-periodic orbits which reduce the basin of attraction of the fixed point. Detailed discussion on it can be found in [52].

6.7 Further Developments

We will provide a brief discussion on further developments in systems with delay, delay feedback and delay coupling in addition to the phenomena discussed above in this chapter.

Recently, generic properties of systems with time-delay that are related to the appearance and stability of periodic solutions are discussed by Yanchuk et al. [67]. In particular they have shown that delay systems generically have families of periodic solutions, which are reappearing for infinitely many delay times and they overlap leading to increasing coexistence of multiple stable as well as unstable solutions. These authors have also reported the stability issue of periodic solutions with large delay by explaining asymptotic properties of the spectrum of characteristic multipliers, which can be split into two parts as pseudocontinuous and strongly unstable [67].

Recent investigations on synchronization has also focused on the synchronization patterns induced by distributed time-delay [68, 69]. In particular, an array of coupled pendulums with randomly distributed distances between any two neighbors are considered and the effect of different distances as distributed time-delays in the coupling interactions is investigated [68]. It has been shown that the distributed delay stabilize the chaotic dynamics of the coupled system and different periodic patterns appears with increase of the range of distributed delay in confirmation with the result of [67]. Further, it has been shown that the distributed delays in chemical coupling can induce a variety of phase-coherent dynamic behaviours in inhibitory coupled bursting Hindmarsh-Rose neurons [69]. It has also been shown that time-delay can induce stabilization of a steady state in network of oscillators [70]. Propagation of synchronization and desynchronization wave fronts in coupled map chains with delayed transmission was investigated [71].

Time-delay feedback has also been investigated largely in inducing spatiotemporal instabilities which plays vital role in controlling pathological activities in neural systems [42]. In particular, it has been shown that short delay beyond a critical threshold may induce spatiotemporal instabilities [72]. Delay induced spatial correlations in one-dimensional stochastic networks with nearest neighbour coupling was also reported [73]. Fundamental design principles are presented for vehicle systems governed by autonomous cruise control devices using delay-induced patterns [74]. It is shown that for any car-following model short-wavelength oscillations can appear due to robotic reaction times, and that there are tradeoffs between the time-delay and the control gains. Recently, it has also been shown that partial time-delay coupling enlarges death island of coupled oscillators [75].

References

1. S. Kim, S.H. Park and C.S. Ryu, Phys. Rev. Lett. **79** 2911 (1997)
2. U. Ernst, K. Pawelzik and T. Geisel, Phys. Rev. E **57** 2150 (1998)
3. D.V. Ramana Reddy, A. Sen, and G.L. Johnston, Phys. Rev. Lett. 85, 3381 (2000)
4. G.C. Sethia, A. Sen, F.M. Atay, Phys. Rev. Lett. **100**, 144102 (2008)
5. J.H. Sheeba, V.K. Chandrasekar, M. Lakshmanan, Phys. Rev. E **79**, 055203(R) (2009); J.H. Sheeba, V.K. Chandrasekar, M. Lakshmanan, Phys. Rev. E **81**, 049906 (2010)
6. M.G. Rosenblum and A.S. Pikovsky, Phys. Rev. Lett. **92** 114102 (2004)
7. S. Boccaletti, J. Kurths, G. Osipov, D.L. Valladares, C.S. Zhou, Phys. Rep. **366**, 1 (2002)
8. A. Arenas, A. Diaz-Guilera, J. Kurths, Y. Moreno, C. Zhou, Phys. Rep. **469**, 93 (2008)
9. A. Clauset, C. Moore, M.E.J. Newman, Nature **453**, 98 (2008)
10. A.T. Winfree, *The Geometry of Biological Time*, (Springer, New York, 1980)
11. Y. Kuramoto, *Chemical Oscillations, Waves, and Turbulence*, (Springer, New York, 1984)
12. K. Wiesenfeld, P. Colet, S.H. Strogatz, Phys. Rev. E, **57**, 1563 (1998)
13. D.L. Turcotte, *Fractals and Choas in Geology and Geophysics* 2nd *edn.*, (Cambridge Univ. Press, Cambridge, 1997)
14. R.M. May, *Stability and Complexity in Model Ecosystems*, (Princeton Univ. Press, Princeton, 1973)
15. S.A. Levin, B.T. Grenfell, A. Hastings, A.S. Parelson, Science, **275**, 334 (1997)
16. M. Arbib (ed.), *The Handbook of Brain Theory and Neural Networks*, (MIT Press, Cambridge, MA, 1995)
17. J. Pantaleone, Phys. Rev. D, **58**, 3002 (1998)
18. F.M. Atay, J. Jost, Phys. Rev. Lett. **92**, 044101 (2004)
19. C. Massoller, A.C. Marti, Phys. Rev. Lett. **94**, 134102 (2005)
20. D.M. Abrams, S.H. Strogatz, Phys. Rev. Lett. **93**, 174102 (2004)
21. Y. Kuramoto, D. Battogtokh, Nonlinear Phenom. Complex Syst. **5**, 380 (2002)
22. D. Tanaka, Y. Kuramoto, Phys. Rev. E **68**, 026219 (2003)
23. S. I. Shima, Y. Kuramoto, Phys. Rev. E **69**, 036213 (2004)
24. D.M. Abrams, R. Mirollo, S.H. Strogatz, and D.A. Wiley, Phys. Rev. Lett. **101**, 084103 (2008)
25. A.T. Winfree, J. Theore. Biol. **16**, 15 (1967)
26. S.H. Strogatz and I. Stewart, Sci. Am. **12** 68 (1993)
27. Y. Kuramoto, Prog. Theor. Phys. **94** 321 (1995)
28. Y. Kuramoto and H. Nakao, Phys. Rev. Lett. **76** 4352 (1996)
29. Y. Kuramoto, D. Battogtokh and H. Nakao, Phys. Rev. Lett. **81** 3543 (1998)
30. Y. Kuramoto, H. Nakao, and D. Battogtokh, Physica A. **288** 244 (2000)
31. Y. Kuramoto, in *Nonlinear Dynamics and Chaos: Where Do We Go from Here?*, ed. by S. J. Hogen et al., (Institute of Physics, Bristol, 2003)
32. Oleh E. Omelćhenko, Y.L. Maisternko, P.A. Tass, Phys. Rev. Lett. **100**, 144105 (2008)
33. E.A. Marten, C.R. Laing, S.H. Strogatz, Phys. Rev. Lett. **104**, 044101 (2010)
34. N. Yousif, X. Liu, Expert Rev. Med. Dev. **4**, 623 (2007)
35. W. Kinzel, A. Englert, G. Reents, M. Zigzag, I. Kanter, Phys. Rev. E **79**, 056207 (2009)
36. C. M. Gonzalez, C. Masoller, M.C. Torrent, J. Garcia-Ojalvo, Euro. Phys. Lett. **79**, 64003 (2007)
37. O. D'Huys, R. Vicente, T. Erneux, J. Danckaert, I. Fischer, Chaos **18**, 037116 (2008)
38. T. Oguchi, H. Nijmeijer, T. Yamamoto, Chaos **18**, 037108 (2008)
39. Y. Kuramoto, *Chemical Oscillations, Waves, and Turbulence*, (Springer, Berlin, 1984)
40. T. Kapitaniak, *Controlling Chaos: Theoretical and Practical Methods in Non-linear Dynamics*, (Academic Press, London, 1996)
41. M. Lakshmanan, K. Murali, *Chaos in Nonlinear Oscillators: Controlling and Synchronization*, (World Scientific,Singapore, 1996)
42. E. Schöll, H.G. Schuster (eds.), *Handbook of Chaos Control* (Wiley-VCH, Weinheim, 2008)
43. T. Erneux, *Applied Delay Differential Equations* (Springer, New York, 2009)

44. B. Balachandran, T. Kalmár-Nagy, E.D. Gilsinn (eds.), *Delay Differential Equations: Recent Advances and New Directions* (Springer, New York, 2009)
45. M.A. Dahlem, F.M. Schneider, E. Schöll, Chaos **18**, 026110 (2008)
46. F.M. Schneider, E. Schöll, M.A. Dahlem, Chaos **19**, 015110 (2009)
47. K.A. Richardson, S.J. Schiff, B.J. Gluckmann, Phys. Rev. Lett. **94**, 028103 (2005)
48. U.B. Barnikol, O.V. Popovych, C. Hauptmann, V. Strum, H.J. Freund, P.A. Tass, Philos. Trans. R. Soc. Lond. Ser. A **366**, 3545 (2008)
49. J. Chiasson, J.J. Loiseau (eds.), *Applications of Time Delay Systems* (Springer, Berlin, 2007)
50. M. Lakshmanan, S. Rajasekar, *Nonlinear Dynamics: Integrability, Chaos and Patterns*, (Springer-Verlag, Berlin, 2003)
51. K. Pyragas, Phys. Lett. A **170**, 421 (1992)
52. R.C. Hinz, P. Hövel, E. Schöll, arXiv:0912.1727v1 (2009)
53. E. Ott, C. Grebogi, Y.A. Yorke, Phys. Rev. Lett. **64**, 1196 (1990)
54. A.G. Balanov, V. Beato, N.B. Janson, H. Engel, E. Schöll, Phys. Rev. E **74**, 016214 (2006)
55. J. Sieber, A. Goneyley-Buelga, S. Neild, D. Wagg, B. Krauskopf, Phys. Rev. Lett. **100**, 244101 (2008)
56. K.B. Blyuss, Y.N. Kyrychko, P. Hövel, E. Schöll, Eurphys. J. B **65**, 571 (2008)
57. V.Z. Tronciu, H.J. Wünsche, M. Wolfrum, M. Radziunas, Phys. Rev. E **73**, 046205 (2006)
58. T. Dahms, P. Hövel, E. Schöll, Phys. Rev. E **78**, 056213 (2008)
59. F.M. Schneider, E. Schöll, M.A. Dahlem, Chaos **19**, 015110 (2009)
60. E. Schöll, *Nonlinear Spatio-temporal Dynamics and Chaos in Semiconductors* (Cambridge University Press, Cambridge, 2001)
61. K. Pyragas, Phys. Lett. A **206**, 323 (1995)
62. H. Nakajima, Y. Ueda, Physica D **111**, 143 (1997)
63. J.E.S. Socolar, J.D. Gauthier, Phys. Rev. E **57**, 6589 (1998)
64. A. Ahlborn, U. Parlitz, Phys. Rev. Lett. **93**, 264101 (2004)
65. B. Fiedler, V. Flunkert, M. Georgi, P. Hövel, E. Schöll, Phys. Rev. Lett. **98**, 114101 (2007)
66. W. Just, B. Fiedler, V. Flunkert, M. Georgi, P. Hövel, E. Schöll, Phys. Rev. E **76**, 026210 (2007)
67. S. Yanchuk, P. Perlikowski, Phys. Rev. E **79**, 046221 (2009)
68. J. Zhou, Z. Liu, Phys. Rev. E **77**, 056213 (2008)
69. X. Liang, M. Tang, M. Dhamala, Z. Liu, Phys. Rev. E **80**, 066202 (2009)
70. K. Konishi, H. Kokame, N. Hara, Phys. Rev. E **81**, 016201 (2010)
71. D. Schmitzer, W. Kinzel, I. Kanter, Phys. Rev. E **80**, 047203 (2009)
72. S. Sen, P. Ghosh, S.S. Riaz, D.S. Ray, Phys. Rev. E **80**, 046212 (2009)
73. A. Pototsky, N.B. Janson, Phys. Rev. E **80**, 066203 (2009)
74. G. Orosz, J. Moehlis, F. Bullo, Phys. Rev. E **81**, 025204(R) (2010)
75. W. Zou, M. Zhan, Phys. Rev. E **80**, 065204(R) (2009)

Chapter 7
Complete Synchronization of Chaotic Oscillations in Coupled Time-Delay Systems

7.1 Introduction

Historically, synchronization phenomenon dates back to the period of Christiaan Huygens (1629–1695), who in 1665 found that two very weakly coupled pendulum clocks, hanging from the same beam, become anti-phase synchronized [1]. Since the early identification of synchronization in coupled chaotic oscillators [2–4], the phenomenon has attracted considerable research activity in different areas of science, and several generalizations and interesting applications have been developed [1, 5–9]. Chaos synchronization is of interest not only from a theoretical point of view but also has potential applications in diverse areas involving physical, chemical, biological, neurological, electrical and fluid mechanical systems. In particular, possible applications of chaos synchronization include secure communication, cryptography, controlling, long term prediction, optimization of nonlinear system performance, modelling brain activity, pattern recognition, and so on [1–18].

In recent years, different kinds of chaos synchronizations, which are characterized by the difference in the degree of correlation between the interacting chaotic dynamical systems, have been identified:

1. Complete (or identical) synchronization refers to the identical evolution of the interacting systems, $Y(t) = X(t)$ [2, 3],
2. Generalized synchronization is observed in coupled nonidentical systems, where there exists some functional relation between the states of the response and the drive systems, that is, $Y(t) = F(X(t))$ [19–21],
3. Phase synchronization means entrainment of phases of the interacting systems, $n\Phi_x - m\Phi_y = const.$ (n and m are integers), while their amplitudes remain chaotic and often uncorrelated [22, 23],
4. Lag synchronization refers to the phenomenon, where the state of the response system lags the state of the drive system with a lag time $\tau > 0$, $Y(t) = X(t - \tau)$ [24–26] and
5. Anticipatory synchronization corresponds to the fact that the state of the response system anticipates the state of the drive system with an anticipating time $\tau > 0$, $Y(t) = X(t + \tau)$ [27–29], etc.

M. Lakshmanan, D.V. Senthilkumar, *Dynamics of Nonlinear Time-Delay Systems*,
Springer Series in Synergetics, DOI 10.1007/978-3-642-14938-2_7,
© Springer-Verlag Berlin Heidelberg 2010

We have provided a brief discussion on the above types of synchronizations along with their characterizations in Appendix B.

Synchronization of chaotic systems with coexisting attractors indicates that the route to complete synchronization is characterized by a sequence of type-I and on-off intermittencies, intermittent phase synchronization, anticipatory synchronization and period-doubling phase synchronization [30]. Transition from one kind of synchronizations to the other, coexistence of different kinds of synchronization in time series and also the nature of transitions have been studied extensively [24–26, 31, 32] in coupled chaotic systems. The role of parameter mismatch in synchronization phenomena is quite versatile and it has also been widely reported in the literature [29, 33–38]. For a critical discussion on the interrelationship between various kinds of synchronizations, we may refer to [39, 40]. There are also attempts to find a unifying framework for defining the overall class of chaotic synchronizations [39–41]. Reviews on the phenomenon of chaos synchronization can be found in [1, 5–7].

One of the most important potential applications of chaos synchronization is secure communication. Several approaches for chaos synchronization and control of chaos with application to secure communication have been the focus of many recent investigations [13, 16, 42–50]. Important milestones in this direction include the first demonstration of synchronizing hyperchaos with a single scalar transmitted signal [51], see for example the comments of L. Pecora in "Physics in Action" section of "Physics World", May 1996 issue under the title "Hyperchaos harnessed" [52]. Successful demonstration of communication with chaotic lasers in the laboratory has been made in 1998, in which erbium-doped fiber ring laser (with delay feedback) was used to produce chaotic light with frequency around 100 MHz [53], see also [54]. An important landmark in this direction is the demonstration of high-speed long-distance communication based on chaos synchronization over a commercial fibre-optic channel. Argyris and colleagues [55] reported the successful transfer of digital information at gigabit rates by chaotically fluctuating laser light travelling over 120 km of a commercial fibre-optic link around Athens, Greece. The scheme used by Argyris et al. exploited time-delayed feedback to generate high-dimensional, high-capacity waveforms at high bandwidths.

It is now an accepted fact that secure communication based on simple low dimensional chaotic systems does not ensure sufficient level of security, as the associated chaotic attractors can be reconstructed with some effort and the hidden message can be retrieved by an eavesdropper [52, 56–58]. Therefore, synchronization of hyperchaotic dynamics has been proposed as an alternative method for improving the security in the communication schemes [13, 51, 59, 60]. However, even here also it was demonstrated that messages masked by a hyperchaotic signal can be extracted by using nonlinear dynamic forecasting as the local dynamics does not reflect more complicated dynamics significantly [56, 57].

One way to overcome this problem is to consider chaos synchronization in high dimensional systems having multiple positive Lyapunov exponents. This increases the security by giving rise to much more complex time series, which are apparently not vulnerable to the unmasking procedures generally. Recently chaotic time-delay systems have been suggested as good candidates for secure communication [51, 61–72], as the time-delay systems are essentially infinite dimensional in nature and are described by delay differential equations, and that they can admit hyperchaotic attractors with large number of positive Lyapunov exponents for suitably chosen nonlinearity.

However, it should be noted that one has to be cautious due to the fact that even in time-delay systems with multiple positive Lyapunov exponents unmasking may be possible. Particularly, this is so if any reconstruction of the dynamics of the system is achieved in some appropriate space even for very high dimensional dynamics as demonstrated by Zhou and Lai [73] in the case of Mackey-Glass equation. Nevertheless, it has been shown that delay time modulation (time-dependent delay) wipes off any imprints of the delay time carved in the time series of a time-delay system and that the reconstructed phase trajectory of the system is not collapsed into a simple manifold [74]. Synchronization, encryption and communication in coupled time-delay systems in the presence of delay time modulation was also reported [75–77].

In addition to the exploitation of the infinite dimensionality nature of the time-delay systems in secure communication, time delay serves as a source of instability and results in many new phenomena that cannot be observed in the absence of delay. Hence time-delay systems are now recognized as veritable black boxes that give rise to novel phenomena such as amplitude death, phase flip bifurcation, Neimark-Sacker type bifurcation, etc. Because of the ease of experimental realization of time-delay systems, these phenomena have also been demonstrated experimentally as discussed in the previous chapters.

In view of the above facts, study of chaos synchronization in coupled time-delay systems has become an active area of research both at the theoretical and experimental levels [37, 38, 41, 74, 76–100]. In recognition of this fact, from this chapter onwards, we will discuss different kinds of synchronizations and their transitions in coupled time-delay systems along with derivations of suitable stability conditions for the asymptotically stable synchronized states. In particular, in this chapter we will demonstrate complete synchronization in coupled time-delay systems with suitable stability condition using Krasvoskii-Lyapunov functional approach. We will show that the same general stability condition is valid for different cases, even for the general situation where all the coefficients of the error equation corresponding to the synchronization manifold are time-dependent (see next section for details). These analytical results are also confirmed by numerical simulation of paradigmatic examples.

7.2 Complete Synchronization in Coupled Time-Delay Systems

As mentioned above, recent studies on synchronization in coupled time-delay systems with or without time-delay coupling is receiving considerable importance both theoretically and experimentally due to the infinite dimensional nature of the underlying systems possessing a very large number of positive Lyapunov exponents as a function of the delay time [62]. In particular, in recent studies on synchronization in coupled time-delay systems, the Krasovskii-Lyapunov theory has been widely used in identifying the stability of the asymptotically stable synchronized states [27, 89, 101, 102].

In this chapter we will show that the same general stability condition resulting from the Krasovskii-Lyapunov theory indeed holds good for the rather general case as well. In particular, we will discuss all the four possible cases that arise due to the nature of the coefficients in the error equation corresponding to the synchronization manifold and show that the same stability condition deduced from the Krasovskii-Lyapunov functional approach is valid for all the cases, subject to certain conditions. We will also confirm these analytical results by numerical analysis using paradigmatic examples.

7.3 Stability Using Krasovskii-Lyapunov Theory

Consider the following linearly coupled scalar time-delay system,

$$\dot{x}(t) = -ax(t) + bf(x(t-\tau)), \tag{7.1a}$$
$$\dot{y}(t) = -ay(t) + bf(y(t-\tau)) + K(t)(x(t) - y(t)), \tag{7.1b}$$

where a and b are positive constants, $\tau > 0$ is the delay-time, $K(t)$ is the coupling function between the drive and the response systems and $f(x)$ is a continuously differentiable function. Now we can deduce the stability condition for complete synchronization of the general unidirectionally coupled time-delay systems (7.1). The time evolution of the difference system with the state variable $\Delta = x(t) - y(t)$ (the error equation corresponding to the complete synchronization manifold of the coupled time-delay system (7.1)) for small values of it can be written as

$$\dot{\Delta} = -(a + K(t))\Delta + bf'(y(t-\tau))\Delta_\tau, \quad \Delta_\tau = \Delta(t-\tau). \tag{7.2}$$

It is to be noted that there arises four cases depending on the nature of the coefficients of the Δ and Δ_τ terms of the above error equation as follows:

1. Both coefficients of the Δ and Δ_τ terms are time-independent.
2. The coefficient of the Δ term is time-independent and that of the Δ_τ term is time-dependent.
3. The coefficient of the Δ term is time-dependent and that of the Δ_τ term is time-independent.
4. Both coefficients of the Δ and Δ_τ terms are time-dependent.

The synchronization manifold of the error equation (7.2) is locally attracting if the origin of this equation is stable. Following the Krasovskii-Lyapunov theory [101], we define a continuous, positive-definite Lyapunov functional of the form

$$V(t) = \frac{1}{2}\Delta^2 + \mu \int_{-\tau}^{0} \Delta^2(t + \theta)d\theta, \qquad V(0) = 0 \qquad (7.3)$$

where μ is an arbitrary positive parameter, $\mu > 0$. The derivative of the functional $V(t)$ along the trajectory of the error equation (7.2),

$$\frac{dV}{dt} = -(a + K(t))\Delta^2 + bf'(y(t - \tau))\Delta\Delta_\tau + \mu\Delta^2 - \mu\Delta_\tau^2, \qquad (7.4)$$

has to be negative to ensure the stability of the solution $\Delta = 0$. The above equation can be written as

$$\frac{dV}{dt} = -\mu\Delta^2\Gamma(X, \mu), \qquad (7.5)$$

where $X = \Delta_\tau/\Delta$ and

$$\Gamma = \left[\left((a + K(t) - \mu)/\mu\right) - \left(bf'(y(t - \tau))/\mu\right)X + X^2\right]. \qquad (7.6)$$

In order to show that $\frac{dV}{dt} < 0$ for all Δ and Δ_τ and so for all X, it is sufficient to show that $\Gamma_{min} > 0$. One can easily check that the absolute minimum of Γ occurs at

$$X = \frac{1}{2\mu}bf'(y(t - \tau)), \qquad (7.7)$$

with Γ_{min} as

$$\Gamma_{min} = \left[4\mu(a + K(t) - \mu) - b^2 f'(y(t - \tau))^2\right]/4\mu^2. \qquad (7.8)$$

Consequently, we have the condition for stability as

$$a + K(t) > \frac{b^2}{4\mu} f'(y(t - \tau))^2 + \mu = \Phi(\mu). \qquad (7.9)$$

Now, $\Phi(\mu)$ as a function of μ for a given $f'(x)$ has an absolute minimum at

$$\mu = (|bf'(y(t - \tau))|)/2, \qquad (7.10)$$

with $\Phi_{min} = |bf'(y(t-\tau))|$. Since $\Phi \geq \Phi_{min} = |bf'(y(t-\tau))|$, from the inequality (7.9), it turns out that a sufficient condition for asymptotic stability is

$$a + K(t) > |bf'(y(t - \tau))|. \qquad (7.11)$$

It is to be noted that since μ is an arbitrary positive parameter due to the definition of the positive definite Lyapunov function (7.3), the above stability condition holds good only when $\mu = (|bf'(y(t - \tau))|)/2$ is a constant, i.e., only when $f'(x)$ is a constant (in other words when the coefficient of Δ_τ term in the error equation (7.2) is time-independent, which corresponds to the cases (1) and (3) discussed above). On the other hand if $f'(x)$ is time-dependent, then μ can be obtained alternatively by rewriting Eq. (7.9) as

$$b^2 f'(y(t - \tau))^2 < 4\mu(a + K(t) - \mu), \tag{7.12}$$
$$= -4\left[\mu - (a + K(t))/2\right]^2 + (a + K(t))^2,$$
$$\equiv \Psi(\mu).$$

Now, $\Psi(\mu)$ as a function of μ for a given $f'(x)$ has an absolute maximum at

$$\mu = (a + K(t))/2, \tag{7.13}$$

with $\Psi_{max} = (a + K(t))^2$. Using this maximum value in the right hand side of (7.12), we obtain the same stability condition as that of (7.11), provided $(a + K(t))/2 > 0$ since $\mu > 0$. Since $a > 0$, this implies $K(t) > -a$, that is coupling function $K(t)$ should be either positive definite or $|K(t)| > a$ if it is negative. In particular for the case 2, since the coefficient of the Δ term in the error equation is time independent (which corresponds to the cases (1) and (2) mentioned above), $K(t) = k > -a$ for all t (k : $const.$).

However, there arises an even more general situation where the coefficients of both the Δ and Δ_τ terms are time dependent (case 4), in which case the arbitrary positive parameter μ in the Lyapunov functional has to be chosen as a positive definite function, $\mu = g(t) > 0$ for all t. In this case, one has to consider the derivative of $\mu = g(t)$ also in the derivative of $V(t)$ as follows,

$$\frac{dV}{dt} = -(a + K(t))\Delta^2 + bf'(y(t - \tau))\Delta\Delta_\tau \tag{7.14}$$
$$+ g(t)\left(\Delta^2 - \Delta_\tau^2\right) + \dot{g}(t)\int_{-\tau}^{0} \Delta^2(t + \theta)d\theta < 0.$$

It is known from the Lyapunov functional that the term $\int_{-\tau}^{0} \Delta^2(t + \theta)d\theta$ is positive definite and let us suppose that $\dot{g}(t) \le 0$ for all t. Then for $\dot{V}(t) < 0$ we need

$$-(a + K(t))\Delta^2 + bf'(y(t - \tau))\Delta\Delta_\tau \tag{7.15}$$
$$+ g(t)\left(\Delta^2 - \Delta_\tau^2\right) < 0,$$

that is,

$$-\left[(a + K(t)) - b^2 f'(y(t - \tau))^2/4g(t) - g(t)\right]\Delta^2 \tag{7.16}$$
$$- g(t)\left[\Delta_\tau - bf'(y(t - \tau))\Delta/2g(t)\right]^2 < 0.$$

The second term in the above equation is positive definite by assumption of $g(t)$ and hence it follows that

$$b^2 f'(y(t - \tau))^2 < 4g(t)(a + K(t) - g(t)), \tag{7.17}$$
$$= -4\left[g(t) - (a + K(t))/2\right]^2 + (a + K(t))^2,$$
$$\equiv \Gamma(g(t)).$$

Consequently we obtain the same stability condition as in Eq. (7.11) with the maximum of Γ, $\Gamma_{max} = (a + K(t))^2$, occurring at $g(t) = (a + K(t))/2 > 0$, along with the condition $\dot{g}(t) = \frac{dK}{dt} \leq 0$ for all t.

Thus, one can show that the same general stability condition, Eq. (7.11), is valid for all the four cases that arise in the error equation (7.2) corresponding to the synchronization manifold of the unidirectionally coupled time-delay systems (subject to the above constraints on $K(t)$ in specific situations).

7.4 Numerical Confirmation

In this section, we will provide numerical confirmation of the above stability analysis for all the four cases using appropriate nonlinear functional forms $f(x)$ and suitable coupling $K(t)$ in the coupled time-delay systems (7.1). For this purpose we will consider the following nonlinear functions $f(x)$:

(1) the piecewise linear function, which has been discussed earlier in Chap. 3,

$$f(x) = \begin{cases} 0, & x \leq -4/3 \\ -1.5x - 2, & -4/3 < x \leq -0.8 \\ x, & -0.8 < x \leq 0.8 \\ -1.5x + 2, & 0.8 < x \leq 4/3 \\ 0, & x > 4/3, \end{cases} \tag{7.18}$$

and (2) the Ikeda model, where

$$f(x) = \sin(x(t - \tau)). \tag{7.19}$$

We have fixed the parameters as $a = 1.0$, $b = 1.2$ and $\tau = 25.0$ for the coupled piecewise linear time-delay system defined by (7.1) and (7.18), for which the uncoupled systems exhibit a hyperchaotic behavior with nine positive Lyapunov exponents [89, 102–104], see Sect. 3.3.4. For the coupled Ikeda systems (7.1) and (7.19), the parameters are chosen as $a = 1.0$, $b = 5.0$ and $\tau = 2.0$ where the uncoupled individual Ikeda systems exhibit a hyperchaotic behavior with three positive Lyapunov exponents [102], see Sect. 4.3.2.

We choose the coupling function $K(t)$ as a square wave function represented as [94]

$$K(t) = \{(t_0, k_1), (t_1, k_2), (t_2, k_1), (t_3, k_2), \ldots\}, \tag{7.20}$$

where $t_j = t_0 + (j - 1)\tau_s$, $j \geq 1$ is the switching instant, $k_1 > 0, k_2 > 0$ with $k_1 \neq k_2$. For constant coupling, $K(t) = k_1 = k_2$. On the other hand, if either $k_1 = 0$ or $k_2 = 0$, then the coupling is called an intermittent coupling/control which is now being widely studied in the literature [105–107].

7.4.1 Case 1

First, we use the piecewise linear function (7.18), and the constant coupling $K(t) = k_1 = k_2$. It is clear from the form of the nonlinear function $f(x)$ and the coupling that both the coefficients of the Δ and Δ_τ terms in the error equation (7.2) are constant (case 1) and consequently μ can be chosen as $\mu = (|bf'(y(t - \tau))|)/2$. The time trajectories of the variables $x(t)$ and $y(t)$ of the coupled piecewise linear time-delay systems (7.1) and (7.18) are shown in Fig. 7.1a indicating complete synchronization between them for the coupling strength $k = k_1 = k_2 = 0.9$ satisfying the stability condition $a + k > bf'(y(t - \tau)) = 1.5b$. Here, the other system parameters are fixed as noted above.

7.4.2 Case 2

Next, we analyse the function $f(x) = \sin(x(t - \tau))$, given by (7.19), of the Ikeda system with constant coupling, which corresponds to the case 2 where the coefficient

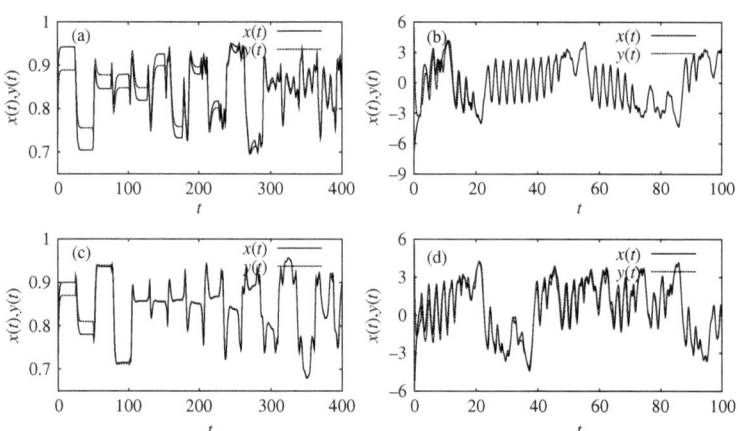

Fig. 7.1 The time trajectory plot of the variables $x(t)$ and $y(t)$ of the coupled time-delay systems (7.1) indicating complete synchronization between them. (**a**) Piecewise linear time-delay system, (7.1) and (7.18), for the parameters $a = 1.0, b = 1.2, \tau = 25.0$ and for the constant coupling $k_1 = k_2 = 0.9$. (**b**) Ikeda time-delay system, (7.1) and (7.19), for the parameters $a = 1.0, b = 5$, $\tau = 2.0$ along with the constant coupling $k_1 = k_2 = 5.0$. (**c**) Piecewise linear time-delay system for the same values of the system parameters as in Fig. 7.1a with the *square wave* coupling rates $k_1 = 0.9$ and $k_2 = 1.0$. (**d**) Ikeda system for the same values of the system parameters and with the *square wave* coupling rates $k_1 = 5.0$ and $k_2 = 6.0$

of the Δ_τ term in the error equation is time-dependent, while that of the Δ term is still time-independent and hence μ can take the form $\mu = (a + K(t))/2$ with $K(t) > 0$. The coupling strength is fixed as $k = k_1 = k_2 = 5.0$ such that the stability condition $a + k > bf'(y(t - \tau)) = b$ is satisfied. The variables $x(t)$ and $y(t)$ of the coupled Ikeda systems, (7.1) and (7.19), are plotted as a function of time in Fig. 7.1b demonstrating complete synchronization between them.

7.4.3 Case 3

Again, we consider the piecewise linear function (7.18), and the same parameter values as in the case 1 but with the square wave coupling $K(t)$ chosen as $k_1 = 0.9$ and $k_2 = 1.0$ such that the stability condition (7.11) is satisfied for all t. The switching instant τ_s between k_1 and k_2 for the square wave coupling rate is fixed as $\tau_s = 1.0$. This situation corresponds to the case 3, where the coefficient of the Δ term in the error equation is time-dependent, while that of the Δ_τ term is time-independent and as a result μ can be fixed as $\mu = (|bf'(y(t - \tau))|)/2$. The time trajectories of the variables $x(t)$ and $y(t)$ are shown in Fig. 7.1c indicating complete synchronization. Note that here $K(t) > 0$ and the stability condition (7.11) is indeed satisfied.

7.4.4 Case 4

Finally for the more general case where both the coefficients of the Δ and Δ_τ terms of the error equation are time-dependent, $\mu = g(t)$ can be given as $g(t) = (a + K(t))/2$ for the chosen form of the square wave coupling $K(t)$ with $K(t) > 0$. Figure 7.1d is plotted for the same values of the system parameters as in Fig. 7.1b with $k_1 = 5.0, k_2 = 6.0$ and $\tau_s = 1.0$ satisfying the stability condition (7.11), indicating complete synchronization between the variables $x(t)$ and $y(t)$.

7.5 Conclusion

In this chapter, asymptotic stability of synchronized state in a unidirectionally coupled general time-delay system is studied using the Krasovskii-Lyapunov theory. It is shown that the same stability condition is valid for all the four cases that arise due to the nature of the coefficients of the Δ and Δ_τ terms in the error equation corresponding to the synchronization manifold. In particular, it is shown that the same general stability condition is valid even for the general case where both the coefficients of the Δ and Δ_τ terms in the error equation are time-dependent, which is of high importance for various applications. We have also numerically confirmed these results using appropriate examples along with suitable coupling configuration.

References

1. A.S. Pikovsky, M.G. Rosenblum, J. Kurths, *Synchronization – A Unified Approach to Nonlinear Science* (Cambridge University Press, Cambridge, 2001)
2. H. Fujisaka, T. Yamada, Prog. Theor. Phys. **69**, 32 (1983)
3. L.M. Pecora, T.L. Carroll, Phys. Rev. Lett. **64**, 821 (1990)
4. J.F. Heagy, L.M. Pecora, T.L. Carroll, Phys. Rev. Lett. **74**, 4185 (1995)
5. S. Boccaletti, J. Kurths, G. Osipov, D.L. Valladares, C.S. Zhou, Phys. Rep. **366**, 1 (2002)
6. J. Kurths (ed.), Special issue on phase synchronization. Int. J. Bifurcat. Chaos **10**, 2289 (2000)
7. L. Pecora (ed.), Special focus issue on chaotic synchronization. Chaos **7**, 520 (1997)
8. M. Lakshmanan, K. Murali, *Chaos in Nonlinear Oscillators: Controlling and Synchronization* (World Scientific, Singapore, 1996)
9. M. Lakshmanan, S. Rajasekar, *Nonlinear Dynamics: Integrability, Chaos and Patterns* (Springer, Berlin, 2003)
10. G. Chen, X. Dong, *From Chaos to Order: Methodologies, Perspectives and Applications* (World Scientific, Singapore, 1998)
11. H.G. Schuster (ed.), *Handbook of Chaos Control* (Wiley-VCH, Weinheim, 1999)
12. S.K. Han, C. Kurrer, Y. Kuramoto, Phys. Rev. Lett. **75**, 3190 (1999)
13. L. Kocarev, U. Parlitz, Phys. Rev. Lett. **74**, 5028 (1995)
14. C. Schafer, M.G. Rosenblum, J. Kurths, H. H. Abel, Nature (London) **392**, 239 (1998)
15. B. Blasius, A. Huppert, L. Stone, Nature (London) **399**, 354 (1999)
16. S. Hayes, C. Grebogy, E. Ott, Phys. Rev. Lett. **70**, 3031 (1993)
17. K. Pyragas, Phys. Lett. A **181**, 203 (1993)
18. A. Kittle, K. Pyragas, R. Richter, Phys. Rev. E **50**, 262 (1994)
19. N.F. Rulkov, M.M. Sushchik, L.S. Tsimring, H.D.I. Abarbanel, Phys. Rev. E **51**, 980 (1995)
20. L. Kocarev, U. Parlitz, Phys. Rev. Lett. **76**, 1816 (1996)
21. R. Brown, Phys. Rev. Lett. **81**, 4835 (1998)
22. M.G. Rosenblum, A.S. Pikovsky, J. Kurths, Phys. Rev. Lett. **76**, 1804 (1996)
23. T. Yalcinkaya, Y.C. Lai, Phys. Rev. Lett. **79**, 3885 (1997)
24. M.G. Rosenblum, A.S. Pikovsky, J. Kurths, Phys. Rev. Lett. **78**, 4193 (1997)
25. S. Rim, I. Kim, P. Kang, Y.J. Park, C.M. Kim, Phys. Rev. E **66**, 015205(R) (2002)
26. M. Zhan, G.W. Wei, C.H. Lai, Phys. Rev. E **65**, 036202 (2002)
27. H.U. Voss, Phys. Rev. E **61**, 5115 (2002)
28. H.U. Voss, Phys. Rev. Lett. **87**, 014102 (2001)
29. C. Masoller, Phys. Rev. Lett. **86**, 2782 (2001)
30. A.N. Pisarchik, R.J. Reategui, J.R. Villalobos-Salazar, J.H. Garcia-Lopez, S. Boccaletti, Phys. Rev. Lett. **96**, 244102 (2006)
31. M. Zhan, Y. Wang, X. Gang, G.W. Wei, C.H. Lai, Phys. Rev. E **68**, 036208 (2003)
32. A. Locquet, F. Rogister, M. Sciamanna, P. Megret, M. Blandel, Phys. Rev. E **64**, 045203(R) (2001)
33. S. Boccaletti, D.L. Valladares, Phys. Rev. E **62**, 7497 (2000)
34. S. Taherion, Y.C. Lai, Phys. Rev. E **59**, R6247 (1999)
35. L. Zhu, Y.C. Lai, Phys. Rev. E **64**, 045205 (2001)
36. A. Locquet, C. Masoller, C.R. Mirasso, Phys. Rev. E **65**, 056205 (2002)
37. E.M. Shahverdiev, S. Sivaprakasam, K.A. Shore, Phys. Rev. E **66**, 037202 (2002)
38. E.M. Shahverdiev, S. Sivaprakasam, K.A. Shore, Phys. Lett. A **292**, 320 (2002)
39. S. Boccaletti, L.M. Pecora, A. Pelaez, Phys. Rev. E **63**, 066219 (2001)
40. R. Brown, L. Kocarev, Chaos **10**, 344 (2000)
41. A.E. Hramov, A.A. Koronovskii, Chaos **14**, 603 (2004)
42. L. Kocarev, K.S. Halle, K. Eckert, L.O. Chua, U. Parlitz, Int. J. Bifurcat. Chaos **2**, 709 (1992)
43. K.M. Cuomo, A.V. Oppenheim, Phys. Rev. Lett. **71**, 65 (1993)
44. S. Hayes, C. Grebogy, E. Ott, A. Mark, Phys. Rev. Lett. **73**, 1781 (1994)
45. J. Garcia-Ojalvo, R. Roy, Phys. Rev. Lett. **86**, 5204 (2001)

46. G.D. VanWiggeren, R. Roy, Phys. Rev. Lett. **88**, 097903 (2002)
47. J. Garcia-Ojalvo, R. Roy, Proc. SPIE **4646**, 525 (2002)
48. B.B. Zhou, R. Roy, Phys. Rev. E **75**, 026205 (2007)
49. R. He, P.G. Vaidya, Phys. Rev. E **57**, 1532 (1998)
50. X. Wang, X. Wu, Y. He, G. Aniwar, Int. J. Mod. Phys. B **22**, 3709 (2008)
51. J.H. Peng, E.J. Ding, M. Ding, W. Yang, Phys. Rev. Lett. **76**, 904 (1996)
52. L. Pecora, Phys. World **9**, 17 (1996)
53. G.D. VanWiggeren, R. Roy, Science **279**, 1198 (1998)
54. L. Pecora, Phys. World **11**, 25 (1998)
55. A. Argyris, D. Syvridis, L. Larger, V. Annovazzi-Lodi, P. Colet, I. Fischer, J. Garcia-Ojalvo, C.R. Mirasso, L. Pesquera, K.A. Shore, Nature **438**, 343 (2005)
56. G. Perez, H. Cerdeira, Phys. Rev. Lett. **74**, 1970 (1995)
57. K.M. Short, A.T. Parker, Phys. Rev. E **58**, 1159 (1998)
58. K. Kaneko, *Theory and Applications of Coupled Map Lattices* (Wiley, New York, 1993)
59. Y.C. Lai, Phys. Rev. E **55**, R4861 (1997)
60. M.K. Ali, J.Q. Fang, Phys. Rev. E **55**, 5285 (1997)
61. M.C. Mackey, L. Glass, Science **197**, 287 (1977)
62. J.D. Farmer, Physica D **4**, 366 (1982)
63. O.E. Rossler, Phys. Lett. A **71**, 155 (1979)
64. P. Grassberger, I. Procaccia, Physica D **9**, 189 (1983)
65. J.P. Goedgebure, L. Larger, H. Porte, Phys. Rev. Lett. **80**, 2249 (1998)
66. L. Yaowen, G. Guangming, Z. Hong, W. Yinghai, G. Liang, Phys. Rev. E **62**, 7898 (2000)
67. V.S. Udaltsov, J.P. Goedgebure, L. Larger, W.T. Rhodes, Phys. Rev. Lett. **86**, 1892 (2001)
68. D. Ghosh, S. Banerjee, A.R. Chowdhury, Euro. Phys. Lett. **80**, 3006 (2008)
69. R. He, P.G. Vaidya, Phys. Rev. E **59**, 4048 (1999)
70. B. Mensour, A. Longtin, Phys. Lett. A **244**, 59 (1998)
71. T. Heil, I. Fischer, W. Elsasser, J. Mulet, C.R. Mirasso, Phys. Rev. Lett. **86**, 795 (2001)
72. J. Fort, V. Mendez, Phys. Rev. Lett. **89**, 178101 (1999)
73. C. Zhou, C.H. Lai, Phys. Rev. E **60**, 320 (1999)
74. W.-H. Kye, M. Choi, S. Rim, M.S. Kurdoglyan, C.-M. Kim, Y.-J. Park, Phys. Rev. E **69**, 055202(R) (2004)
75. W.-H. Kye, M. Choi, M.S. Kurdoglyan, C.-M. Kim, Y.-J. Park, Phys. Rev. E **70**, 046211 (2004)
76. W.-H. Kye, M. Choi, C.-M. Kim, Y.-J. Park, Phys. Rev. E **71**, 045202(R) (2005)
77. D.V. Senthilkumar, M. Lakshmanan, Chaos **17**, 013112 (2007)
78. K. Pyragas, Phys. Rev. E **58**, 3067 (1998)
79. M. Zhan, X. Wang, X. Gong, G. W. Wei, C.H. Lai, Phys. Rev. E **68**, 036208 (2003)
80. M.-Y. Kim, C. Sramek, A. Uchida, R. Roy, Phys. Rev. E **74**, 016211 (2006)
81. S. Sano, A. Uchida, S. Yoshimori, R. Roy, Phys. Rev. E **75**, 016207 (2007)
82. M.J. Bunner, W. Just, Phys. Rev. E **58**, R4072 (1998)
83. S. Boccaletti, D.L. Valladares, J. Kurths, D. Maza, H. Mancini, Phys. Rev. E **61**, 3712 (2000)
84. E.M. Shahverdiev, K.A. Shore, Phys. Rev. E **71**, 016201 (2005)
85. A.E. Hramov, A.A. Koronovskii, Europhys. Lett. **79**, 169 (2005)
86. A.E. Hramov, A.A. Koronovskii, P.V. Popov, Phys. Rev. E **72**, 037201 (2005)
87. A.E. Hramov, A.A. Koronovskii, Physica D **206**, 252 (2005)
88. E.M. Shahverdiev, R. A. Nuriev, R.H. Hashimov, K.A. Shore, Chaos Solit. Fract. **25**, 325 (2005)
89. D.V. Senthilkumar, M. Lakshmanan, Phys. Rev. E **71**, 016211 (2005)
90. D.V. Senthilkumar, M. Lakshmanan, J. Kurths, Phys. Rev. E **74**, 035205(R) (2006)
91. D.V. Senthilkumar, M. Lakshmanan, J. Phys. Conf. Ser. **23**, 300 (2005)
92. D.V. Senthilkumar, M. Lakshmanan, Phys. Rev. E **76**, 066210 (2007)
93. S. Zhou, H. Li, Z. Wu, Phys. Rev. E **75**, 037203 (2007)
94. M. Chen, J. Kurths, Phys. Rev. E **76**, 036212 (2007)
95. T. Huang, C. Li, X. Liu, Chaos **76**, 033122 (2008)

96. W.-H. Kye, M. Choi, M.-W. Kim, S.-Y. Lee, S. Rim, C.-M. Kim, Y.-J. Park, Phys. Lett. A **322**, 338 (2004)
97. E.M. Shahverdiev, S. Sivaprakasam, K.A. Shore, Phys. Rev. E **66**, 017204 (2002)
98. E.M. Shahverdiev, K.A. Shore, Phys. Rev. E **71**, 016201 (2005)
99. S. Sivaprakasam, E.M. Shahverdiev, K.A. Shore, Phys. Rev. Lett. **87**, 154101 (2001)
100. E.M. Shahverdiev, S. Sivaprakasam, K.A. Shore, Phys. Rev. E **66**, 017206 (2002)
101. N.N. Krasovskii, *Stability of Motion* (Stanford University Press, Stanford, 1963)
102. D.V. Senthilkumar, J. Kurths, M. Lakshmanan, Chaos **19**, 023107 (2009)
103. M. Lakshmanan, R. Sahadevan (Eds.), *Proceedings of the Third National Conference on Nonlinear Systems and Dynamics* (Allied Publishers, Chennai, 2006), p. 202
104. D.V. Senthilkumar, M. Lakshmanan, Int. J. Bifurcat. Chaos **15**, 2985 (2005)
105. M. Zochowski, Physics D **145**, 181 (2000)
106. C.D. Li, X.F. Liao, T.W. Huang, Chaos **17**, 013103 (2007)
107. T.W. Huang, C.D. Li, X.Z. Liu, Chaos **18**, 033122 (2008)

Chapter 8
Transition from Anticipatory to Lag Synchronization via Complete Synchronization

8.1 Introduction

In this chapter we will consider chaos synchronization of two single scalar piecewise-linear time-delay systems studied in Chaps. 3 and 7 with unidirectional coupling between them and having two different time-delays: one in the coupling term and the other in the individual systems, namely feedback delay. We deduce [1] the corresponding stability condition for synchronization following Krasovskii-Lyapunov theory as in the previous chapter for complete synchronization, and demonstrate that there exist transitions between three different kinds of direct, and their inverse synchronizations, namely anticipatory, complete and lag synchronizations, as a function of the time-delay parameter in the coupling. To characterize the existence of anticipatory and lag synchronizations, we plot the similarity function $S(\tau)$. We then also demonstrate that when one of the system parameters is varied, the onset of exact anticipatory/complete/lag synchronization from the desynchronized state is preceded by a region of approximate synchronized state. We also show that the latter is characterized by a transition from on-off intermittency to a periodic structure in the laminar phase distribution, as suggested in the work of Zhan et al. [2] for the case of lag synchronization in coupled Rössler systems.

8.2 Coupled System and the General Stability Condition

Now let us consider the following general unidirectionally coupled drive $x_1(t)$ and response $x_2(t)$ systems with two different time-delays τ_1 and τ_2 as feedback and coupling time-delays, respectively,

$$\dot{x}_1(t) = -ax_1(t) + b_1 f(x_1(t - \tau_1)), \tag{8.1a}$$
$$\dot{x}_2(t) = -ax_2(t) + b_2 f(x_2(t - \tau_1)) + b_3 f(x_1(t - \tau_2)), \tag{8.1b}$$

where b_1, b_2 and b_3 are constants, $a > 0$, and $f(x)$ is a continuously differentiable (or even a continuous) function. The parameters are so chosen that both the systems evolve chaotically. Now we can deduce the stability condition for chaos

M. Lakshmanan, D.V. Senthilkumar, *Dynamics of Nonlinear Time-Delay Systems*, Springer Series in Synergetics, DOI 10.1007/978-3-642-14938-2_8, © Springer-Verlag Berlin Heidelberg 2010

synchronization for the two time-delay systems, Eqs. (8.1a) and (8.1b), in the presence of the delay coupling $b_3 f(x_1(t - \tau_2))$. Designating $x_1(t - \tau)$ as $x_{1\tau}$, so that $x_{1\tau_2-\tau_1} = x_1(t - (\tau_2 - \tau_1))$, the time evolution of the difference system with the state variable $\Delta = x_{1\tau_2-\tau_1} - x_2$ can be written for small values of Δ by using the evolution Eqs. (8.1) as

$$\dot{\Delta} = -a\Delta + (b_2 + b_3 - b_1)f(x_1(t - \tau_2)) + b_2 f'(x_1(t - \tau_2))\Delta_{\tau_1}, \quad \Delta_\tau = \Delta(t - \tau). \tag{8.2}$$

Here $f'(x) = \frac{df}{dx}$. From the definition of Δ, one may immediately note that the error variable Δ corresponds to anticipatory synchronization when $\tau_2 < \tau_1$, identical synchronization for $\tau_2 = \tau_1$ and lag synchronization when $\tau_2 > \tau_1$ (see also below).

The above evolution equation (8.2) corresponding to the error variable of the synchronization manifold is inhomogeneous and so it is difficult to analyse it analytically. Nevertheless, the evolution equation can be written as a homogeneous equation,

$$\dot{\Delta} = -a\Delta + b_2 f'(x_1(t - \tau_2))\Delta_{\tau_1}, \tag{8.3}$$

for the specific choice of the parameters,

$$b_1 = b_2 + b_3. \tag{8.4}$$

The synchronization manifold $\Delta = x_{1\tau_2-\tau_1} - x_2 = 0$ (as well as the inverse synchronization manifold $\Delta = x_{1\tau_2-\tau_1} + x_2 = 0$, which will be discussed in the later Sect. 8.3.4) corresponds to the following distinct cases:

1. Anticipatory synchronization occurs when $\tau_2 < \tau_1$ with $x_2(t) = x_1(t - \hat{\tau})$; $\hat{\tau} = \tau_2 - \tau_1 < 0$, where the state of the response system anticipates exactly the state of the drive system in a synchronized manner with an anticipating time $|\hat{\tau}|$ (whereas in the case of inverse anticipatory synchronization, the state of the response system anticipates the inverse state of the drive system, that is, $x_2(t) = -x_1(t - \hat{\tau})$).
2. Complete synchronization results when $\tau_2 = \tau_1$ with $x_2(t) = x_1(t)$; $\hat{\tau} = \tau_2 - \tau_1 = 0$, where the state of the response system evolves exactly identical with the state of the drive system (whereas in the case of inverse complete synchronization, the state of the response system evolves in a synchronized manner to the inverse state of the drive system, that is, $x_2(t) = -x_1(t)$).
3. Lag synchronization occurs when $\tau_2 > \tau_1$ with $x_2(t) = x_1(t - \hat{\tau})$; $\hat{\tau} = \tau_2 - \tau_1 > 0$, where the state of the response system lags the state of the drive system in a synchronized manner with a lag time $\hat{\tau}$ (whereas in the case of inverse lag synchronization, the state of the response system lags the inverse state of the drive system, that is, $x_2(t) = -x_1(t - \hat{\tau})$).

The synchronization manifold corresponding to Eq. (8.3) is locally attracting if the origin of this equation is stable. Following Krasovskii-Lyapunov functional approach [3, 4], discussed earlier in Chap. 7, we define a positive definite Lyapunov functional of the form

$$V(t) = \frac{1}{2}\Delta^2 + \mu \int_{-\tau_1}^{0} \Delta^2(t+\theta)d\theta, \tag{8.5}$$

where μ is an arbitrary positive parameter, $\mu > 0$. Note that $V(t)$ approaches zero as $\Delta \to 0$.

To estimate a sufficient condition for the stability of the solution $\Delta = 0$, we again require (as in Chap. 7) the derivative of the functional $V(t)$ along the trajectory of Eq. (8.3),

$$\frac{dV}{dt} = -a\Delta^2 + b_2 f'(x_1(t-\tau_2))\Delta\Delta_{\tau_1} + \mu\Delta^2 - \mu\Delta_{\tau_1}^2, \tag{8.6}$$

to be negative. The requirement that $\frac{dV}{dt} < 0$, for all Δ and Δ_τ, results in the condition for stability (see Sect. 7.3) as

$$a > \frac{b_2^2}{4\mu} f'(x_1(t-\tau_2))^2 + \mu = \Phi(\mu). \tag{8.7}$$

Again $\Phi(\mu)$ as a function of μ for a given $f'(x)$ has an absolute minimum at $\mu = (|b_2 f'(x_1(t-\tau_2))|)/2$ with $\Phi_{min} = |b_2 f'(x_1(t-\tau_2))|$. Since $\Phi \geq \Phi_{min} = |b_2 f'(x_1(t-\tau_2))|$, from the inequality (8.7), it turns out that sufficient condition for asymptotic stability is

$$a > |b_2 f'(x_1(t-\tau_2))| \tag{8.8}$$

along with the condition (8.4) on the parameters b_1, b_2 and b_3.

The above condition indeed corresponds to the stability condition for exact anticipatory, identical as well as lag synchronizations for suitable values of the coupling delay τ_2. In the following, we will demonstrate the transition from anticipatory to lag synchronization via complete synchronization as the coupling delay τ_2 is varied from $\tau_2 < \tau_1$ to $\tau_2 > \tau_1$, subject to the stability condition (8.8) with the parametric restriction $b_1 = b_2 + b_3$ in three typical cases: (1) coupled piecewise linear, (2) Mackey-Glass and (3) Ikeda time-delay systems.

8.3 Coupled Piecewise Linear Time-Delay System and Stability Condition: Transition from Anticipatory to Lag Synchronization

As a specific example, we first consider the coupled piecewise linear time-delay system studied in detail in Chap. 3, given by Eq. (8.1) along with the nonlinear function $f(x)$ specified by the piecewise linear form,

$$f(x) = \begin{cases} 0, & x \le -4/3 \\ -1.5x - 2, & -4/3 < x \le -0.8 \\ x, & -0.8 < x \le 0.8 \\ -1.5x + 2, & 0.8 < x \le 4/3 \\ 0, & x > 4/3. \end{cases} \tag{8.9}$$

Therefore

$$|f'(x_1(t - \tau_2))| = \begin{cases} 1.5, & 0.8 \le |x_1| \le \frac{4}{3} \\ 1.0, & |x_1| < 0.8. \end{cases} \tag{8.10}$$

Note that the region $|x_1| > 4/3$ is outside the dynamics of the present system. Consequently the stability condition (8.8) becomes $a > 1.5|b_2| > |b_2|$ along with the parametric restriction $b_1 = b_2 + b_3$.

Thus one can take $a > |b_2|$ as a less stringent condition for (8.8) to be valid, while

$$a > 1.5|b_2| \tag{8.11}$$

as the most general condition specified by (8.8) for asymptotic stability of the synchronized state $\Delta = 0$ of the coupled piecewise linear time-delay systems.

8.3.1 Anticipatory Synchronization for $\tau_2 < \tau_1$

To start with, we first consider the transition to anticipatory synchronization in the coupled system (8.1) with the nonlinear function $f(x)$ taken as in (8.9). We have fixed the value of the feedback time-delay τ_1 at $\tau_1 = 25.0$ while the other parameters are fixed as $a = 0.16, b_1 = 0.2, b_2 = 0.1, b_3 = 0.1$ and the time-delay in the coupling, τ_2, is treated as the control parameter. With the above mentioned stability condition (8.11) and with the coupling delay τ_2 being less than the feedback delay τ_1, one can observe the transition to anticipatory synchronization. The time trajectory plot is shown in Fig. 8.1a depicting anticipatory synchronization, for the specific value of $\tau_2 = 20$ with the anticipating time equal to that of the difference between the feedback and the coupling delays, that is, $|\tau| = |\tau_2 - \tau_1|$. The time-shifted plot Fig. 8.1b, $x_1(t - \tau), \tau < 0$ Vs $x_2(t)$, shows a concentrated diagonal line confirming the existence of anticipatory synchronization (In all the numerical studies in this monograph sufficiently large number of transients have been left out before presenting our figures).

Rosenblum et al. [5] have introduced the notion of similarity function $S_l(\tau)$ for characterizing the lag synchronization as a time averaged difference between the variables x_1 and x_2 (with mean values being subtracted) taken with the time shift τ,

$$S_l^2(\tau) = \frac{\langle [x_2(t + \tau) - x_1(t)]^2 \rangle}{[\langle x_1^2(t) \rangle \langle x_2^2(t) \rangle]^{1/2}}, \tau > 0, \tag{8.12}$$

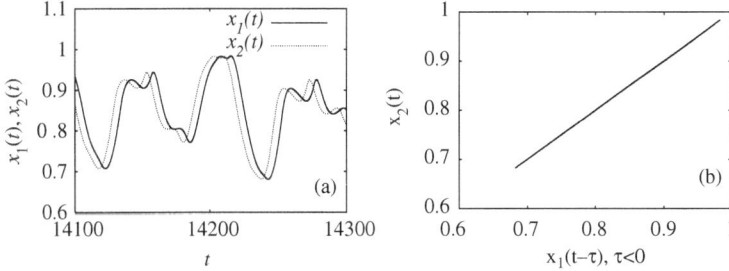

Fig. 8.1 Exact anticipatory synchronization for the parameter values $a = 0.16$, $b_1 = 0.2$, $b_2 = 0.1$, $b_3 = 0.1$, $\tau_1 = 25.0$ and $\tau_2 = 20.0$. (**a**) Time series plot of $x_1(t)$ and $x_2(t)$, (**b**) Synchronization manifold between $x_1(t - \tau)$ and $x_2(t)$, $\tau = \tau_2 - \tau_1$, $\tau < 0$. The response $x_2(t)$ anticipates the drive $x_1(t)$ with a time shift of $\tau = -5.0$

where, $\langle x \rangle$ means time average over the variable x. If the signals $x_1(t)$ and $x_2(t)$ are independent, the difference between them is of the same order as the signals themselves. If $x_1(t) = x_2(t)$, as in the case of complete synchronization, the similarity function reaches a minimum $S(\tau) = 0$ for $\tau = 0$. But for the case of nonzero value of time shift τ, if $S_l(\tau) = 0$, then there exists a time shift τ between the two signals $x_1(t)$ and $x_2(t)$ such that $x_2(t) = x_1(t - \tau)$, $\tau > 0$, demonstrating lag synchronization.

In the present study, we have used the same similarity function $S_l(\tau)$ to characterize anticipatory synchronization with negative time shift, $\tau < 0$, instead of the positive time shift, $\tau > 0$, in Eq. (8.12). In other words, one may define the similarity function for anticipatory synchronization as

$$S_a^2(\tau) = \frac{\langle [x_1(t - \tau) - x_2(t)]^2 \rangle}{\left[\langle x_1^2(t) \rangle \langle x_2^2(t) \rangle \right]^{1/2}}, \tau < 0 \tag{8.13}$$

Then the minimum of $S_a(\tau)$, that is $S_a(\tau) = 0$, indicates that there exists a time shift $-\tau$ between the two signals $x_1(t)$ and $x_2(t)$ such that $x_2(t) = x_1(t - \tau)$, $\tau < 0$, demonstrating anticipatory synchronization. Figure 8.2 shows the similarity function $S_a(\tau)$ as a function of the coupling delay τ_2 for four different values of b_2, the parameter whose value determines the stability condition given by Eq. (8.11), while satisfying the parametric condition $b_1 = b_2 + b_3$. Curves 1 and 2 are plotted for the values of $b_2 = 0.18(> a = 0.16 > a/1.5)$ and $b_2 = 0.16(= a > a/1.5)$, respectively, where the minimum values of $S_a(\tau)$ is found to be greater than zero, indicating that there is no exact time shift between the two signals $x_1(t)$ and $x_2(t)$. Note that in both the cases the stringent stability condition (8.11) and the less stringent condition $a > |b_2|$ are violated. Curve 3 corresponds to the value of $b_2 = 0.15$ (which is less than a but greater than $a/1.5$), where the minimum value of $S_a(\tau)$ is almost zero, but not exactly zero (as may be seen in the inset of Fig. 8.2), indicating an approximate anticipatory synchronization $x_1(t - \tau) \approx x_2(t)$, $\tau < 0$. On the other hand the curve 4 is plotted for the value of $b_2 = 0.1(< a/1.5)$, satisfying the general

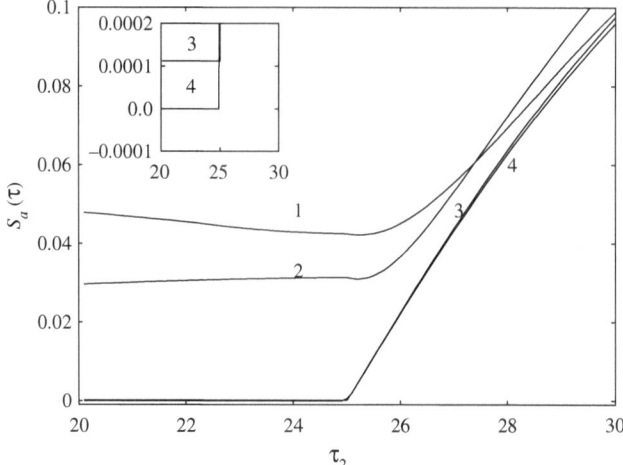

Fig. 8.2 Similarity function $S_a(\tau)$ for different values of b_2, the other system parameters are $a = 0.16$, $b_1 = 0.2$ and $\tau_1 = 25.0$. (*Curve 1*: $b_2 = 0.18$, $b_3 = 0.02$, *Curve 2*: $b_2 = 0.16$, $b_3 = 0.04$, *Curve 3*: $b_2 = 0.15$, $b_3 = 0.05$ and *Curve 4*: $b_2 = 0.1$, $b_3 = 0.1$)

stability criterion, Eq. (8.11). It shows that the minimum of $S_a(\tau) = 0$, thereby indicating that there exists an exact time shift between the two signals demonstrating anticipatory synchronization. The anticipating time is found to be equal to the difference between the coupling and feedback delay times, that is, $|\tau| = |\tau_2 - \tau_1|$. Note that $S_a(\tau) = 0$ for all values of $\tau_2 < \tau_1$, indicating anticipatory synchronization for a range of delay coupling. A further significance is that the anticipating time $|\tau| = |\tau_2 - \tau_1|$ is an adjustable quantity as long as $\tau_2 < \tau_1$, which can be tuned suitably to satisfy experimental situations.

Next, we show that the emergence of exact anticipatory synchronization is preceded by a region of approximate anticipatory synchronization, which is associated with the transition from on-off intermittency to a periodic structure in the laminar phase distribution [2] as a function of the parameter b_2. First we choose the value of b_2 as $b_2 = 0.17$ (with $b_1 = 0.2$ and $b_3 = 0.03$), above the value of $a = 0.16$, such that the general stability criterion, Eq. (8.11), as well the less stringent condition $a > |b_2|$ are violated. Figure 8.3a shows the difference of $x_1(t-\tau)-x_2(t)$, $\tau < 0$ Vs t, exhibiting a typical feature of on-off intermittency [6, 7] with the *off* state near the laminar phase and the *on* state showing random bursts. In Fig. 8.3b, $x_1(t-\tau)$, $\tau < 0$, is plotted against $x_2(t)$, where the distribution is scattered around the diagonal. To analyze the statistical feature associated with the irregular motion, we calculated the distribution of laminar phases $\Lambda(t)$ with amplitude less than a threshold value $\Delta = 0.005$ as was done in the statistical analysis of intermittency [6, 7], where the power law behavior of mean laminar length is calculated as a function of control parameter. A universal asymptotic $-\frac{3}{2}$ power law distribution is observed in Fig. 8.3c, which is quite typical for on-off intermittency.

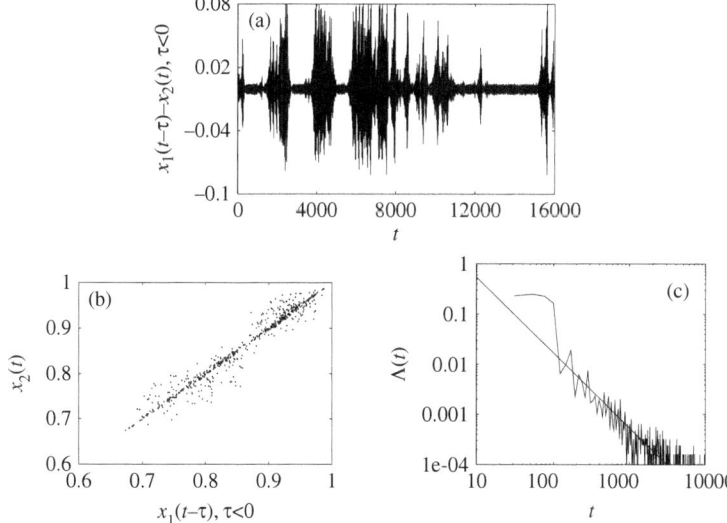

Fig. 8.3 (**a**) The time series $x_1(t - \tau) - x_2(t)$, $\tau < 0$ for $b_2 = 0.17$ and $b_3 = 0.03$ with all other parameters as in Fig. 8.1 (so that the stability condition is violated for anticipatory synchronization), (**b**) Projection of $x_1(t - \tau)$, $\tau < 0$ Vs $x_2(t)$ and (**c**) The statistical distribution of laminar phase satisfying $-\frac{3}{2}$ power law scaling

Now, we choose the value of $b_2 = 0.15$, below the value of $a = 0.16$ so that the less stringent condition $a > |b_2|$ is satisfied while the general stability criterion given by Eq. (8.11) is violated and we carry out the same analysis as above. In Fig. 8.4a, the difference of $x_1(t - \tau) - x_2(t)$, $\tau < 0$, is plotted against time t, which is more regular and is much smaller in amplitude but not exactly zero, thereby implying an approximate anticipatory synchronization $x_1(t - \tau) \approx x_2(t)$, $\tau < 0$. Figure 8.4b shows the plot of $x_1(t - \tau)$, $\tau < 0$ vs $x_2(t)$, where the distribution is localized entirely on the diagonal, but not sharply on it. Earlier we noted that for this case the minimum of similarity function $S_a(\tau)$ (Curve 3, inset of Fig. 8.2) is nearly zero, but not exactly zero. The distribution of laminar phase $\Lambda(t)$ is plotted in Fig. 8.4c as for the Fig. 8.3c. It shows a periodic structure in the distribution of laminar phase, where the peaks occur approximately at $t = nT$, $n = 1, 2, ...$, where T is of the order of the period of the lowest periodic orbit of the uncoupled system (8.1a). It should be remembered that the periodic behavior is associated with the statistical analysis, while the signals remain chaotic. Finally for the case $b_2 = 0.1 (< a/1.5)$, which satisfies the stringent stability criterion (8.11), and where the similarity function vanishes exactly (Curve 4 in Fig. 8.2), exact anticipatory synchronization occurs as confirmed in Fig. 8.1. Thus we find that the transition to exact anticipatory synchronization precedes a region of approximate anticipatory synchronization from desynchronized state as the parameter b_2 changes. We have also demonstrated that the emergence of this approximate anticipatory synchronization from the desyn-

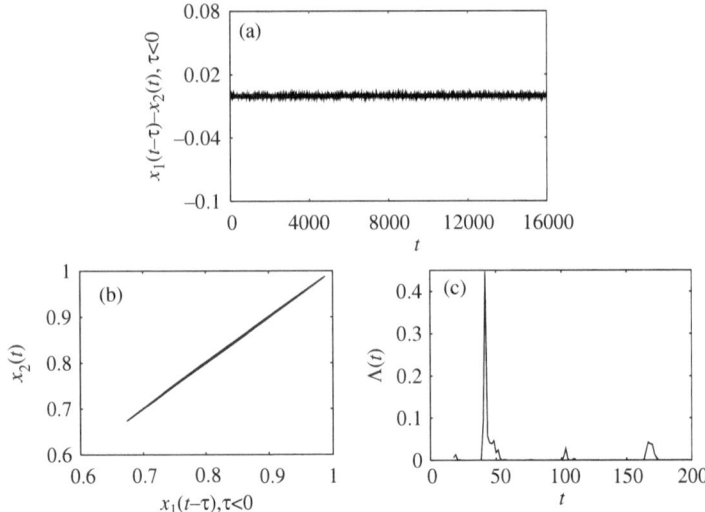

Fig. 8.4 (**a**) The time series $x_1(t - \tau) - x_2(t)$, $\tau < 0$ for $b_2 = 0.15$ and $b_3 = 0.05$ with all other parameters fixed as in Fig. 8.1 (so that the less stringent condition $a > |b_2|$ is satisfied while (8.11) is violated), (**b**) Projection of $x_1(t - \tau)$, $\tau < 0$ Vs $x_2(t)$ and (**c**) The statistical distribution of laminar phase showing a periodic structure

chronized state is characterized by the transition of on-off intermittency to periodic structure in the laminar phase distribution.

8.3.2 Complete Synchronization for $\tau_2 = \tau_1$

Complete synchronization follows the anticipatory synchronization as the value of the coupling time-delay τ_2 is increased to equal the feedback time-delay τ_1, from a lower value. With $\tau_2 = \tau_1$, the same stability criterion, Eq. (8.11), holds good for this case of complete synchronization as well with the same condition $b_1 = b_2 + b_3$. Figure 8.5a shows the time trajectory plot of $x_1(t)$ and $x_2(t)$, exhibiting synchronized evolution between them, which is also confirmed by the entirely localized diagonal line of $x_1(t)$ Vs $x_2(t)$ as shown in Fig. 8.5b. As in the case of anticipatory synchronization, we have found that the transition to complete synchronization precedes a region of approximate complete synchronization ($x_1(t) \approx x_2(t)$) from the desynchronized state as the parameter b_2 varies. Here also we have identified that the emergence of approximate complete synchronization for the case $\tau_2 = \tau_1$ is associated with a transition from on-off intermittency to a periodic structure in the laminar phase distribution as a function of the parameter b_2. In the next subsection we will discuss the existence of lag synchronization for the values of τ_2 greater than τ_1.

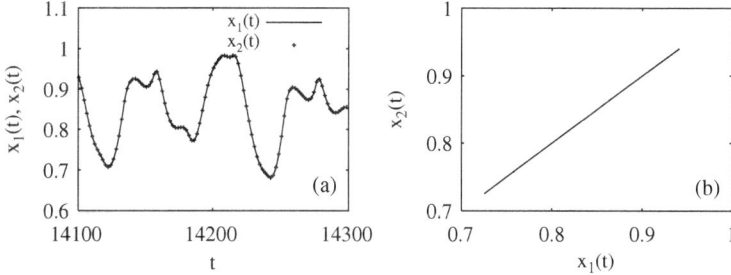

Fig. 8.5 Exact complete synchronization for the parameter values $a = 0.16, b_1 = 0.2, b_2 = 0.1, b_3 = 0.1, \tau_1 = 25.0$ and $\tau_2 = 25.0$. Here the general stability criterion (8.11) is satisfied. (**a**) Time series plot of $x_1(t)$ and $x_2(t)$ and (**b**) Synchronization manifold between $x_1(t)$ and $x_2(t)$. The response $x_2(t)$ follows identically the drive $x_1(t)$ without any time shift

8.3.3 Lag Synchronization for $\tau_2 > \tau_1$

For the coupling delay τ_2 greater than the feedback delay τ_1, we find that the system (8.1) exhibits exact lag synchronization provided the parameters satisfy the stringent stability criterion (8.11), with the lag time equal to the difference between the coupling and feedback delay times. Figure 8.6a shows the plot of $x_1(t)$ and $x_2(t)$ Vs time t, where the response system lags the state of the drive system with constant lag time $\tau = |\tau_2 - \tau_1|$. Figure 8.6b shows the time-shifted plot of $x_1(t - \tau), \tau > 0$ and $x_2(t)$. However, in the region of less stringent stability condition, $1.5|b_2| < a < |b_2|$, approximate lag synchronization occurs as in the cases of anticipatory and complete synchronizations.

We have also calculated the similarity function $S_l(\tau)$ from Eq. (8.12) to characterize the lag synchronization. Figure 8.7 shows the similarity function $S_l(\tau)$ Vs coupling delay τ_2 for four different values of b_2. Curves 1 and 2 show the similarity

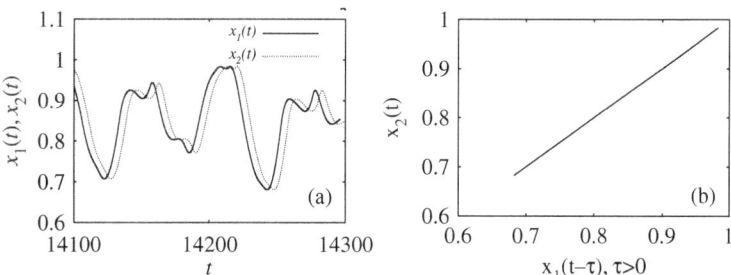

Fig. 8.6 Exact lag synchronization for the parameter values $a = 0.16, b_1 = 0.2, b_2 = 0.1, b_3 = 0.1, \tau_1 = 25.0$ and $\tau_2 = 30.0$. Here the general stability criterion (8.11) is satisfied. (**a**) Time series plot of $x_1(t)$ and $x_2(t)$, (**b**) Synchronization manifold between $x_1(t - \tau), \tau > 0$ and $x_2(t)$. The response $x_2(t)$ lags the drive $x_1(t)$ with a time shift of $\tau = 5.0$

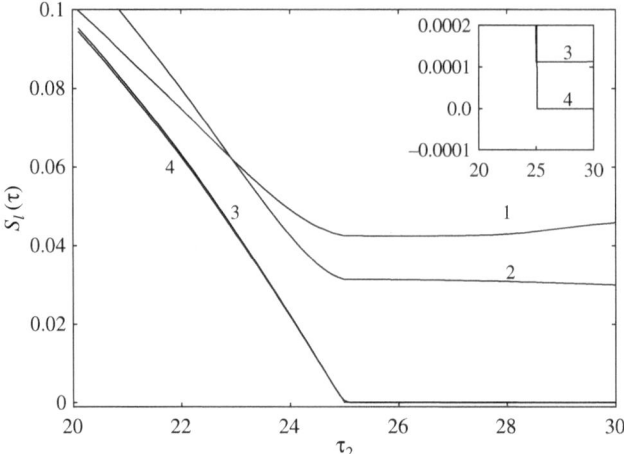

Fig. 8.7 Similarity function $S_l(\tau)$ for different values of b_2, the other system parameters are $a = 0.16$, $b_1 = 0.2$ and $\tau_1 = 25.0$. (*Curve* 1: $b_2 = 0.18$, $b_3 = 0.02$, *Curve* 2: $b_2 = 0.16$, $b_3 = 0.04$, *Curve* 3: $b_2 = 0.15$, $b_3 = 0.05$ and *Curve* 4: $b_2 = 0.1$, $b_3 = 0.1$)

function $S_l(\tau)$ for the values of $b_2 = 0.18$ and 0.16, respectively. The minimum of the similarity function $S_l(\tau)$ occurs for values of $S_l(\tau) > 0$ and hence there is a lack of exact lag time between the drive and response signals indicating asynchronization. Curve 3 corresponds to the value of $b_2 = 0.15$ (which is less than a but greater than $a/1.5$), where the minimum values of $S_l(\tau)$ is almost zero, but not exactly zero (as may be seen in the inset of Fig. 8.7), so that $x_1(t - \tau) \approx x_2(t)$, $\tau > 0$. However for the value of $b_3 = 0.1$, for which the general condition (8.11) is satisfied, the minimum of similarity function becomes exactly zero (Curve 4) indicating that there is an exact time shift (Fig. 8.6) between drive and response signals $x_1(t)$ and $x_2(t)$, respectively, confirming the occurrence of lag synchronization.

We have also confirmed that as in the case of anticipatory synchronization, when the parameter b_2 varies, the onset of exact lag synchronization is preceded by a region of approximate lag synchronization, which is characterized by a transition from on-off intermittency of the desynchronized state to a periodic structure in the laminar phase distribution. For the value of $b_2 = 0.17$ (which violates the stability condition (8.11) as well as the less stringent condition $a > |b_2|$), Fig. 8.8a shows the difference of $x_1(t - \tau) - x_2(t)$, $\tau > 0$, Vs time t, exhibiting a typical on-off intermittency. In Fig. 8.8b, $x_1(t - \tau)$, $\tau > 0$, is plotted against $x_2(t)$, where the distribution is not concentrated along the diagonal. In Fig. 8.8a, the laminar phase distribution $\Lambda(t)$ is characterized by an exponential $-\frac{3}{2}$ power law behavior as shown in Fig. 8.8c. In order to show that there is a transition from on-off intermittency to periodic behavior in the laminar phase distribution corresponding to approximate lag synchronization, we have changed the value of b_2 from 0.17 to 0.15, (so that the less stringent condition $a > |b_2|$ is satisfied but not the general condition (8.11)), and examined the nature of the laminar phase distribution $\Lambda(t)$.

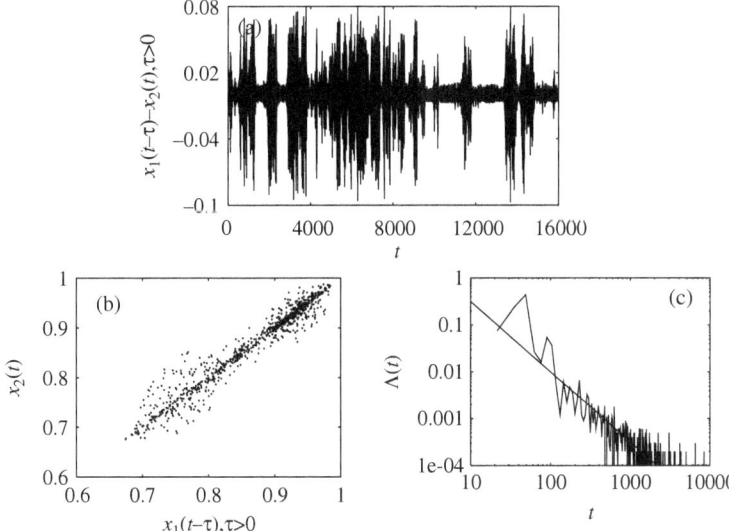

Fig. 8.8 (**a**) The time series $x_1(t - \tau) - x_2(t)$, $\tau > 0$ for $b_2 = 0.17$ and $b_3 = 0.03$ with all other parameters as in Fig. 8.6 (so that the stability condition is violated), (**b**) Projection of $x_1(t-\tau)$, $\tau > 0$ Vs $x_2(t)$ and (**c**) The statistical distribution of laminar phase satisfying $-\frac{3}{2}$ power law scaling

The difference between $x_1(t - \tau)$, $\tau > 0$, and $x_2(t)$ is shown as a function of time t in Fig. 8.9a, where there is only a laminar phase present for the threshold value $\Delta = 0.002$ without any intermittent burst. The corresponding laminar phase distribution $\Lambda(t)$ is again characterized by a periodic structure as shown in Fig. 8.9c. As in the case of approximate anticipatory synchronization, here also the peaks occur approximately at $t = nT$, $n = 1, 2, ...$, where T is roughly of the order of the period of the lowest periodic orbit of the uncoupled system (8.1a). Time-shifted plot $x_1(t - \tau)$, $\tau > 0$, Vs $x_2(t)$ is shown in Fig. 8.9b, where the distribution is concentrated along but not exactly on the diagonal line confirming the onset of approximate lag synchronization. As noted previously that for this case the minimum of similarity function $S_l(\tau)$ is nearly zero but not exactly zero (Curve 3, inset of Fig. 8.7). Finally for $b_2 = 0.1$, which satisfies the general stability criterion (8.11), we have exact lag synchronization as demonstrated in Figs. 8.6 and 8.7. Thus we find that as the parameter b_2 varies the transition to exact lag synchronization precedes a region of approximate lag synchronization from desynchronized state, where the latter is characterized by the transition from on-off intermittency to periodic structure in the laminar phase distribution.

8.3.4 Inverse Synchronizations

Next, we study the existence of inverse synchronization in coupled piecewise linear systems. Transition between inverse anticipatory and inverse lag via complete

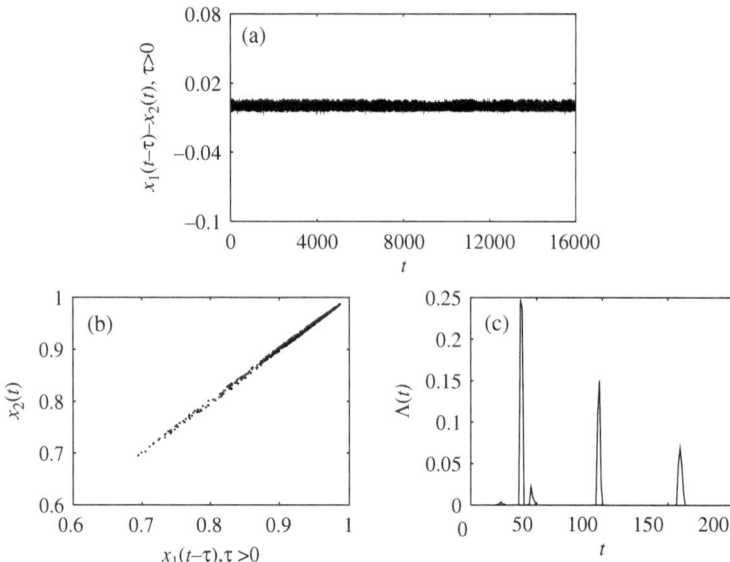

Fig. 8.9 (a) The time series $x_1(t - \tau) - x_2(t)$, $\tau > 0$ for $b_2 = 0.15$ and $b_3 = 0.05$ so that the less stringent condition $a > |b_2|$ is satisfied while (8.11) is violated, (b) Projection of $x_1(t - \tau)$, $\tau > 0$ Vs $x_2(t)$ and (c) The statistical distribution of laminar phase showing periodic structure

inverse synchronization is demonstrated in this section as a function of the coupling delay in the coupled piecewise linear time-delay systems with inhibitory coupling represented by

$$\dot{x}_1(t) = -ax_1(t) + b_1 f(x_1(t - \tau_1)), \tag{8.14a}$$
$$\dot{x}_2(t) = -ax_2(t) + b_2 f(x_2(t - \tau_1)) - b_3 f(x_1(t - \tau_2)), \tag{8.14b}$$

where the functional form $f(x)$ is the same as given by Eq. (8.9).

Importance of inhibitory or repulsive couplings are well acknowledged in biological systems. It is a well established fact that couplings between neurons are both excitatory and inhibitory [8]. Ecological webs typically have both positive and negative connections between their components [9, 10]. Coupled lasers with negative couplings have also been widely studied [11]. The well known Swift-Hohenberg and Kuramoto-Sivashinsky equations have such a term [12]. Currently, it has also been realized that a large class of natural networks also have inhibitory interactions among the interacting units [13, 14].

Now, the time evolution of the difference system with the state variable $\Delta = x_1{}_{\tau_2 - \tau_1} + x_2$, where $x_1{}_{\tau_2 - \tau_1} = x_1(t - (\tau_2 - \tau_1))$, corresponds to the following distinct cases:

1. Inverse anticipatory synchronization occurs when $\tau_2 < \tau_1$ with $x_2(t) = -x_1(t - \hat{\tau})$; $\hat{\tau} = \tau_2 - \tau_1 < 0$, where the state of the response system anticipates the

inverse state of the drive system in a synchronized manner with the anticipating time $\hat{\tau}$ (whereas in the case of direct anticipatory synchronization, the state of the response system anticipates exactly the state of the drive system, that is, $x_2(t) = x_1(t - \hat{\tau})$).

2. Inverse complete synchronization results when $\tau_2 = \tau_1$ with $x_2(t) = -x_1(t); \hat{\tau} = \tau_2 - \tau_1 = 0$, where the state of the response system evolves in a synchronized manner with the inverse state of the drive system (whereas in the case of complete synchronization, the state of the response system evolves exactly identical to the state of the drive system, that is, $x_2(t) = x_1(t)$).

3. Inverse lag synchronization occurs when $\tau_2 > \tau_1$ with $x_2(t) = -x_1(t - \hat{\tau}); \hat{\tau} = \tau_2 - \tau_1 > 0$, where the state of the response system lags the inverse state of the drive system in a synchronized manner with the lag time $\hat{\tau}$ (whereas in the case of direct lag synchronization, the state of the response system lags exactly the state of the drive system, that is, $x_2(t) = x_1(t - \hat{\tau})$).

By following the stability analysis using Krasovskii-Lyapunov functional approach as was done in Sect. 8.2 for direct synchronization, one can obtain the same asymptotic stability condition given by Eq. (8.8) along with the parametric condition $b_1 = b_2 + b_3$ in the present case also. Now again from the form of the piecewise linear function $f(x)$, one can have $a > |b_2|$ as a less stringent condition and $a > 1.5|b_2|$ as the more general condition specified by (8.8) for asymptotic stability of the inverse synchronized state $\Delta = 0$ as discussed in Sect. 8.2 for direct synchronization.

Further, it is interesting to note that if one substitutes $x_2 \to \hat{x}_2 = -x_2$ in Eq. (8.14), then the coupling becomes excitatory for the choice of functional forms we have chosen. This is exactly the case we have studied in Sect. 8.2, where direct anticipatory, complete and lag synchronizations exist as a function of the coupling delay. However, one cannot obtain inverse (anticipatory, complete and lag) synchronization with excitatory coupling or direct (anticipatory, complete and lag) synchronization with inhibitory coupling for the chosen form of the unidirectional nonlinear coupling because of the nature of the parametric relation between b_1, b_2 and b_3 and the stability condition (8.8).

In the following, we will demonstrate the existence of inverse anticipatory, inverse complete and inverse lag synchronizations as a function of the coupling strength for fixed values of the other parameters.

8.3.4.1   Inverse Anticipatory Synchronization

We have chosen the same values for all the parameters as discussed in the case of direct anticipatory synchronization in Sect. 8.3.1 for $\tau_2 < \tau_1$. Time trajectories of the state variables $x_1(t)$, $x_2(t)$ and $-x_2(t)$ of the coupled piecewise linear time-delay system (8.14) with inhibitory coupling is shown in Fig. 8.10 for the parameter values $a = 0.16, b_1 = 0.2, b_2 = 0.1, b_3 = 0.1, \tau_1 = 25.0$ and $\tau_2 = 20.0$, satisfying the general stability criterion (8.11). It is clear from the Fig. 8.10 that the response system $x_2(t)$ anticipates inversely the state of the drive system $x_1(t)$. To view this

Fig. 8.10 Time series plot of the variables $x_1(t)$, $x_2(t)$ and $-x_2(t)$ of the coupled piecewise linear time-delay system (8.14) depicting exact inverse anticipatory synchronization for the same parameter values as in Fig. 8.1

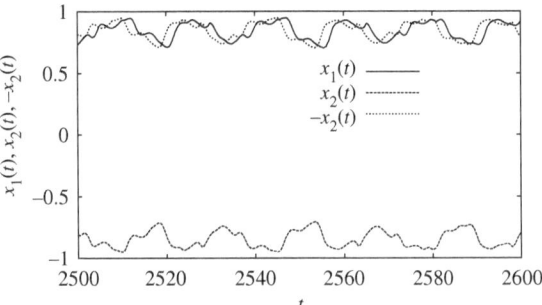

clearly in Fig. 8.10, we have also plotted the inverse of the response variable $x_2(t)$, that is $-x_2(t)$.

It is to be noted that all the other dynamical behaviours observed in Sect. 8.3.1, namely transition from approximate anticipatory to exact anticipatory synchronization as function of the parameter b_2 and their characterization by the transition of on-off intermittency to a periodic structure in the laminar phase distribution can be observed in this case of inverse anticipatory synchronization also for the same parameter values.

8.3.4.2 Complete Inverse Synchronization

Existence of exact complete inverse synchronization in the coupled piecewise linear time-delay system (8.14) with inhibitory coupling is shown in Fig. 8.11 for the same values of all the parameters as in Fig. 8.5 in Sect. 8.3.2.

8.3.4.3 Inverse Lag Synchronization

Again we have fixed the same values for all the parameters as in the Sect. 8.3.3, where the existence of exact lag synchronizations has been shown for the param-

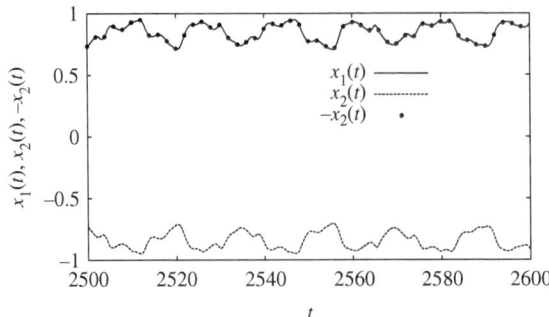

Fig. 8.11 Time series plot of the variables $x_1(t)$, $x_2(t)$ and $-x_2(t)$ of coupled piecewise linear time-delay system (8.14) depicting exact complete inverse synchronization for the same parameter values as in Fig. 8.5

Fig. 8.12 Time series plot of the variables $x_1(t)$, $x_2(t)$ and $-x_2(t)$ of coupled piecewise linear time-delay system (8.14) depicting exact inverse lag synchronization for the same parameter values as in Fig. 8.6

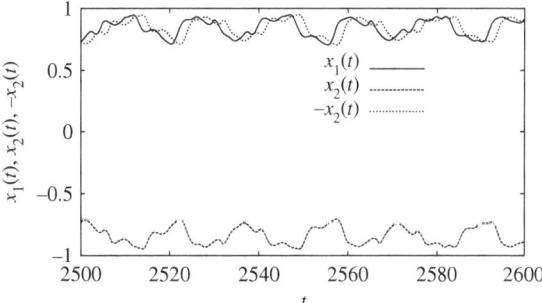

eter values satisfying the general stability condition (8.11). Time series plot of $x_1(t)$, $x_2(t)$ and $-x_2(t)$ depicting exact inverse lag synchronization is shown in Fig. 8.12. All the other dynamical transitions can also be observed in this case of inverse lag synchronization as well which are not presented here for avoiding repetition.

8.4 Transition from Anticipatory to Lag via Complete Synchronization: Mackey-Glass and Ikeda Systems

In this section, we will discuss briefly the generality of the above results on different kinds of synchronizations and their transitions in two other prototype delay models, which are widely studied in the literature. In particular, we will consider the Mackey-Glass [15, 16] and Ikeda [17–20] time-delay systems to bring out the existence of the above results.

Let us now consider the unidirectionally coupled drive $x_1(t)$ and response $x_2(t)$ systems with two different time-delays τ_1 and τ_2 as feedback and coupling time-delays, respectively, as in Eq. (8.1), with the following functional forms for the nonlinearity function,

$$f(x) = \frac{x(t - \tau)}{(1.0 + x(t - \tau)^c)} \qquad (8.15)$$

for Mackey-Glass system [15, 16] and

$$f(x) = sin(x(t - \tau)) \qquad (8.16)$$

for Ikeda system [17–20].

Now, we apply the stability criterion (8.8) deduced for an arbitrary nonlinear function $f(x)$ in Eq. (8.1) to the present examples. We note that (1) for the Mackey-Glass system

$$f'(x) = \frac{1 + (1 - c)x(t - \tau)^c}{(1 + x(t - \tau)^c)^2} \qquad (8.17)$$

and (2) for the Ikeda system

$$f'(x) = \cos(x(t - \tau)). \tag{8.18}$$

As the derivatives, $f'(x)$, themselves now depend on the instantaneous value of x for both the systems, it is not possible to pinpoint analytically the exact range of the parameter a at which the general stability condition (8.8), $a > |b_2 f'(x_1(t - \tau_2))|$, is satisfied unlike the earlier case of piecewise linear time-delay system (8.9). Nevertheless, one is able to find out numerically the value of a for which the asymptotic stability condition is satisfied in both the cases of coupled Mackey-Glass and Ikeda time-delay systems. However, it is also possible to find the value of $f'(x_{max})$ by identifying the maximal value of $x(t)$ numerically from the corresponding attractors and then find the values of the control parameter b_2 for which the stability condition,

$$a > b_2 |f'(x_{max})|, \tag{8.19}$$

is satisfied. In the following, we will demonstrate the transition from anticipatory to lag synchronization via complete synchronization as the coupling delay τ_2 is varied from $\tau_2 < \tau_1$ to $\tau_2 > \tau_1$, using the above stability criterion.

8.4.1 Anticipatory Synchronization for $\tau_2 < \tau_1$

We first demonstrate the transition to anticipatory synchronization from the desynchronized state in the coupled system (8.1) along with the functional form (8.15) for the Mackey-Glass system as the value of the coupling strength b_3 is varied. After this, we discuss the corresponding results for the coupled Ikeda system with the functional form (8.16).

8.4.1.1 Coupled Mackey-Glass Systems

We have fixed the value of the feedback time-delay τ_1 at $\tau_1 = 30.0$ and the coupling delay τ_2 at $\tau_2 = 25$ while the other parameters are fixed as $a = 0.1, b_1 = 0.2$ and $c = 10$. Hyperchaotic attractor of the uncoupled Mackey-Glass system for the chosen value of the parameters is shown in Fig. 8.13a, from which one can recognize that the maximum value of the state variable $x(t)$ does not exceed $x_{max} = 1.4$. Consequently, one can find the values of the parameter b_2 for which the stability condition $a > b_2|f'(x_{max} = 1.4)|$ with $|f'(x_{max} = 1.4)| = 0.2895$ is satisfied. For the chosen value of the parameter $a = 0.1$ the stability condition for asymptotic stability is satisfied for the values of $b_2 < 0.35$.

However, it is to be noted that according to the parametric condition $b_1 = b_2 + b_3$ and the chosen value of the parameter $b_1 = 0.2$, the value of b_2 should be less than 0.2. For any value of $b_2 > 0.2$, the value of the other parameter b_3 becomes negative, and as a consequence the coupling in Eq. (8.14) becomes excitatory and hence inverse synchronization cannot be realized as discussed in the Sect. 8.3.4.

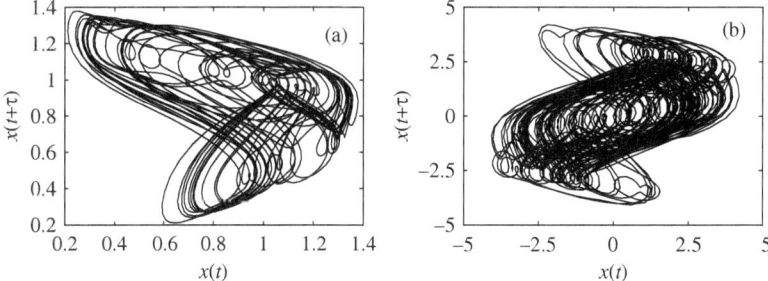

Fig. 8.13 (**a**) Hyperchaotic attractor of the Mackey-Glass time-delay system (Eq. (8.1) with the functional form as in Eq. (8.15)) for the parameter values $a = 0.1$, $b = 0.2$, $c = 10.0$ and $\tau = 30$ and (**b**) Hyperchaotic attractor of the Ikeda time-delay system (Eq. (8.1) with the functional form as in Eq. (8.16)) for the parameter values $a = 1.0$, $b = 5.0$ and $\tau = 4.0$

Numerical simulation indicates that one cannot obtain synchronized state for all the values of $b_2 < 0.35$ as expected, instead one can obtain stable synchronized state for $b_2 < 0.14$ consistent with the parametric condition $b_1 = b_2 + b_3$, for which the stability condition is satisfied.

The time trajectory plot of the variables $x_1(t)$ and $x_2(t)$ is shown in Fig. 8.14a depicting the existence of anticipatory synchronization for the value of the control parameter $b_2 = 0.14$ satisfying the stability condition with the anticipating time equal to that of the difference between the feedback and coupling delays, that is, $\tau_a = |\tau_2 - \tau_1|$. The time-shifted plot, Fig. 8.14b, of $x_1(t - \tau)$ Vs $x_2(t)$, $\tau < 0$ shows a sharp diagonal line confirming the existence of anticipatory synchronization. Upon decreasing the value of b_2 from $b_2 = 0.2$ consistent with the parametric condition $b_1 = b_2 + b_3$, one can find that there is a transition from desynchronized state to approximate anticipatory synchronization at the value of $b_2 = 0.145$ and then to exact anticipatory synchronization state at $b_2 \leq 0.14$. It is to be noted that the transition from approximate to exact anticipatory synchronized state is also charac-

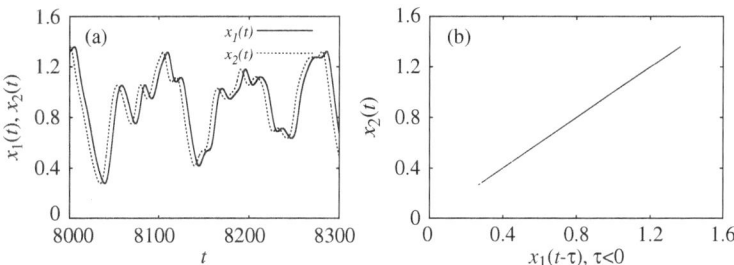

Fig. 8.14 Exact anticipatory synchronization for the parameter values $a = 0.1$, $b_1 = 0.2$, $b_2 = 0.14$, $b_3 = 0.06$, $\tau_1 = 30.0$ and $\tau_2 = 25.0$ of the Mackey-Glass system. (**a**) Time series plot of $x_1(t)$ and $x_2(t)$, (**b**) Synchronization manifold between $x_1(t - \tau)$ and $x_2(t)$, $\tau = \tau_2 - \tau_1$. The response $x_2(t)$ anticipates the drive $x_1(t)$ with a time shift of $\tau = -5.0$

terized by transition from on-off intermittency to periodic structures in the laminar phase distribution as shown for the case of piecewise linear time-delay system.

Now let us characterize the emergence of anticipatory state from asynchronous state using the notion of similarity function (8.13) for anticipatory synchronization S_a introduced in Sect. 8.3.1. Similarity function S_a for anticipatory synchronization is plotted in Fig. 8.15 for three different values of the control parameter b_2 consistent with the parametric condition (8.4). Curve 3 corresponds to the value of $b_2 = 0.15$ at which the coupled system (8.1) (with the functional form (8.15) of Mackey-Glass system) is in an asynchronous state. As the value of the parameter b_2 is decreased from $b_2 = 0.15$, transition towards approximate anticipatory synchronization is observed and it is shown for the value of $b_2 = 0.145$ in curve 2 (see the inset of Fig. 8.15). Further decrease in the value of b_2 leads the coupled system (8.1) to exhibit exact anticipatory synchronization (curve 1 is plotted for $b_2 = 0.14$).

8.4.1.2 Coupled Ikeda Systems

Now, we will point out the existence of anticipatory synchronization in the Ikeda system, Eq. (8.1) along with the functional form (8.16). We have fixed the value of the parameters as $\tau_1 = 4, \tau_2 = 3.0, a = 1.0, b_1 = 5$. Hyperchaotic attractor of the individual Ikeda time-delay system for the chosen parameter values is shown in Fig. 8.13b. It is evident from this figure that the maximum value of $x(t)$ does not exceed $x_{max} = 5$. As a consequence the stability condition (8.8) can be written as $a > b_2 \cos(5)$ with $f'(x_{max}) = \cos(5) = 0.2836$ and one can obtain asymptotically stable synchronized state for $b_2 < 2.88$, for which the stability condition is satisfied.

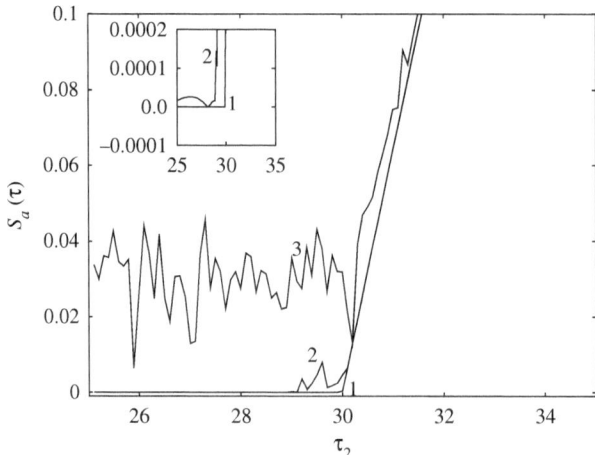

Fig. 8.15 Similarity function $S_a(\tau)$ as a function of coupling delay τ_2 for different values of b_3 in the case of the Mackey-Glass system, the other system parameters are $a = 0.1, b_1 = 0.2$ and $\tau_1 = 30.0$. (Curve 1: $b_2 = 0.14, b_3 = 0.06$, Curve 2: $b_2 = 0.145, b_3 = 0.055$, and Curve 3: $b_2 = 0.15, b_3 = 0.05$).

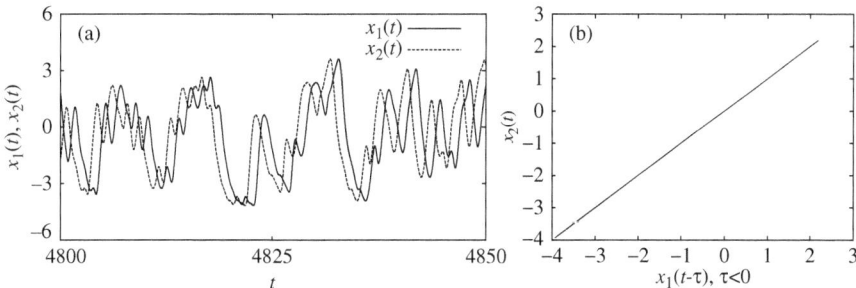

Fig. 8.16 Exact anticipatory synchronization for the parameter values $a = 1.0, b_1 = 5, b_2 = 2.8, b_3 = 2.2, \tau_1 = 4.0$ and $\tau_2 = 3.0$ of the Ikeda system. (a) Time series plot of $x_1(t)$ and $x_2(t)$, (b) Synchronization manifold between $x_1(t - \tau)$ and $x_2(t), \tau = \tau_2 - \tau_1$. The response $x_2(t)$ anticipates the drive $x_1(t)$ with a time shift of $\tau = -1.0$

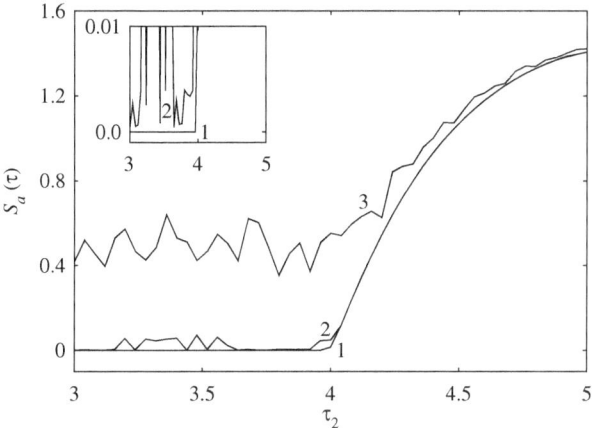

Fig. 8.17 Similarity function $S_a(\tau)$ as a function of coupling delay τ_2 for different values of b_3 in the case of the Ikeda system, the other system parameters are $a = 1.0, b_1 = 5$ and $\tau_1 = 4.0$. (*Curve 1: $b_2 = 2.8, b_3 = 2.2$, Curve 2: $b_2 = 2.9, b_3 = 2.1$, and Curve 3: $b_2 = 3.0, b_3 = 2.0$*)

Exact anticipatory synchronization is shown in Fig. 8.16 for the value of the parameter $b_2 = 2.8$. The state of drive $x_1(t)$ being anticipated by that of the response $x_2(t)$ is shown in Fig. 8.16a and the corresponding time-shifted plot is shown in Fig. 8.16b. Similarity function, $S_a(\tau)$, for the Ikeda system (Fig. 8.17) is also plotted for three different values of b_2 as a function of the coupling delay τ_2. Curve 3 in Fig. 8.17 is plotted for $b_2 = 3.0$ which corresponds to desynchronized state. Approximate anticipatory synchronized state (curve 2) is shown for the value of $b_2 = 2.9$, for which the value of $S_a(\tau)$ is close to zero but not exactly equal to zero as seen in the inset of Fig. 8.17, while that corresponds to exact anticipatory synchronization (curve 1) is equal to zero as seen in the inset for $b_2 = 2.8$. Thus as the parameter b_2 is decreased from $b_2 = 3.0$, one can observe that there is a transition from asynchronous to approximate state and then to exact anticipatory synchronization.

8.4.2 Complete Synchronization for $\tau_2 = \tau_1$

Complete synchronization follows the anticipatory synchronization when the value of the coupling time-delay τ_2 equals the feedback time-delay τ_1, when τ_2 is increased from a lower value. With $\tau_2 = \tau_1$, the same stability criterion, Eq. (8.8), holds good for this case of complete synchronization as well with the same condition $b_1 = b_2 + b_3$.

Figure 8.18 shows the existence of complete synchronization in the Mackey-Glass system when the value of the coupling delay $\tau_2 = 30$ equals that of the feedback delay $\tau_1 = 30$. Complete synchronous evolution of both the drive, $x_1(t)$, and the response, $x_2(t)$, systems is shown in Fig. 8.18a for the same value of the other parameters as in case of anticipatory synchronization, while the entirely localized diagonal line of $x_1(t)$ and $x_2(t)$ confirms the existence of complete synchronization as depicted in Fig. 8.18b. Similarly the existence of complete synchronization in Ikeda system is shown in Fig. 8.19 for the value of coupling delay $\tau_2 = \tau_1 = 4$ for the same values of the parameters as in Fig. 8.16.

8.4.3 Lag Synchronization for $\tau_2 > \tau_1$

For the value of the coupling delay τ_2 greater than that of the feedback delay τ_1, we find that the system (8.1) exhibits lag synchronization with the lag time equal to the difference between the coupling and the feedback delay times. The same stability condition (8.8) also holds good in this case of lag synchronization as well for $\tau_2 > \tau_1$ as discussed in Sect. 8.2. Time evolution of both the drive, $x_1(t)$, and the response, $x_2(t)$, of the Mackey-Glass system depicting the exact lag synchronization is shown

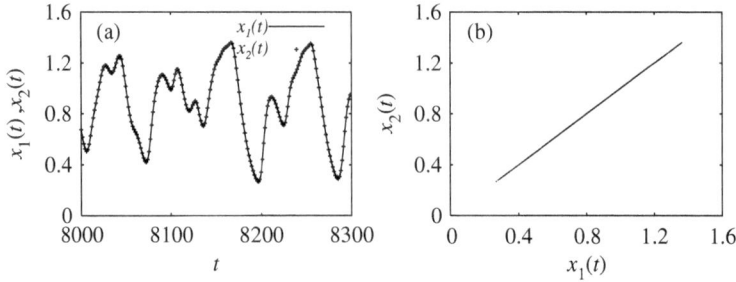

Fig. 8.18 Exact complete synchronization for the parameter values $a = 0.1, b_1 = 0.2, b_2 = 0.14, b_3 = 0.06, \tau_1 = 30.0$ and $\tau_2 = 30.0$ of Mackey-Glass system. (**a**) Time series plot of $x_1(t)$ and $x_2(t)$ and (**b**) Synchronization manifold between $x_1(t)$ and $x_2(t)$. The response $x_2(t)$ follows identically the drive $x_1(t)$ without any time shift

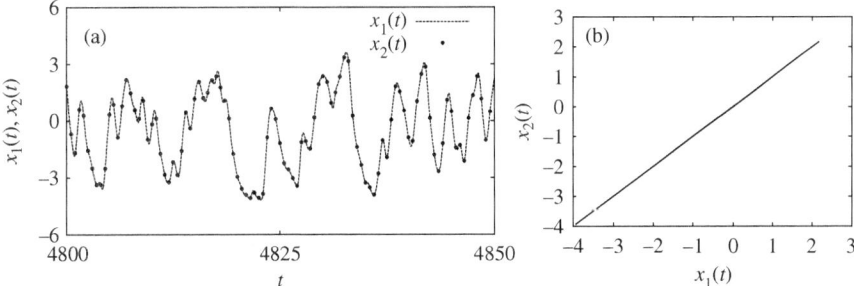

Fig. 8.19 Exact complete synchronization for the parameter values $a = 1.0, b_1 = 5.0, b_2 = 2.8, b_3 = 2.2, \tau_1 = 4.0$ and $\tau_2 = 4.0$ of Ikeda system. (**a**) Time series plot of $x_1(t)$ and $x_2(t)$ and (**b**) Synchronization manifold between $x_1(t)$ and $x_2(t)$. The response $x_2(t)$ follows identically the drive $x_1(t)$ without any time shift

in Fig. 8.20a while the corresponding time-shifted plot is depicted in Fig. 8.20b for the same value of the parameters as in Fig. 8.14. Similarity function S_l for lag synchronization (8.12) is plotted (Fig. 8.21) as a function of the coupling delay τ_2 for three different values of the coupling strength b_3 as discussed in the case of anticipatory synchronization. As the coupling strength is increased from zero, the coupled system (8.1) shows transition from asynchronous state (curve 3 for the value of the coupling strength $b_3 = 0.05$) to exact lag synchronized state (curve 1 for $b_3 = 0.06$) preceded by approximate lag synchronized state (curve 2 for $b_3 = 0.055$). The value of S_l is exactly equal to zero for the exact lag synchronized state as seen in the inset of Fig. 8.21 while that corresponds to approximate lag synchronization has finite value close to zero.

Similarly, the time traces of the drive, $x_1(t)$, and response, $x_2(t)$, systems of the Ikeda system are plotted in Fig. 8.22a along with the corresponding time shifted plot in Fig. 8.22b. The existence of lag synchronization in Ikeda system is also characterized using similarity function as shown in Fig. 8.23. Asynchronous state is represented by the curve 3 for the value of the parameter $b_2 = 3.0$ whereas

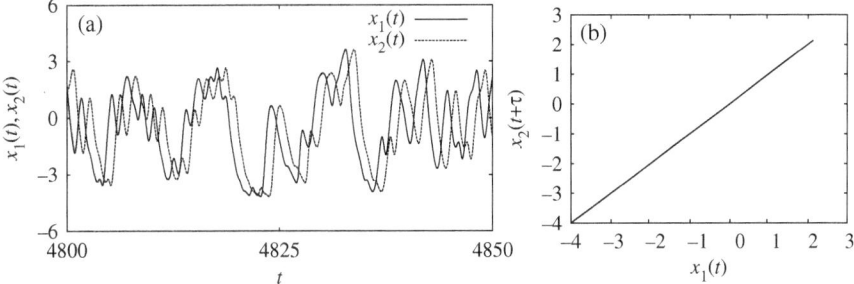

Fig. 8.20 Exact lag synchronization for the parameter values $a = 0.1, b_1 = 0.2, b_2 = 0.14, b_3 = 0.06, \tau_1 = 30.0$ and $\tau_2 = 35.0$ of the Mackey-Glass system. (**a**) Time series plot of $x_1(t)$ and $x_2(t)$, (**b**) Synchronization manifold between $x_1(t)$ and $x_2(t + \tau), \tau = \tau_2 - \tau_1$

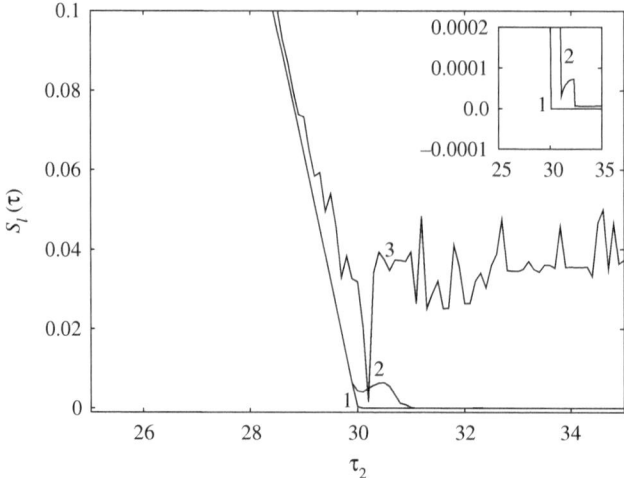

Fig. 8.21 Similarity function $S_l(\tau)$ as a function of coupling delay τ_2 for different values of b_3 in the case of the Mackey-Glass system, the other system parameters are $a = 0.1$, $b_1 = 0.2$ and $\tau_1 = 30.0$. (*Curve 1*: $b_2 = 0.14$, $b_3 = 0.06$, *Curve 2*: $b_2 = 0.145$, $b_3 = 0.055$, and *Curve 3*: $b_2 = 0.15$, $b_3 = 0.05$)

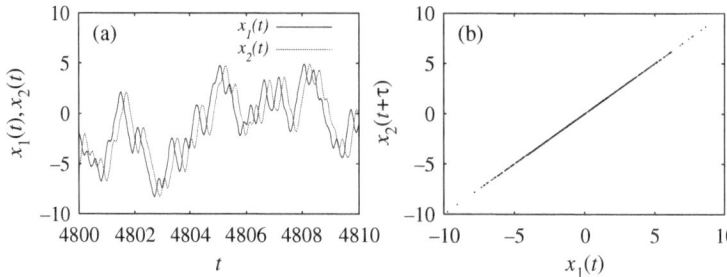

Fig. 8.22 Exact lag synchronization for the parameter values $a = 1.0$, $b_1 = 5.0$, $b_2 = 2.8$, $b_3 = 2.2$, $\tau_1 = 4.0$ and $\tau_2 = 5.0$ of the Ikeda system. (**a**) Time series plot of $x_1(t)$ and $x_2(t)$, (**b**) Synchronization manifold between $x_1(t)$ and $x_2(t + \tau)$, $\tau = \tau_2 - \tau_1$

approximate lag synchronized state is shown by curve 2 for $b_2 = 2.9$. Exact lag synchronization corresponds to the curve 1 plotted for the value of $b_2 = 2.8$. It is clear from the inset of Fig. 8.23 that the similarity function S_l corresponding to the approximate lag synchronization fluctuates around zero while that of exact lag synchronization is exactly equal to zero.

It is to be noted that as in the case of the coupled piecewise linear time-delay system (8.1) with (9.9) the emergence of approximate synchronization (anticipatory/complete/lag) is also associated with a transition from on-off intermittency to periodic structure in the laminar phase distribution for the appropriate values of the parameters in both the examples of the coupled Mackey-Glass and the coupled Ikeda systems.

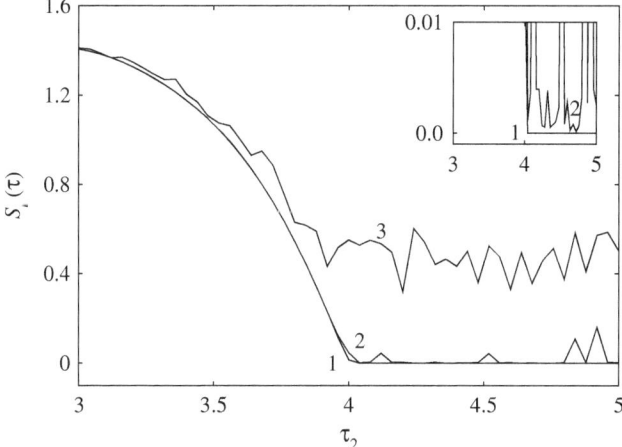

Fig. 8.23 Similarity function $S_l(\tau)$ as a function of coupling delay τ_2 for different values of b_2 in the case of the Ikeda system, the other system parameters are $a = 1.0$, $b_1 = 5.0$ and $\tau_1 = 4.0$. (*Curve* 1: $b_2 = 2.8$, $b_3 = 2.2$, *Curve* 2: $b_2 = 2.9$, $b_3 = 2.1$, and *Curve* 3: $b_2 = 3.0$, $b_3 = 2.0$)

8.5 Inverse Synchronizations: Mackey-Glass and Ikeda Systems

Finally, we will briefly demonstrate the existence of inverse (anticipatory,complete and lag) synchronizations in the coupled time-delay systems with inhibitory coupling. Consider the same coupled systems as in Eq. (8.14) but with the functional form (8.15) corresponding to the Mackey-Glass systems and (8.16) for the Ikeda systems. We have chosen the parameter values as $a = 0.1$, $b_1 = 0.2$, $c = 10.0$, $b_2 = 0.14$, $b_3 = 0.06$ and $\tau = 30.0$ for the coupled Mackey-Glass time-delay systems and $a = 1.0$, $b_1 = 5$, $b_2 = 2.8$, $b_3 = 2.2$ and $\tau = 4.0$ for the coupled Ikeda time-delay systems satisfying their appropriate stability conditions for asymptotic stability of the synchronized states. We have fixed the same values for all the parameters as in the case of exact anticipatory, complete and lag synchronizations as in Sect. 8.4 for both the systems and hence all the discussions in Sect. 8.4 corresponding to the stability analysis and dynamical transitions are also valid here. In the following, we will just point out the existence of different types of inverse synchronizations as a function of the coupling delay τ_2 for both the systems for the above values of the parameters. The time trajectory plots of the variables $x_1(t)$, $x_2(t)$ and $-x_2(t)$ are plotted in the following figures for both the Mackey-Glass and Ikeda systems for different values of the coupling delay τ_2. The inverse of the variable of the response system $x_2(t)$, that is $-x_2(t)$, is shown in the figures in order to clearly visualize the inverse synchronizations for the corresponding values of the parameters. Inverse anticipatory synchronization is shown in Fig. 8.24a, b for the values of the coupling delay $\tau_2 = 25 < \tau_1 = 30$ and $\tau_2 = 3.0 < \tau_1 = 4.0$, respectively, corresponding to the coupled Mackey-Glass and Ikeda time-delay systems. Complete

Fig. 8.24 Time series plots of the variables $x_1(t)$, $x_2(t)$ and $-x_2(t)$ of coupled time-delay systems depicting inverse anticipatory synchronization. (**a**) Mackey-Glass systems and (**b**) Ikeda systems. The value of the parameters are given in the text

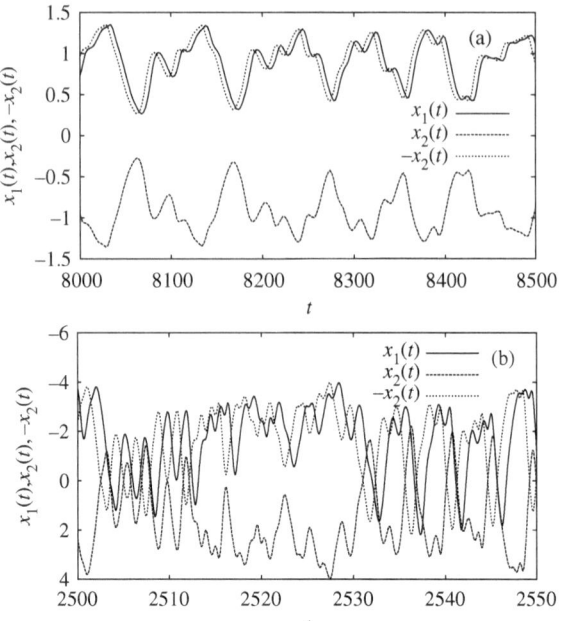

Fig. 8.25 Time series plots of the variables $x_1(t)$, $x_2(t)$ and $-x_2(t)$ of coupled time-delay systems depicting complete inverse synchronization. (**a**) Mackey-Glass systems and (**b**) Ikeda systems

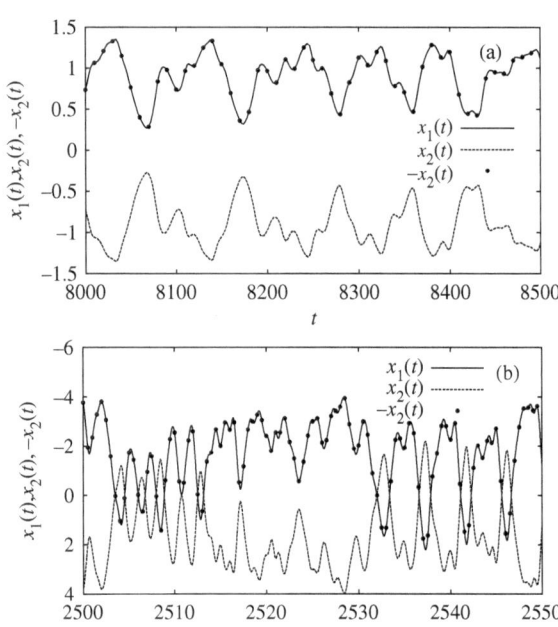

Fig. 8.26 Time series plots of the variables $x_1(t)$, $x_2(t)$ and $-x_2(t)$ of coupled time-delay systems depicting inverse lag synchronization. (**a**) Mackey-Glass systems and (**b**) Ikeda systems

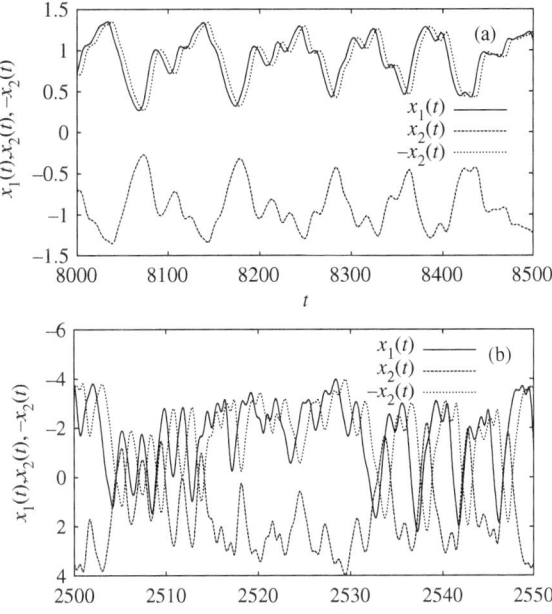

inverse synchronizations for both the systems are plotted in Fig. 8.25a, b for the values of the coupling delay $\tau_2 = \tau_1 = 30$ and $\tau_2 = \tau_1 = 4.0$, respectively. The existence of inverse lag synchronization in coupled Mackey-Glass systems for $\tau_2 = 35 > \tau_1 = 30$ and in coupled Ikeda systems for $\tau_2 = 5.0 > \tau_1 = 4.0$ are shown in Fig. 8.26a, b, respectively. These results clearly demonstrate the existence of inverse synchronizations in coupled Mackey-Glass systems and in coupled Ikeda systems also as in the case of piecewise linear time-delay systems studied in Sect. 8.3. In fact, these results are also corroborated using similarity function, probability of synchronization and transitions in the spectrum of Lyapunov exponents of the coupled piecewise linear and Ikeda time-delay systems [21].

References

1. D.V. Senthilkumar, M. Lakshmanan, Phys. Rev. E **71**, 016211 (2005)
2. M. Zhan, G.W. Wei, C.H. Lai, Phys. Rev. E **65**, 036202 (2002)
3. K. Pyragas, Phys. Rev. E **58**, 3067 (1998)
4. N.N. Krasovskii, *Stability of Motion* (Stanford University Press, Stanford, CA, 1963)
5. M.G. Rosenblum, A.S. Pikovsky, J. Kurths, Phys. Rev. Lett. **76**, 1804 (1996)
6. N. Platt, S.M. Hammel, J.F. Heagy, Phys. Rev. Lett. **72**, 3498 (1994)
7. J.F. Heagy, N. Platt, S.M. Hammel, Phys. Rev. E **49**, 1140 (1994)
8. A.V. Rangan, D. Cai, Phys. Rev. Lett. **96**, 178101 (2006)
9. X. Chen, J.E. Cohen, J. Theor. Biol. **212**, 223 (2001)
10. S. Sinha, S. Sinha, Phys. Rev. E **71**, 020902(R) (2005)
11. T.W. Carr, M.Y. Taylor, I.B. Schwartz, Physica D **213**, 152 (2006)

12. M.C. Cross, P.C. Hohenberg, Rev. Mod. Phys. **65**, 851 (1993)
13. S. Boccaletti, V. Latora, Y. Morenoy, M. Chavez, D.U. Hwang, Phys. Rep. **424**, 175 (2006)
14. A. Arenas, A. Diaz-Guilera, J. Kurths, Y. Moreno, C.S. Zhou, Phys. Rep. **469**, 93 (2008)
15. M.C. Mackey, L. Glass, Science **197**, 287 (1977)
16. M.C. Mackey, L. Glass, *From Clocks to Chaos, The Rhythms of Life* (Princeton University Press, Princeton, NJ, 1988)
17. K. Ikeda, H. Daido, O. Akimoto, Phys. Rev. Lett. **45**, 709 (1980)
18. K. Ikeda, K. Matsumoto, Physica D **29**, 223 (1987)
19. C. Masoller, D.H. Zanette, Physica A **300**, 359 (2001)
20. M. Le Berre, E. Ressayre, A. Tallet, H.M. Gibbs, D.L. Kaplan, M.H. Rose, Phys. Rev. A **35**, 3020 (1987)
21. D.V. Senthilkumar, J. Kurths, M. Lakshmanan, Chaos **19**, 023107 (2009)

Chapter 9
Intermittency Transition to Generalized Synchronization

9.1 Introduction

One of the interesting synchronization behaviors of unidirectionally coupled chaotic systems is the generalized synchronization (GS), which was conceptually introduced in [1]. Generalized synchronization is observed in coupled nonidentical systems, where there exists some functional relationship between the drive $X(t)$ and the response $Y(t)$ systems, that is, $Y(t) = F(X(t))$. With GS, all the response systems coupled to the drive lose their intrinsic chaoticity (sensitivity to initial conditions) under the same driving and follow the same trajectory. Hence the presence of GS can be detected using the so called auxiliary system approach [2], where an additional system (auxiliary system) identical to the response system is coupled to the drive in a similar fashion. Auxiliary system approach is particularly appealing since it can be implemented directly in an experiment and, in addition, this method allows one to utilize analytical approaches for studying GS. However, one has to be aware that if there are multiple basins of attraction for the coupled drive-response system, then the auxiliary system approach can fail.

Generalized synchronization (GS) has been well studied and understood in systems with few degrees of freedom and for discrete maps [1–7]. The concept of GS has also been extended to spatially extended chaotic systems such as coupled Ginzburg-Landau equations [8]. Recently, the terminology intermittent generalized synchronization (IGS) [9] was introduced in diffusively coupled Rössler systems in analogy with intermittent lag synchronization (ILS) [10, 11] and intermittent phase synchronization (IPS) [12–14], and also verified experimentally in coupled Chua's circuits. Very recently, it has been shown [15] that transition to intermittent chaotic synchronization (in the case of complete synchronization) is characteristically distinct for geometrically different chaotic attractors. In particular, it has been shown that for phase-coherent chaotic attractors (Rössler attractor) the transition occurs immediately as soon as the coupling strength is increased from zero and for non-phase-coherent attractors (Lorenz attractor), the transition occurs slowly as the coupling strength is increased from zero.

As noted earlier, time-delay systems form an important class of dynamical systems and recently they are receiving central importance in investigating various

M. Lakshmanan, D.V. Senthilkumar, *Dynamics of Nonlinear Time-Delay Systems*,
Springer Series in Synergetics, DOI 10.1007/978-3-642-14938-2_9,
© Springer-Verlag Berlin Heidelberg 2010

types of chaotic synchronizations, in view of their infinite dimensional nature and feasibility of experimental realization [16–19]. While the concept of GS has been well established in low dimensional systems, it has not yet been studied in detail in coupled time-delay systems and only very few recent studies have dealt with GS in time-delay systems [16, 17]. In particular, the mechanism of onset of GS in coupled time-delay systems and its characteristic properties have not yet been clearly understood.

In this chapter, we present some of the characteristic properties associated with the nature of transition to GS from an asynchronous state in unidirectionally coupled piecewise linear time-delay systems exhibiting highly non-phase-coherent hyperchaotic attractors [18], and also in the coupled Mackey-Glass and Ikeda systems. We find that the onset of GS is preceded by an on-off intermittency mechanism from the desynchronized state. We have also identified that the intermittency transition to GS exhibits characteristically distinct behaviors for different coupling schemes. In particular, the intermittency transition occurs in a broad range of coupling strength for error feedback coupling configurations and in a narrow range of coupling strength for direct feedback coupling configurations, beyond certain threshold value of the coupling strength. In addition, the intermittent dynamics is characterized by periodic bursts away from the temporal synchronized state with period equal to the delay time of the response system in the case of broad range intermittency transition whereas it is characterized by random time intervals in the case of narrow range intermittency transition. We have also confirmed these dynamical behaviors in both linear and nonlinear coupling configurations. We have analyzed these transitions analytically using Krasvoskii-Lyapunov functional approach and numerically by the probability of synchronization and by the subLyapunov exponents. We have also addressed the reason behind these transitions using periodic orbit theory. The robustness of these transitions with the system parameters in both the linear and nonlinear, error feedback and direct feedback coupling configurations are also studied.

9.2 Broad Range (Slow/Delayed) Intermittency Transition to GS for Linear Error Feedback Coupling of the Form $(x_1(t) - x_2(t))$

To be specific, we first consider the following unidirectional, linearly coupled systems with drive $x_1(t)$, response $x_2(t)$ and an auxiliary $x_3(t)$,

$$\dot{x}_1(t) = -ax_1(t) + b_1 f(x_1(t - \tau_1)), \tag{9.1a}$$

$$\dot{x}_2(t) = -ax_2(t) + b_2 f(x_2(t - \tau_2)) + b_3(x_1(t) - x_2(t)), \tag{9.1b}$$

$$\dot{x}_3(t) = -ax_3(t) + b_2 f(x_3(t - \tau_2)) + b_3(x_1(t) - x_3(t)), \tag{9.1c}$$

where b_1, b_2 and b_3 are constant parameters, and τ_1 and τ_2 are constant delay parameters (Dynamics of individual systems have already discussed in Chap. 3). Note that when $b_1 \neq b_2$ or $\tau_1 \neq \tau_2$ or both, corresponding to parameter mismatches, we have unidirectionally coupled nonidentical systems (Eqs. (9.1a) and (9.1b)), while the auxiliary system is given by (9.1c) and $f(x)$ is the odd piecewise linear function (3.2). The coupling in (9.1b) may be also called a linear error feedback coupling.

For simplicity, we have chosen $b_1 = b_2$ so that the time-delays τ_1 and τ_2 alone introduce a simple form of parameter mismatch between the drive $x_1(t)$ and the response $x_2(t)$. We have chosen the values of parameters as $a = 1.0, b_1 = b_2 = 1.2, \tau_1 = 20$ and $\tau_2 = 25$. For this parametric choice, in the absence of coupling, all the three systems (9.1) evolve independently and exhibit hyperchaotic attractors, which is confirmed by the existence of multiple positive Lyapunov exponents (Fig. 3.8).

9.3 Stability Condition

With GS, as all the response systems under the same driving follow the same trajectory, it is sufficient to identify the existence condition for establishment of complete synchronization (CS) between the response $x_2(t)$ and the auxiliary $x_3(t)$ systems in order to achieve GS between the drive $x_1(t)$ and the response $x_2(t)$ systems.

Now, for CS to occur between the response $x_2(t)$ and the auxiliary $x_3(t)$ variables, we consider the time evolution of the difference system with the state variable $\Delta = x_3(t) - x_2(t)$. It can be written for small values of Δ as

$$\dot{\Delta} = -(a + b_3)\Delta + b_2 f'(x_2(t - \tau_2))\Delta_{\tau_2}, \tag{9.2}$$

where for the odd piecewise linear function (3.2) we have

$$f'(x) = \begin{cases} -1.5, & -4/3 < x \leq -0.8 \\ 1, & -0.8 < x \leq 0.8 \\ -1.5, & 0.8 < x \leq 4/3. \end{cases} \tag{9.3}$$

The synchronization manifold, $x_2(t) = x_3(t)$, is locally attracting if the origin, $\Delta = 0$ is stable. Following Krasovskii-Lyapunov functional approach (discussed in the previous chapters), we define a positive definite Lyapunov functional of the form [19–21]

$$V(t) = \frac{1}{2}\Delta^2 + \mu \int_{-\tau_2}^{0} \Delta^2(t + \theta)d\theta, \tag{9.4}$$

where μ is an arbitrary positive parameter, $\mu > 0$. The solution of Eq. (9.2), namely $\Delta = 0$, is stable if the derivative of the functional along the trajectory of Eq. (9.2) is negative. This negativity condition is satisfied if $b_3 + a > \frac{b_2^2 f'^2(x_2(t-\tau_2))}{4\mu} + \mu$, from which it turns out that a sufficient condition for asymptotic stability is

$$a + b_3 > \left| b_2 f'(x_2(t - \tau_2)) \right|. \tag{9.5}$$

Now from the form of the piecewise linear function $f(x)$ given by Eq. (3.2), we have

$$\left| f'(x_2(t - \tau_2)) \right| = \begin{cases} 1.5, & 0.8 \le |x_2| \le \frac{4}{3} \\ 1.0, & |x_2| < 0.8. \end{cases} \tag{9.6}$$

Consequently the stability condition (9.5) becomes $a + b_3 > |1.5b_2| > |b_2|$. Thus one can take

$$a + b_3 > |b_2| \tag{9.7}$$

as the less stringent condition for (9.5) to be valid, while

$$a + b_3 > |1.5b_2| \tag{9.8}$$

can be considered as the most general condition specified by (9.5) for asymptotic stability of the synchronized state $\Delta = 0$.

9.4 Approximate (Intermittent) Generalized Synchronization

In order to understand the mechanism of transition to the synchronized state, it will be important to follow the dynamics from the parameter values at which the less stringent condition is satisfied. Figure 9.1a shows the approximate GS (which may also be termed as intermittent generalized synchronization (IGS) in analogy with the concept of intermittent lag synchronization (ILS)) between the drive $x_1(t)$ and the response $x_2(t)$ systems, whereas Fig. 9.1b shows the approximate CS between the response $x_2(t)$ and the auxiliary $x_3(t)$ systems for the values of the parameters $a = 1.0, b_1 = b_2 = 1.2, \tau_1 = 20, \tau_2 = 25$ and $b_3 = 0.4$ satisfying the less stringent condition (9.7). Perfect GS and perfect CS are shown in Figs. 9.1c, d, respectively, for $b_3 = 0.9$ satisfying the general stability condition (9.8). Time traces of the difference $x_2(t) - x_3(t)$ corresponding to approximate CS (Fig. 9.1b) are shown in Fig. 9.2, which show periodic bursts with period between two consecutive bursts approximately equal to the time-delay of the response system $t \approx 25$ when "on" states of amplitude greater than $|0.01|$ are considered. Figure 9.2b shows an enlarged (in x scale) part of Fig. 9.2a to view the bursts at periodic intervals when bursts of larger amplitudes ($\Delta > |0.01|$) are considered, while Fig. 9.2c is an enlarged (in y

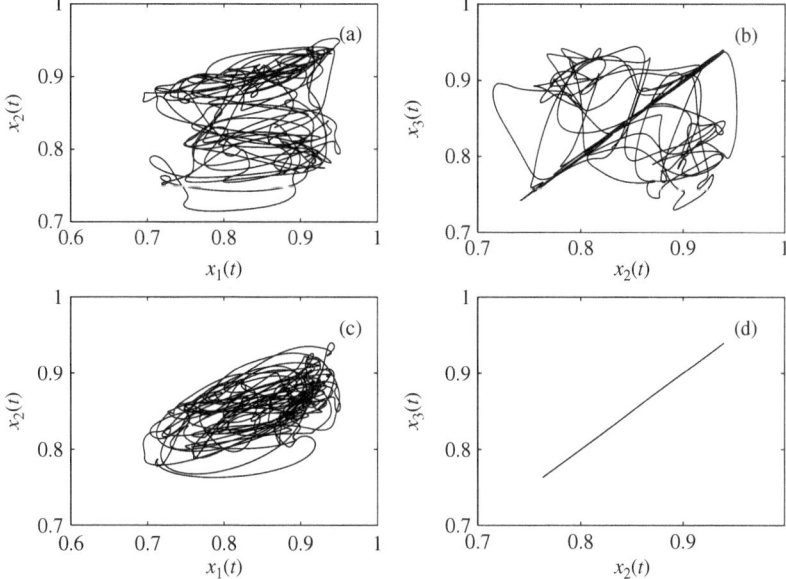

Fig. 9.1 Dynamics in the phase space of the systems (9.1). (**a**) and (**b**) Approximate GS and CS, respectively, for the value of the coupling strength $b_3 = 0.4$. (**c**) and (**d**) Perfect GS and CS, respectively, for the value of the coupling strength $b_3 = 0.9$

scale) version of Fig. 9.2b to show random bursts when bursts of smaller amplitude, $\Delta < |0.01|$, are considered.

Usually the intermittent dynamics is characterized by the entrainment of the dynamical variables in random time intervals of finite duration [22, 23]. But from Fig. 9.2b, it is evident that the intermittent dynamics displays periodic bursts from the synchronous state with period approximately equal to the delay time of the response system, when amplitudes of the state variable $|\Delta| = |x_3(t) - x_2(t)| > 0.01$ are considered, for the values of the coupling strength at which the less stringent stability condition (9.7) is satisfied (It is to be noted that such periodic bursts of period approximately equal to the time-delay of the response system has also been observed by Zhan et al. [16], where the authors discussed relation between two modes of synchronization, namely, CS and GS in unidirectionally coupled Mackey-Glass systems). The statistical features associated with the intermittent dynamics is analyzed by calculating the distribution of laminar phases $\Lambda(t)$ with amplitude less than a threshold value of Δ. A universal asymptotic power law distribution $\Lambda(t) \propto t^{-\alpha}$ is observed for the threshold value $\Delta = 0.0001$ with the value of the exponent $\alpha = 1.5$ as shown in Fig. 9.3, which is quite typical for on-off intermittency. Note that $-3/2$ power law is observed for the intermittent dynamics shown in Fig. 9.2 for laminar phases $\Lambda(t)$ with amplitude less than $\Delta = 0.0001$ (as an illustrative example), which is also evident from Fig. 9.2c, while periodic bursts are observed for "on" state of amplitude greater than $|0.01|$.

Fig. 9.2 The intermittent
dynamics of the response
$x_2(t)$ and auxiliary $x_3(t)$
systems for the value of the
coupling strength $b_3 = 0.4$.
(**a**) Time traces of the
difference $x_2(t) - x_3(t)$
corresponding to Fig. 9.1b,
(**b**) Enlarged in x scale to
show bursts at periodic
intervals when bursts of
larger amplitudes $\Delta > |0.01|$
are considered and (**c**)
Enlarged in y scale to show
random bursts when bursts of
smaller amplitudes
$\Delta < |0.01|$ are considered

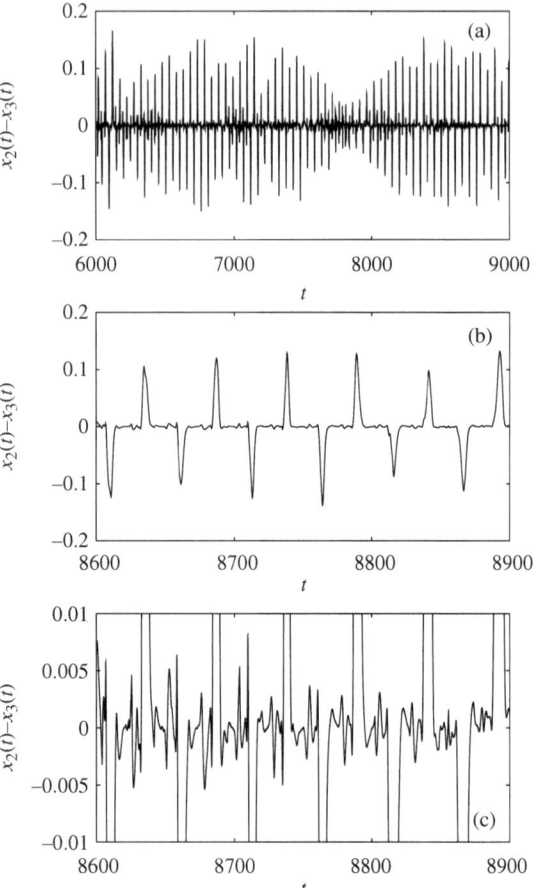

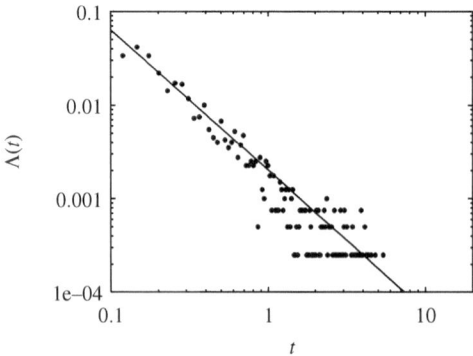

Fig. 9.3 The statistical
distribution of laminar phases
for the Fig. 9.2

9.5 Characterization of IGS

Now we characterize the intermittency transition to GS by using (i) the notion of the probability of synchronization $\Phi(b_3)$ as a function of the coupling strength b_3 [15], which can be defined as the fraction of time during which $|x_2(t) - x_3(t)| < \varepsilon$ occurs, where ε is a small but arbitrary threshold, and (ii) from the changes in the sign of *sub*Lyapunov exponents (which are nothing but the Lyapunov exponents of the subsystem) in the spectrum of Lyapunov exponents of the coupled time-delay systems. Figure 9.4a shows the probability of synchronization $\Phi(b_3)$ as a function of the coupling strength b_3 calculated from the variables of the response $x_2(t)$ and the auxiliary $x_3(t)$ systems for CS to occur between them. For the range of $b_3 \in (0, 0.39)$, there is an absence of any entrainment between the systems resulting in an asynchronous behavior and the probability of synchronization $\Phi(b_3)$ is practically zero in this region. However, starting from the value of $b_3 = 0.39$ and above, there appear oscillations in the value of the probability of synchronization $\Phi(b_3)$ between zero and some finite values less than unity, exhibiting intermittency transition to GS in the range of $b_3 \in (0.4, 0.62)$ for which the less stringent stability condition (9.7) is satisfied. Beyond $b_3 = 0.62$, $\Phi(b_3)$ attains unit value indicating perfect GS. Note that the above intermittency transition occurs in a rather wide range of the coupling strength (this can also be termed as slow or delayed intermittency transition in analogy with the terminology used in [15]), which has also been confirmed from the transition of successive largest *sub*Lyapunov exponents in the corresponding range of the coupling strength.

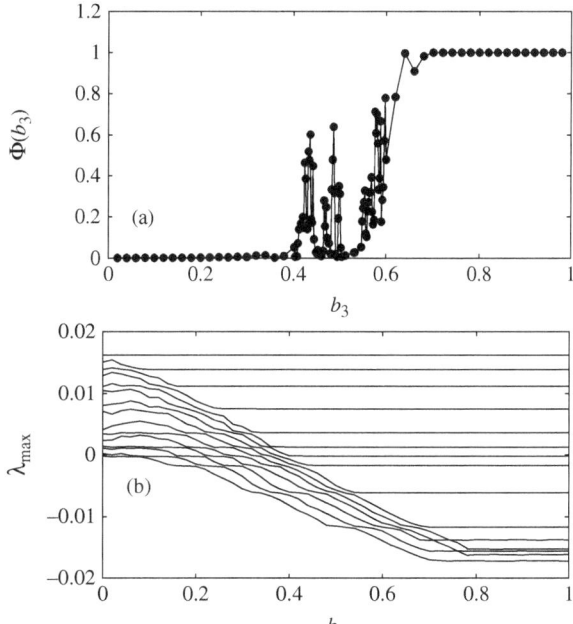

Fig. 9.4 (**a**) The probability of synchronization $\Phi(b_3)$ between the response $x_2(t)$ and the auxiliary $x_3(t)$ systems and (**b**) Largest Lyapunov exponents of the coupled drive $x_1(t)$ and response $x_2(t)$ systems (9.1a) and (9.1b)

The spectrum containing the first fifteen largest Lyapunov exponents λ_{max} of the coupled drive $x_1(t)$ and response $x_2(t)$ systems is shown in Fig. 9.4b. From the general stability condition (9.5), it is evident that for the chosen value of the parameter $a = 1.0$, the less stringent stability condition (9.7) is satisfied for values of coupling strength $b_3 > 0.2$. Correspondingly, the least positive subLyapunov exponent of the response system (9.1b) gradually becomes negative from $b_3 > 0.2$. Subsequently, the remaining positive subLyapunov exponents gradually become negative and attain saturation in the range of $b_3 \in (0.2, 0.8)$. This is in accordance with the fact that the less stringent stability condition is satisfied only in the corresponding range of coupling strength b_3. This is a strong indication of the broad range intermittency (IGS) transition to GS. For $b_3 > 0.8$, the general stability condition (9.8) is satisfied, where one can observe perfect GS as is evidenced by both the probability of synchronization approaching unit value and by the negative saturation of subLyapunov exponents, calculated between the drive and response systems. The inference is that the correlation between the oscillations of the systems eventually becomes stronger with the strength of the coupling, and this is indicated by the successive transition of subLyapunov exponents to negative values.

It is a well established fact that a chaotic attractor can be considered as a pool of infinitely many unstable periodic orbits of all periods. Synchronization between two coupled systems is said to be asymptotically stable, if all the unstable periodic orbits of the response system are stabilized in the transverse direction of the synchronization manifold. Consequently, all the trajectories transverse to the synchronization manifold converge for suitable values of coupling strength and this is reflected in the negative values of the transverse Lyapunov exponents (subLyapunov exponents) upon synchronization [15]. From our results, we find that the subLyapunov exponents gradually become negative in a broad range of coupling strength b_3 after certain threshold value and this is in accordance with gradual stabilization of unstable periodic orbits of the response system in the complex synchronization manifold. Unfortunately, methods for locating UPO's and calculating their transverse Lyapunov exponents have not been well established for time-delay systems and hence a quantitative proof for the gradual stabilization of UPO's has not been given here. However, the gradual stabilization of UPO's along with their transverse Lyapunov exponent in the range of intermittency transition have been reported for the case of coupled Rössler and Lorenz systems in [15]. It is also to be noted that the broad range intermitteny transition in the case of error feedback coupling configuration is due to the fact that the strength of the coupling b_3 contributes only less significantly to stabilize the UPO's as the error $x_1(t) - x_2(t)$ gradually becomes smaller from the transition regime after certain threshold value of the coupling strength.

The robustness of the intermittency transition in a broad range of coupling strength with the system parameter $b_2 \in (1.1, 1.6)$ and with the coupling delay $\tau_2 \in (10, 20)$ has also been confirmed. Figure 9.5a shows the 3-dimensional plot of the probability of synchronization as a function of the system parameter b_2 and the coupling strength b_3, while Fig. 9.5b shows the 3-dimensional plot of $\Phi(b_3)$ as

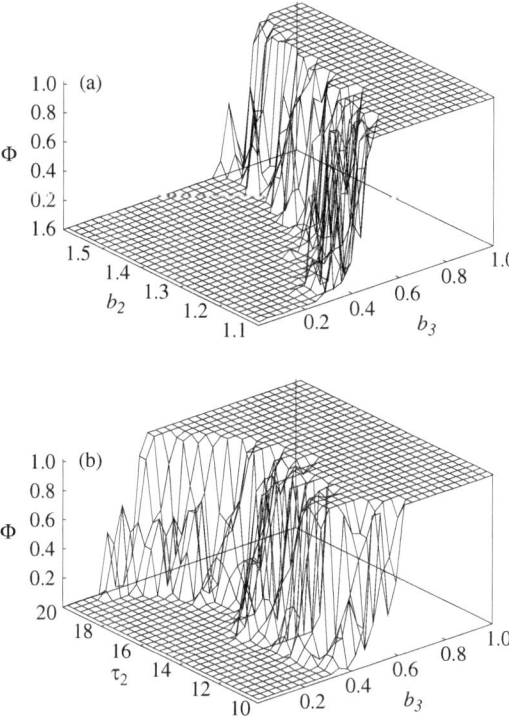

Fig. 9.5 The probability of synchronization $\Phi(b_3)$ in 3-dimensional plots (**a**) as a function of the system parameter b_2 and the coupling strength b_3 and (**b**) as a function of the coupling delay τ_2 and the coupling strength b_3, exhibiting broad range intermittency transition to GS for linear error feedback coupling

a function of the coupling delay τ_2 and the coupling strength b_3. The above figures (Fig. 9.5) clearly reveal the broad range intermittency transition to GS in the case of linear error feedback coupling scheme.

9.6 Narrow Range (Immediate) Intermittency Transition to GS for Linear Direct Feedback Coupling of the Form $x_1(t)$

To illustrate the narrow range intermittency transition to GS, we consider the unidirectional linear direct feedback coupling of the form

$$\dot{x}_1(t) = -ax_1(t) + b_1 f(x_1(t - \tau_1)), \tag{9.9a}$$

$$\dot{x}_2(t) = -ax_2(t) + b_2 f(x_2(t - \tau_2)) + b_3 x_1(t), \tag{9.9b}$$

$$\dot{x}_3(t) = -ax_3(t) + b_2 f(x_3(t - \tau_2)) + b_3 x_1(t), \tag{9.9c}$$

where $f(x)$ is of the same odd piecewise linear form as in Eq. (3.2). Assuming the same values of the parameters as before and proceeding in the same way as in the previous case, one can obtain a sufficient condition for asymptotically stable CS between the response $x_2(t)$, Eq. (9.9b), and the auxiliary $x_3(t)$, Eq. (9.9c), systems as

$$a > \left| b_2 f'(x_2(t - \tau_2)) \right|. \tag{9.10}$$

It is to be noted that the above stability condition holds good only for the case when coupling is present, that is $b_3 \neq 0$. When there is no coupling ($b_3 = 0$), by definition, there will be a desynchronized chaotic state. As soon as the value of the coupling strength is increased from zero, the stability condition (9.10) always lead to synchronized state even for very feeble values of b_3 for parameters satisfying the stability condition, as it is independent of the coupling strength b_3. As the values of the parameters satisfying the stability condition (9.10) rapidly leads to immediate transition to synchronized state as soon as the coupling is switched on, it is difficult to identify the possible transitions to synchronized state. In addition, as the stability condition is independent of the coupling strength b_3, one is not able to explore the dynamical transitions as a function of coupling strength for the parameter values satisfying the stability condition (9.10). Hence we study the synchronization transition by choosing the parameters violating the stability condition as $a = 1.0, b_1 = 1.2, b_2 = 1.1, \tau_1 = \tau_2 = 20$ and varying the coupling strength b_3 in order to identify the mechanism of synchronization transition. Here, in this case b_1 and b_2 alone introduce the parameter mismatch while $\tau_1 = \tau_2$ (It may be added that the qualitative nature of the dynamical transitions remain the same even when the mismatch is either in time delays alone, that is $\tau_1 \neq \tau_2, b_1 = b_2$ or in both the system parameters and time delays, $b_1 \neq b_2, \tau_1 \neq \tau_2$, as confirmed below in the three dimensional plots of Fig. 9.10).

As b_3 is varied from zero, transition from desynchronized state to approximate GS occurs for $b_3 > 0.6$. Approximate GS (IGS) between the drive $x_1(t)$, Eq. (9.9a), and the response $x_2(t)$, Eq. (9.9b), systems is shown in Fig. 9.6a whereas the approximate CS between the response $x_2(t)$, Eq. (9.9b), and the auxiliary $x_3(t)$, Eq. (9.9c), systems is shown in Fig. 9.6b for the value of the coupling strength $b_3 = 0.64$. Perfect GS and CS are shown in Fig. 9.6c, d, respectively, for the value of the coupling strength $b_3 = 0.8$. The intermittent dynamics at the transition regime corresponding to the value of the coupling strength $b_3 = 0.64$ is shown in Fig. 9.7, in which Fig. 9.7b shows the enlarged part of Fig. 9.7a. It is clear from this figure that the intermittent dynamics displays intermittent bursts at random time intervals. The statistical distribution of the laminar phases again shows a universal asymptotic -1.5 power law behavior for the threshold value $|\Delta| = 0.0001$, which is typical for on-off intermittency transitions, as shown in Fig. 9.8.

Now we characterize the intermittency transition to GS in the present case, again by using the notion of probability of synchronization $\Phi(b_3)$ and from the changes in the sign of subLyapunov exponents of the coupled system. The probability of synchronization is shown in Fig. 9.9a as a function of the coupling strength, again

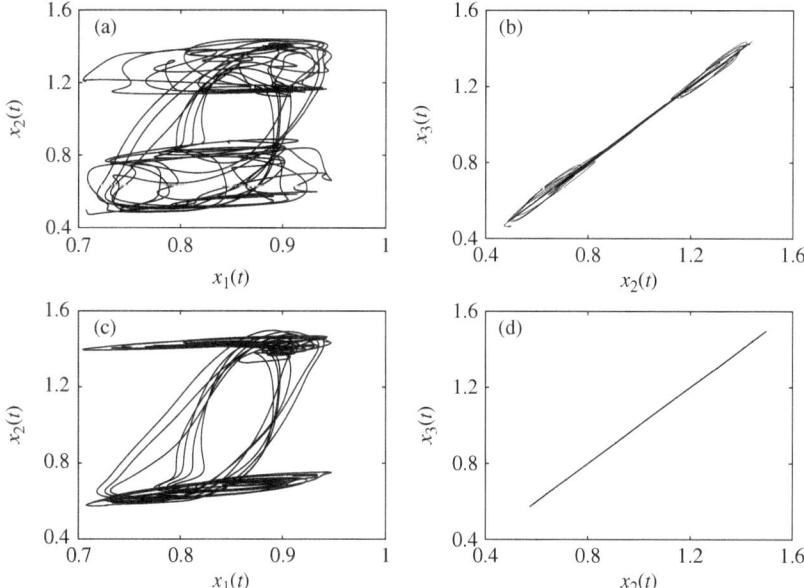

Fig. 9.6 Dynamics in the phase space of the systems (9.9). (**a**) and (**b**) Approximate GS and CS, respectively, for the value of the coupling strength $b_3 = 0.64$. (**c**) and (**d**) Perfect GS and CS, respectively, for the value of the coupling strength $b_3 = 0.8$

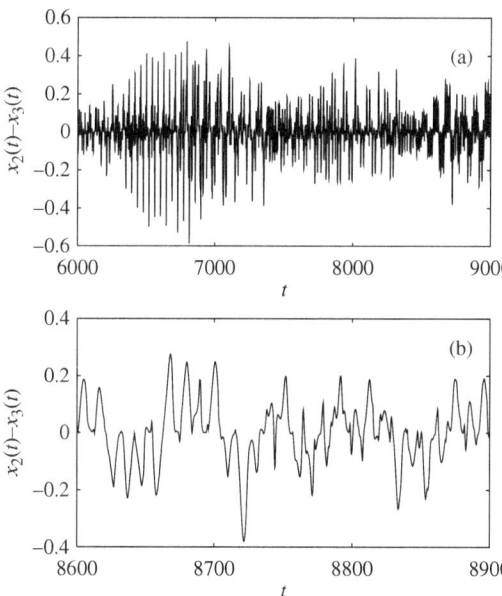

Fig. 9.7 The intermittent dynamics of the response $x_2(t)$ (9.9b) and auxiliary $x_3(t)$ (9.9c) systems for the value of the coupling strength $b_3 = 0.64$. (**a**) and (**b**) Time traces of the difference $x_2(t) - x_3(t)$ corresponding to Fig. 9.6b

Fig. 9.8 The statistical distribution of laminar phases for the Fig. 9.7

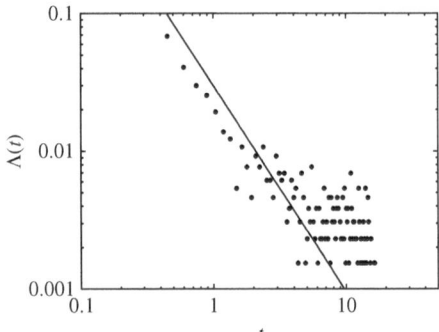

Fig. 9.9 (**a**) The probability of synchronization $\Phi(b_3)$ between the response $x_2(t)$ (9.9b) and the auxiliary $x_3(t)$ (9.9c) systems and (**b**) Largest Lyapunov exponents of the coupled drive $x_1(t)$ and response $x_2(t)$ systems (9.9a) and (9.9b)

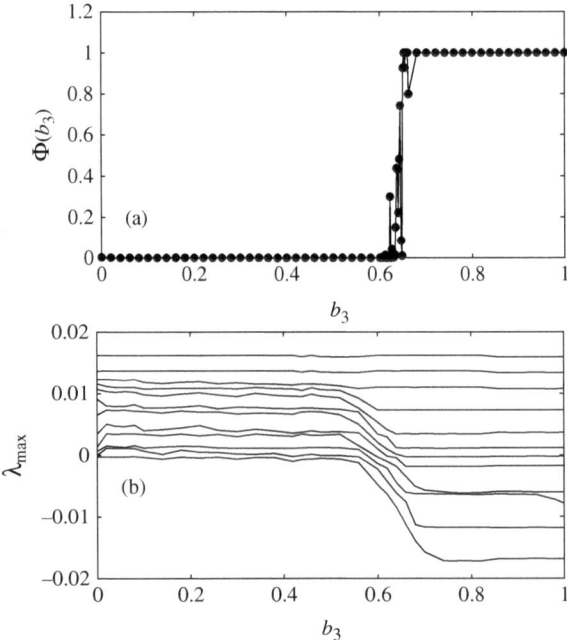

calculated from the response $x_2(t)$ and the auxiliary $x_3(t)$ systems, Eqs. (9.9b) and (9.9c), respectively, which remains zero in the range of $b_3 \in (0, 0.60)$ and oscillates between its maximum and minimum values in a narrow range of $b_3 \in (0.60, 0.68)$ confirming the existence of approximate CS in the latter range. Above $b_3 = 0.68$ the probability of synchronization acquires its maximum value depicting perfect CS between the response $x_2(t)$ and the auxiliary $x_3(t)$ systems. Correspondingly there exists perfect GS between the drive $x_1(t)$ and the response $x_2(t)$ systems. Fig. 9.9b shows the first twelve maximal Lyapunov exponents of the coupled drive $x_1(t)$ and the response $x_2(t)$ systems. The least positive subLyapunov exponent of the response system starts to become negative from $b_3 > 0.60$. Subsequently, all the other positive *sub*Lyapunov exponents become negative and reach saturation in a rather narrow range of $b_3 \in (0.60, 0.68)$. Thus the narrow range intermittency (IGS) transition (this can also be termed as immediate intermittency transition in analogy with the terminology used in [15]) is confirmed from both the probability of synchronization, calculated from the response and the auxiliary systems, and negative saturation of *sub*Lyapunov exponents, calculated from the drive and the response systems.

As discussed in the previous section, the narrow range intermittency transition is in accordance with the stabilization of all the unstable periodic orbits of the response system in a narrow range as a function of the coupling strength b_3 and this is reflected in the immediate transition of all the *sub*Lyapunov exponents (Fig. 9.9b) to negative values. It is also to be noted that the narrow range intermitteny transition in the case of direct feedback coupling configuration is due to the fact that the strength of the coupling b_3 contributes significantly proportional to the strength of the signal $x_1(t)$ to stabilize all the UPO's immediately at the transition regime after certain threshold value of the coupling strength.

The robustness of the intermittency transition in a narrow range of the coupling strength b_3 for a range of values of the parameter $b_2 \in (1.1, 1.6)$ and the delay $\tau_1 = \tau_2 \in (10, 20)$ is shown in Fig. 9.10. The 3-dimensional plot of the probability of synchronization as a function of the system parameter b_2 and the coupling strength

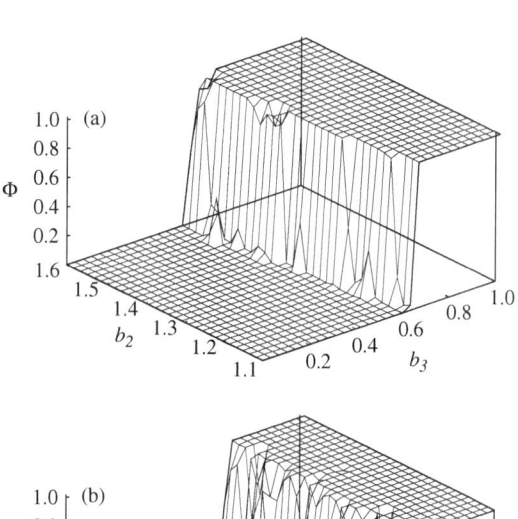

Fig. 9.10 The probability of synchronization $\Phi(b_3)$ in 3-dimensional plots (**a**) as a function of the system parameter b_2 and the coupling strength b_3 and (**b**) as a function of the coupling delay τ_2 and the coupling strength b_3, exhibiting narrow range intermittency transition to GS for linear direct feedback coupling

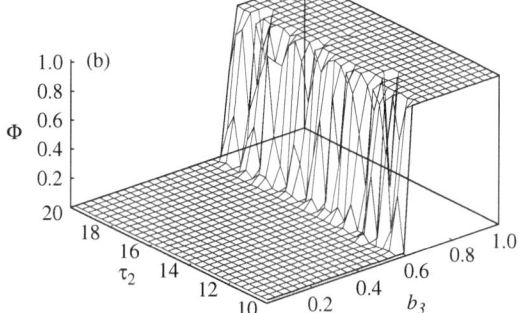

b_3 is shown in Fig. 9.10a, while Fig. 9.10b shows the 3-dimensional plot of $\Phi(b_3)$ as a function of the coupling delay τ_2 and the coupling strength b_3.

9.7 Broad Range Intermittency Transition to GS for Nonlinear Error Feedback Coupling of the Form $(f(x_1(t - \tau_2)) - f(x_2(t - \tau_2)))$

Next we demonstrate the existence of the above types of distinct characteristic transitions for nonlinear coupling configurations as well. For this purpose, we consider the unidirectional nonlinear error feedback coupling of the form

$$\dot{x}_1(t) = - ax_1(t) + b_1 f(x_1(t - \tau_1)), \tag{9.11a}$$

$$\dot{x}_2(t) = - ax_2(t) + b_2 f(x_2(t - \tau_2)) + b_3(f(x_1(t - \tau_2)) - f(x_2(t - \tau_2))), \tag{9.11b}$$

$$\dot{x}_3(t) = - ax_3(t) + b_2 f(x_3(t - \tau_2)) + b_3(f(x_1(t - \tau_2)) - f(x_3(t - \tau_2))), \tag{9.11c}$$

where $f(x)$ is again of the same piecewise linear form as in Eq. (3.2). The parameters are now fixed as $a = 1.0, b_1 = b_2 = 1.2, \tau_1 = 20$ and the coupling delay $\tau_2 = 25$, where the delays alone form the parameter mismatch between the drive and response systems in Eqs. (9.11). Following Krasovskii-Lyapunov theory, for complete synchronization so that the manifold $\Delta = x_3(t) - x_2(t)$ between the response $x_2(t)$ and the auxiliary $x_3(t)$ approaches zero, one can obtain the stability condition as

$$a > \left| (b_2 - b_3) f'(x_2(t - \tau_2)) \right|. \tag{9.12}$$

Consequently from the form of the piecewise linear function (3.2), the stability condition (9.12) becomes $a > |1.5(b_2 - b_3)| > |(b_2 - b_3)|$. Thus one can take

$$a > |b_2 - b_3| \tag{9.13}$$

as less stringent condition for (9.12) to be valid, while

$$a > |1.5b_2 - 1.5b_3| \tag{9.14}$$

can be considered as the most general condition specified by (9.12) for asymptotic stability of the synchronized state $\Delta = x_2(t) - x_3(t) = 0$. For the chosen values of the parameters, the less stringent stability condition (9.13) is satisfied for the values of the coupling strength $b_3 \in (0.2, 0.535)$ and the general stability condition (9.14) is satisfied for $b_3 > 0.535$.

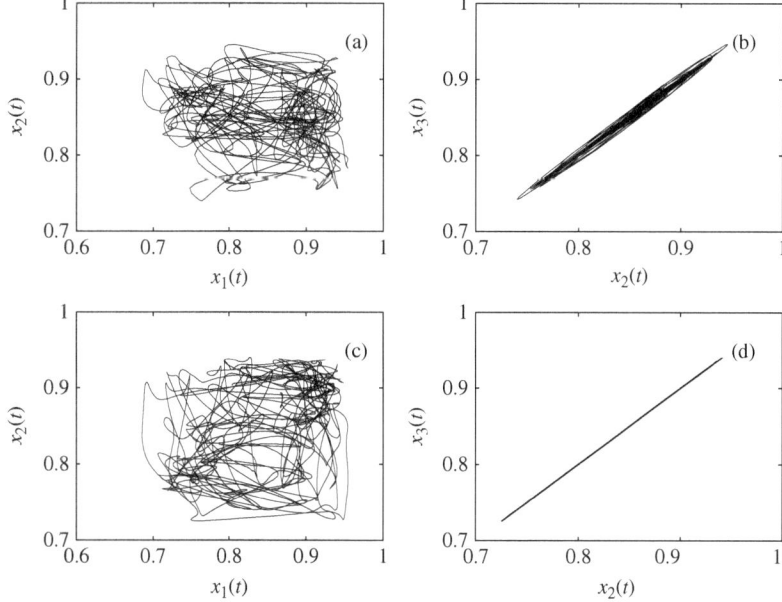

Fig. 9.11 Dynamics in the phase space of the systems (9.11). (**a**) and (**b**) Approximate GS and CS, respectively, for the value of the coupling strength $b_3 = 0.37$. (**c**) and (**d**) Perfect GS and CS, respectively, for the value of the coupling strength $b_3 = 0.8$

As the coupling strength is increased from zero, approximate GS occurs from $b_3 > 0.2$. Figure 9.11a shows the approximate GS (IGS) between the drive $x_1(t)$ (Eq. (9.11a)) and the response $x_2(t)$ (Eq. (9.11b)) systems for the value of the coupling strength $b_3 = 0.37$, while the approximate CS between the response $x_2(t)$ (Eq. (9.11b)) and the auxiliary $x_3(t)$ (Eq. (9.11c)) systems is shown in Fig. 9.11b. Perfect GS and perfect CS are shown in Fig. 9.11c,d respectively for $b_3 = 0.8$. The intermittent dynamics exhibited by the coupled systems at the transition regime is shown in Fig. 9.12, which shows bursts at the period approximately equal to the delay time of the response system $x_2(t)$ for bursts of amplitude greater than $|0.01|$ (Fig. 9.12b). The statistical distribution of the laminar phases away from the intermittent bursts shows an asymptotic -1.5 power law behavior for the threshold value $\Delta = 0.0001$ (see Fig. 9.12c), typical for on-off intermittency, which is shown in Fig. 9.13.

Now, the intermittency transition is again characterized using the probability of synchronization and the *sub*Lyapunov exponents as in the previous cases. Figure 9.14a shows the probability of synchronization $\Phi(b_3)$, the value of which remains zero in the range $b_3 \in (0, 0.2)$ due to the fact that there lacks any entrainment between the response $x_2(t)$ and the auxiliary $x_3(t)$ systems, whereas it fluctuates between the two extreme values in a rather broad range of the coupling strength $b_3 \in (0.2, 0.42)$, depicting the existence of intermittency transition in the corresponding range of b_3. Perfect CS exists for $b_3 > 0.42$ as evidenced from

Fig. 9.12 The intermittent
dynamics of the response
$x_2(t)$ (9.11b) and auxiliary
$x_3(t)$ (9.11c) systems for the
value of the coupling strength
$b_3 = 0.37$ for nonlinear error
feedback coupling. (**a**) Time
traces of the difference
$x_2(t) - x_3(t)$ corresponding
to Fig. 9.11b, (**b**) Enlarged in
x scale to show bursts at
periodic intervals when bursts
of larger amplitudes
$\Delta > |0.01|$ are considered
and (**c**) Enlarged in y scale to
show random bursts when
bursts of smaller amplitudes
$\Delta < |0.01|$ are considered

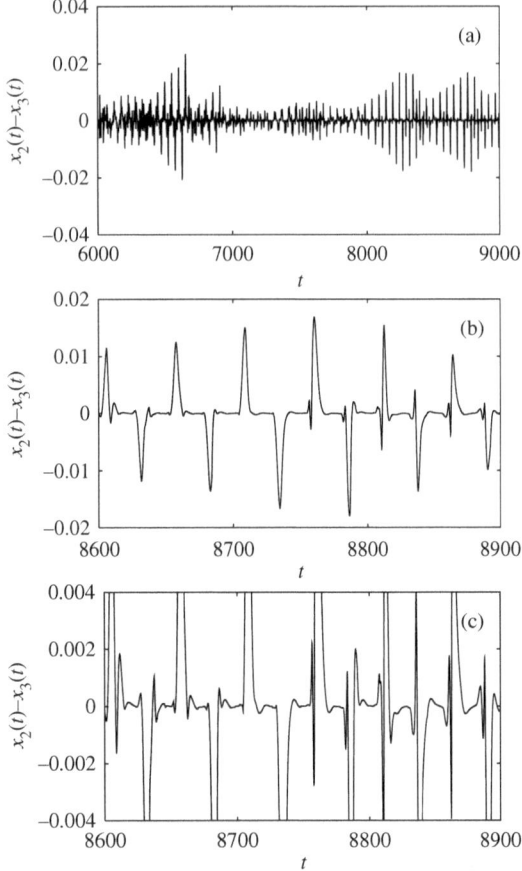

Fig. 9.13 The statistical
distribution of laminar phases
for the Fig. 9.12

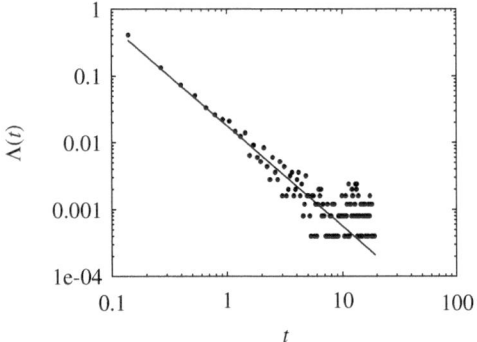

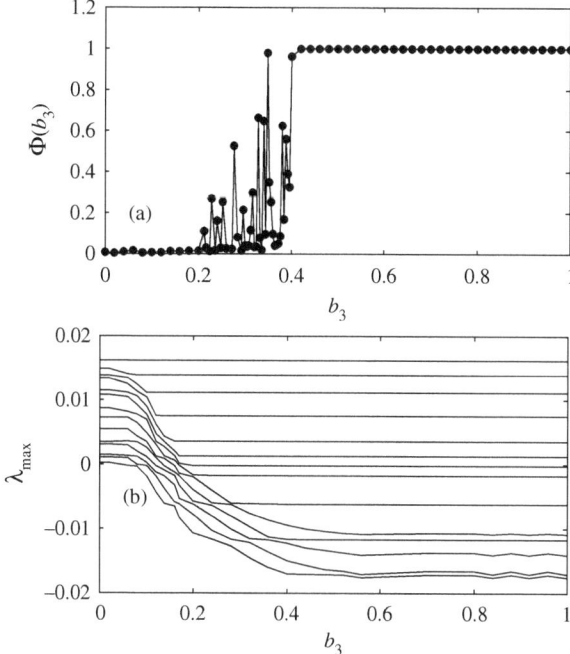

Fig. 9.14 (**a**) The probability of synchronization $\Phi(b_3)$ between the response $x_2(t)$ (9.11b) and the auxiliary $x_3(t)$ (9.11c) systems and (**b**) Largest Lyapunov exponents of the coupled drive $x_1(t)$ and response $x_2(t)$ systems (9.11a) and (9.11b)

the maximum value of $\Phi(b_3)$. Correspondingly there exists perfect GS between the drive $x_1(t)$ and the response $x_2(t)$ systems. Figure 9.14b shows the transition of *sub*Lyapunov exponents of the spectrum of Lyapunov exponents of the coupled drive $x_1(t)$ (Eq. (9.11a)) and the response $x_2(t)$ (Eq. (9.11b)) systems. The *sub*Lyapunov exponents become negative in the range $b_3 \in (0.2, 0.42)$ confirming the broad range intermittency (IGS) transition in a rather wide range of the coupling strength and this is again due to the gradual stabilization of the unstable periodic orbits of the response systems because of the less significant contribution of the coupling strength b_3 as the error becomes gradually smaller from the transition regime beyond certain threshold value of the coupling strength as discussed in Sect. 9.5. The robustness of the intermittency transition with the system parameter b_2 and the coupling delay τ_2 as a function of coupling strength b_3 is shown as 3-dimensional plots in Fig. 9.15.

9.8 Narrow Range Intermittency Transition to GS for Nonlinear Direct Feedback Coupling of the Form $f(x_1(t - \tau_2))$

Now we consider the unidirectional nonlinear coupling of the form

Fig. 9.15 The probability of synchronization $\Phi(b_3)$ in 3-dimensional plots (**a**) as a function of the system parameter b_2 and the coupling strength b_3 and (**b**) as a function of the coupling delay τ_2 and the coupling strength b_3 for the case of nonlinear error feedback coupling configuration given by Eq. (9.11), showing broad range intermittency transition

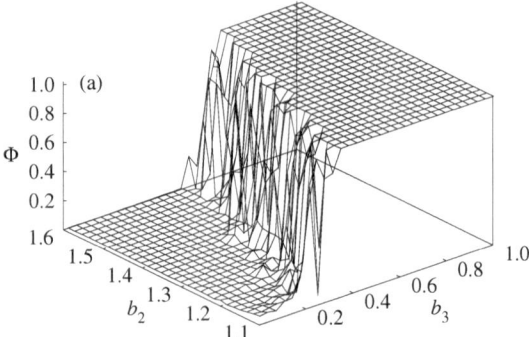

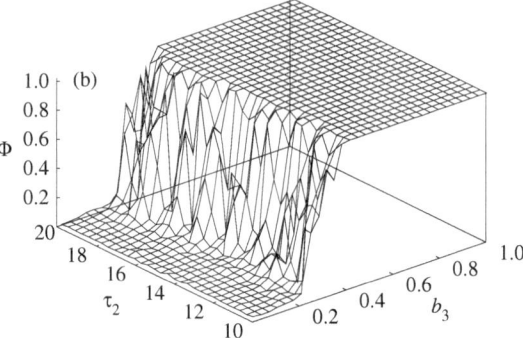

$$\dot{x}_1(t) = -ax_1(t) + b_1 f(x_1(t-\tau_1)), \tag{9.15a}$$

$$\dot{x}_2(t) = -ax_2(t) + b_2 f(x_2(t-\tau_2)) + b_3 f(x_1(t-\tau_2)), \tag{9.15b}$$

$$\dot{x}_3(t) = -ax_3(t) + b_2 f(x_3(t-\tau_2)) + b_3 f(x_1(t-\tau_2)). \tag{9.15c}$$

Choosing the values of the parameters as in the previous case and following Krasvoskii-Lyapunov functional approach for the asymptotically stable synchronized state $\Delta = x_3(t) - x_2(t) = 0$, one can obtain a sufficient condition for asymptotic stability for complete synchronization of the response $x_2(t)$ and the auxiliary $x_3(t)$ systems as

$$a > \left| b_2 f'(x_2(t-\tau_2)) \right|. \tag{9.16}$$

The above stability condition rapidly leads to immediate transition to synchronized state even for very feeble values of the coupling strength b_3 for the parameter values satisfying the stability condition (9.16) as the stability condition is independent of b_3 as in the previous linear coupling case (Sect. 9.6). Hence it is difficult to identify the possible dynamical transitions to synchronized state as a function of the coupling strength b_3. So we study the synchronization transition as a function of the

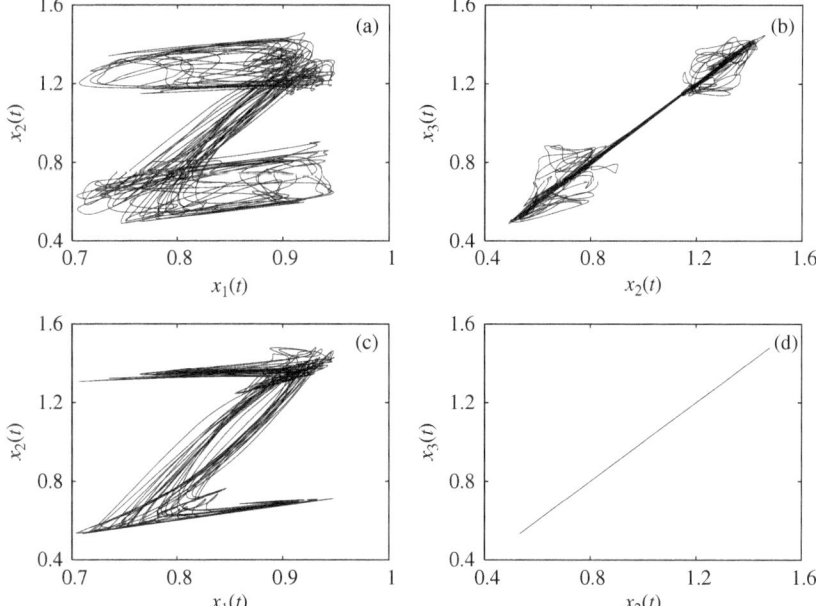

Fig. 9.16 Dynamics in the phase space of the systems (9.15). (**a, b**) Approximate GS and CS for the value of the coupling strength $b_3 = 0.78$ and (**c, d**) Perfect GS and CS for the value of the coupling strength $b_3 = 0.9$

coupling strength b_3 by choosing the parameters violating the stability condition as $a = 1.0$, $b_1 = 1.2$, $b_2 = 1.1$ and $\tau_1 = \tau_2 = 15$.

As b_3 is varied from zero for the above values of the parameters, transition from desynchronized state to approximate GS occurs for $b_3 > 0.74$. The approximate GS (IGS) between the drive $x_1(t)$ and the response $x_2(t)$ variables described by Eqs.(9.15a) and (9.15b) is shown in Fig. 9.16a, whereas Fig. 9.16b shows the approximate CS between the response $x_2(t)$ and the auxiliary $x_3(t)$ variables (Eqs. (9.15a) and (9.15c)) for the value of the coupling strength $b_3 = 0.78$. Perfect GS and perfect CS are shown in Fig. 9.16c, d respectively for $b_3 = 0.9$. Time traces of the difference $x_2(t) - x_3(t)$ corresponding to approximate CS (Fig. 9.16b) is shown in Fig. 9.17, which shows intermittent dynamics with the entrainment of the dynamical variables in random time intervals of finite duration. Fig. 9.17b shows the enlarged picture of part of Fig. 9.17a. The statistical distribution of the laminar phases again shows a universal asymptotic -1.5 power law behavior for the threshold value $|\Delta| = 0.0001$, typical for on-off intermittency, as shown in Fig. 9.18.

As in the previous cases, now we characterize the intermittency transition to GS using the notion of probability of synchronization $\Phi(b_3)$ and from the changes in the sign of subLyapunov exponents in the spectrum of Lyapunov exponents of the coupled time-delay systems. Figure 9.19a shows the probability of synchronization $\Phi(b_3)$ as a function of the coupling strength b_3 calculated from the response

Fig. 9.17 The intermittent
dynamics of the response
$x_2(t)$ (9.15b) and auxiliary
$x_3(t)$ (9.15c) systems for the
value of the coupling strength
$b_3 = 0.78$. (**a**) and (**b**) Time
traces of the difference
$x_2(t) - x_3(t)$ corresponding
to Fig. 9.16b

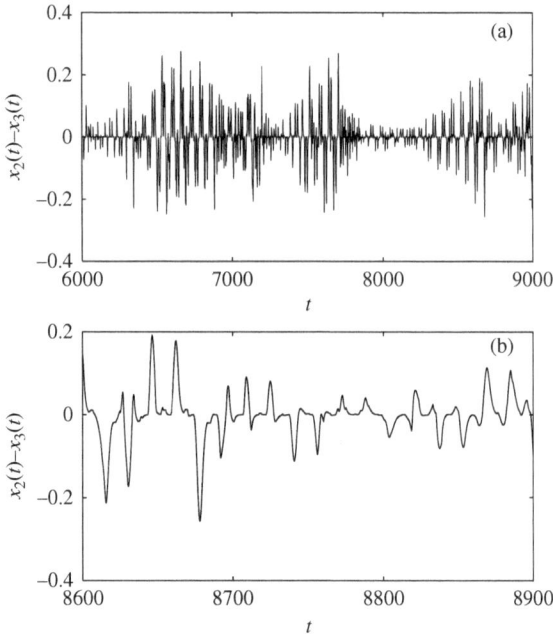

Fig. 9.18 The statistical
distribution of laminar phases
for the Fig. 9.17

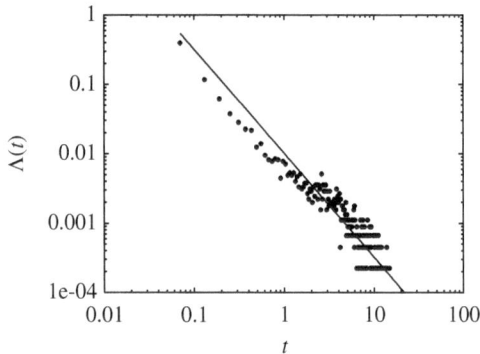

$x_2(t)$ and the auxiliary $x_3(t)$ variables (Eqs. (9.15b) and (9.15c)) for CS between
them. In the range of $b_3 \in (0, 0.74)$, the probability of synchronization remains
approximately zero. Upon increasing the value of b_3, $\Phi(b_3)$ oscillates in the narrow range of $b_3 \in (0.74, 0.78)$ depicting the existence of intermittency transition. This narrow range transition is also confirmed from the transition of successive largest *sub*Lyapunov exponents. The spectrum of the first nine largest Lyapunov exponents λ_{max} of the coupled drive $x_1(t)$ and response $x_2(t)$ variables (Eqs.
(9.15a) and (9.15b)) is shown in Fig. 9.19b. It is also evident from the spectrum
that the *sub*Lyapunov exponents suddenly become negative in the narrow range of
$b_3 \in (0.74, 0.78)$, and then reach saturation values for $b_3 > 0.78$. This confirms the
narrow range intermittency (IGS) transition to GS. This is also in accordance with

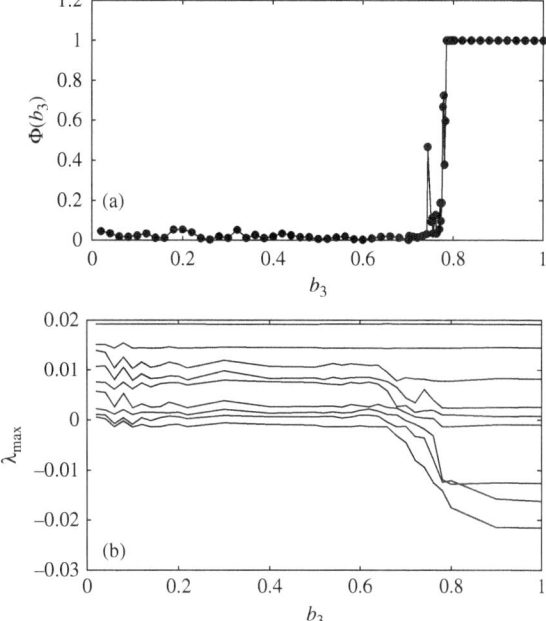

Fig. 9.19 (**a**) The probability of synchronization $\Phi(b_3)$ between the response $x_2(t)$ (9.15b) and the auxiliary $x_3(t)$ (9.15c) systems and (**b**) Largest Lyapunov exponents of the coupled drive $x_1(t)$ and response $x_2(t)$ systems (9.15a) and (9.15b)

the immediate stabilization of all the UPO's of the response system as discussed in Sect. 9.6.

The robustness of the transition with the system parameter b_2 and the delay time τ_2 as a function of coupling strength b_3 is shown as 3-dimensional plots in Fig. 9.20.

9.9 Intermittency Transition to Generalized Synchronization: Mackey-Glass & Ikeda Systems

In this section, we will show that the intermittency transition to GS exhibits characteristically distinct behaviors (as before) in both the coupled Mackey-Glass and the coupled Ikeda time-delay systems for different kinds of couplings. In particular, intermittency transition occurs in a broad range of coupling strength for error feedback coupling configurations and in a narrow range of coupling strength for direct feedback coupling configurations, beyond a certain threshold value of the coupling strength. In addition, the intermittent dynamics is characterized by periodic bursts away from the temporal synchronized state with period equal to the delay time of the response system in the case of broad range intermittency transition whereas it is characterized by random time intervals in the case of narrow range intermittency transition. These transitions have also been investigated analytically using Krasvoskii-Lyapunov functional approach and numerically by the probability of synchronization. The robustness of these transitions with the system parameters in both the error feedback and direct feedback coupling configurations are also shown.

9.9.1 Broad Range Intermittency Transition to GS

To be specific, we first consider the unidirectional, linearly coupled systems of the form Eq. (9.1) with the drive $x_1(t)$, response $x_2(t)$, and auxiliary $x_3(t)$, variables where the nonlinear function $f(x)$ is now of the form (8.15) for the Mackey-Glass system and (8.16) for the Ikeda system.

In Sect. 9.2, we have chosen the parameter mismatch in the time-delays τ_1 and τ_2 alone for the sake of simplicity. Now we have chosen the parameter mismatch in both the parameters b_1 and b_2, and τ_1 and τ_2, that is $b_1 \neq b_2$ and $\tau_1 \neq \tau_2$ to show the generality of the obtained results.

The stability condition (9.5) obtained in Sect. 9.3 for the existence of GS holds good in the above cases of coupled Mackey-Glass and Ikeda systems as well, except that the derivatives of the function $f(x)$ have now different forms. From the stability condition (9.5), it is not possible to identify analytically the regions of approximate and exact GS unlike the case of piecewise linear time-delay system. However, it is also possible find the value of $f'(x_{max})$ by identifying the maximal value of $x(t)$ numerically from the corresponding attractors as discussed in Sect. 8.4 of Chap. 8. From the value of $f'(x_{max})$, the value of parameters b_2 and b_3 satisfying the stability condition $a + b_3 > |b_2 f'(x_2(t - \tau_2))|$ for fixed value of a can be determined. Nevertheless, one can numerically vary the coupling strength b_3 for fixed values of

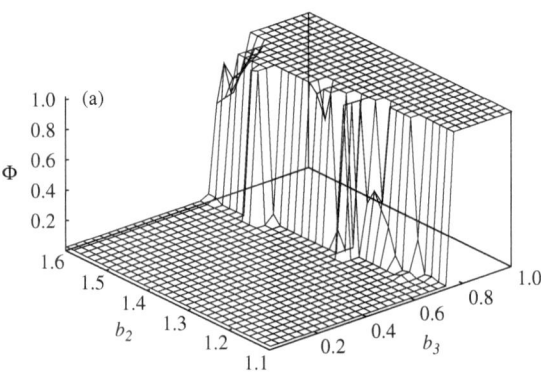

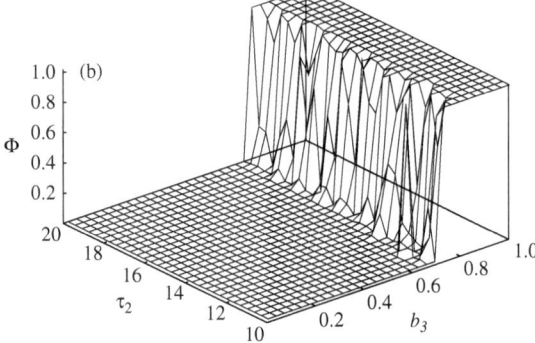

Fig. 9.20 The probability of synchronization $\Phi(b_3)$ in 3-dimensional plots (**a**) as a function of the system parameter b_2 and the coupling strength b_3 and (**b**) as a function of the coupling delay τ_2 and the coupling strength b_3 for the case of nonlinear direct feedback coupling configuration given by Eq. (9.15) showing narrow range intermittency transition to GS

all other parameters and identify the regions of intermittency transition to GS. To start with, we will discuss the results for coupled Mackey-Glass systems which is then followed by the results of coupled Ikeda systems.

9.9.1.1 Coupled Mackey-Glass System

In our analysis, we have fixed the values of the parameters in Eq. (9.1) as $a = 0.1$, $b_1 = 0.2, b_2 = 0.198, \tau_1 = 25, \tau_2 = 30$ and varied the coupling strength b_3 which is now treated as the control parameter. As the value of b_3 is increased from zero numerically, intermittency transition (approximate GS) appears for $b_3 > 0.06$ and complete GS appears for $b_3 > 0.094$. Approximate GS between the drive $x_1(t)$ and the response $x_2(t)$ is shown in Fig. 9.21a for the value of the coupling strength $b_3 = 0.084$ whereas approximate CS between the response $x_2(t)$ and the auxiliary $x_3(t)$ systems is shown in Fig. 9.21b. Perfect GS and perfect CS are shown in Fig. 9.21c, d respectively for $b_3 = 0.1$. The time trace of the difference variable $\Delta = x_2(t) - x_3(t)$ is plotted in Fig. 9.22 corresponding to Fig. 9.21b. It is seen from Fig. 9.22 that the intermittent dynamics displays bursts at periodic intervals of time equal (approximately) to the delay time of the response system when "on" states of amplitude greater than certain threshold value for Δ is considered. It is to be noted that clear distinction between the on and off states contributing to periodic bursts cannot be made from the Fig. 9.22 for the chosen value of the time-delay. However, similar intermittent dynamics with bursts at periodic intervals is also observed in the coupled Mackey-Glass system with the same coupling configurations as in Eq. (9.1) for large delays [16].

The statistical features associated with the intermittent dynamics is again analyzed by calculating the distribution of laminar phases $\Lambda(t)$ with amplitude less than a threshold value of Δ. A universal asymptotic power law distribution $\Lambda(t) \propto t^{-\alpha}$ is observed for the threshold value $|\Delta| = 0.0001$ with the value of the exponent $\alpha = -1.5$ as shown in Fig. 9.23, which is quite typical for on-off intermittency. Figure 9.24 shows the probability of synchronization $\Phi(b_3)$ for the above parameter values of the coupled Mackey-Glass system. For the range of $b_3 \in (0, 0.06)$, the value of $\Phi(b_3)$ is nearly zero due to the absence of any correlation between the response and the auxiliary systems. However, for $b_3 > 0.06$ there appears oscillations in the value of the probability of synchronization $\Phi(b_3)$ between zero and some finite values less than unity, exhibiting intermittency transition to GS in the range of the coupling strength $b_3 \in (0.06, 0.094)$. Above $b_3 > 0.094$, the value of probability of synchronization attains unit value indicating perfect GS. It is to be noted that the intermittency transition occurs here also in a rather broad range of the coupling strength.

The robustness of the intermittency transition in a broad range of coupling strength with the system parameter $b_2 \in (0.196, 0.204)$ and with the coupling delay $\tau_2 \in (25, 30)$ has also been confirmed. Figure 9.25a shows the 3-dimensional plot of the probability of synchronization as a function of the system parameter b_2 and the coupling strength b_3, while Fig. 9.25b shows the 3-dimensional plot of $\Phi(b_3)$ as a function of the coupling delay τ_2 and the coupling strength b_3. The above figures

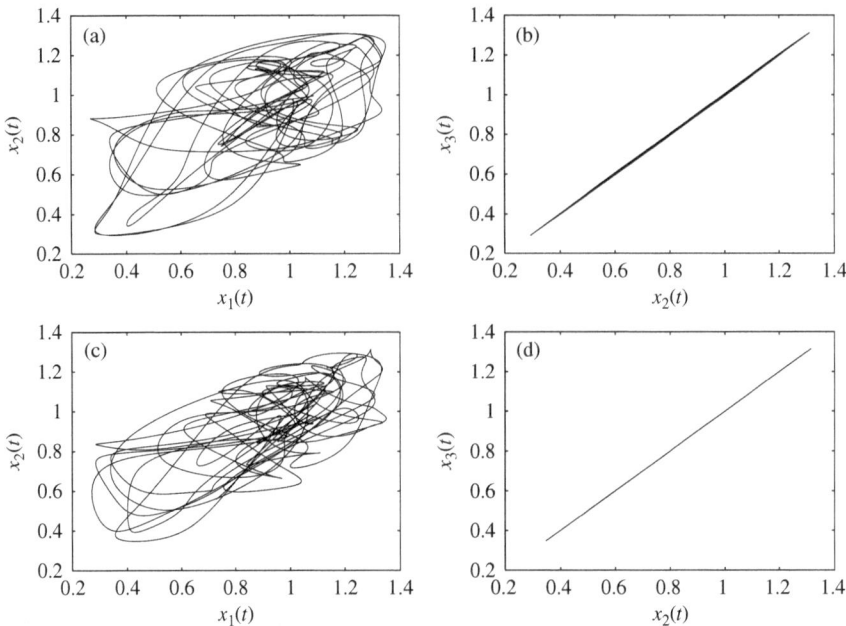

Fig. 9.21 Dynamics in the phase space of the systems (9.1) for the functional form of the Mackey-Glass system (8.15). (**a**) and (**b**) Approximate GS and CS, respectively, for the value of the coupling strength $b_3 = 0.084$. (**c**) and (**d**) Perfect GS and CS, respectively, for the value of the coupling strength $b_3 = 0.1$

(Fig. 9.25) clearly reveal the broad range intermittency transition to GS in the case of linear error feedback coupling scheme.

9.9.1.2 Coupled Ikeda Systems

Next we will demonstrate the existence of broad range intermittency transition to GS in the coupled Ikeda system (see Eq.(7.19)) for the linear error feedback coupling configuration of the form given in Eq. (9.1b). We have fixed the values of the parameters at $a = 1.0, b_1 = 20, b_2 = 22, \tau_1 = 10, \tau_2 = 5$ and varied the coupling strength b_3 as the control parameter. As the value of b_3 is increased from zero, again intermittency transition (approximate GS) is observed in a broad range of $b_3 \in (8, 9.4)$, and for $b_3 > 9.4$ perfect GS is observed. Approximate GS and approximate CS are shown in Fig. 9.26a, b, respectively, for the value of the coupling strength $b_3 = 9$ whereas perfect GS and perfect CS are plotted in Fig. 9.26c, d, respectively, for $b_3 = 10$. The time trace of the difference variable $\Delta = x_2(t) - x_3(t)$ corresponding to Fig. 9.26b is shown in Fig. 9.27. It is evident from Fig. 9.27b, which is an enlarged part of Fig. 9.27a, that the intermittent dynamics displays bursts at periodic intervals with period approximately equal to the delay time of the response system $t \approx \tau_2 = 5$ when bursts of larger amplitudes are considered. The statistical features associated with the intermittent dynamics

Fig. 9.22 The intermittent dynamics of the response $x_2(t)$ and auxiliary $x_3(t)$ systems for the value of the coupling strength $b_3 = 0.084$. (**a**) Time traces of the difference $x_2(t) - x_3(t)$ corresponding to Fig. 9.21b and (**b**) Enlarged in x scale to show bursts at periodic intervals when bursts of larger amplitudes are considered

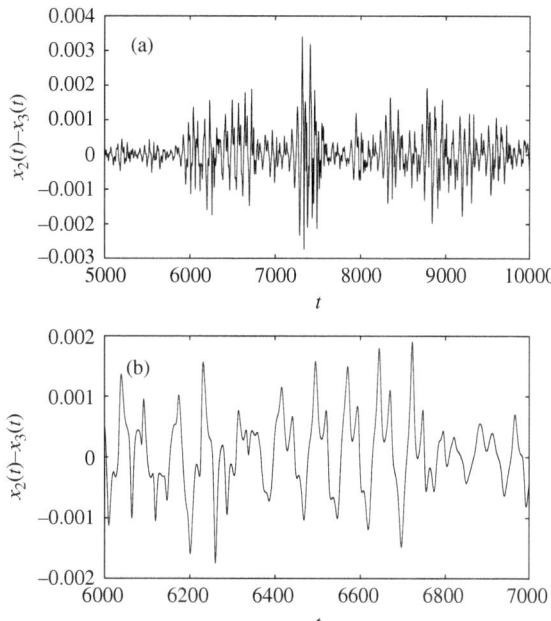

in Fig. 9.27 shows a universal asymptotic -1.5 power law behavior for bursts of amplitude $|\Delta| > 0.0001$ as shown in Fig. 9.28. The intermittency transition is also characterized by the probability of synchronization $\Phi(b_3)$ as shown in Fig. 9.29. Upto the threshold value $b_3 = 8$, the value of $\Phi(b_3)$ is zero corresponding to the asynchronous state of the response and auxiliary systems. There exist oscillations in the value of $\Phi(b_3)$ in the broad range of $b_3 \in (8, 9.4)$, confirming the intermittency transition to GS. Beyond $b_3 > 9.4$, the probability of synchronization $\Phi(b_3)$ reaches unit value showing the existence of perfect GS between the drive and the response systems.

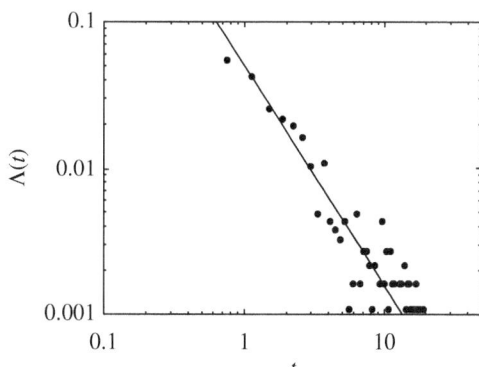

Fig. 9.23 The statistical distribution of laminar phases for the Fig. 9.22

Fig. 9.24 The probability of synchronization $\Phi(b_3)$ between the response $x_2(t)$ and the auxiliary $x_3(t)$ systems for the coupled Mackey-Glass systems

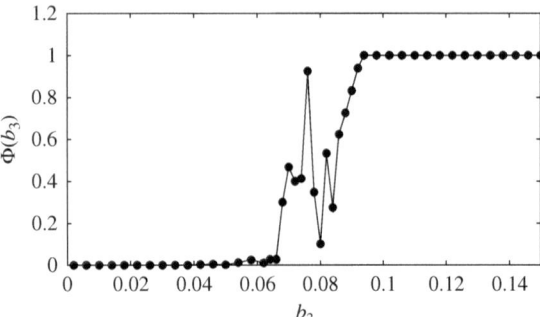

The robustness of the intermittency transition in a broad range of coupling strength with the system parameter $b_2 \in (20, 25)$ and with the coupling delay $\tau_2 \in (5, 10)$ has also been confirmed from Fig. 9.30. The 3-dimensional plot of the probability of synchronization as a function of the system parameter b_2 and the coupling strength b_3 is shown in Fig. 9.30a, while Fig. 9.30b shows the 3-dimensional plot of $\Phi(b_3)$ as a function of the coupling delay τ_2 and the coupling strength b_3. The above figures (Fig. 9.30) clearly reveal the broad range intermittency transition to GS in the case of linear error feedback coupling scheme.

9.9.2 Narrow Range Intermittency Transition to GS

To illustrate the narrow range intermittency transition to GS in coupled Mackey-Glass and Ikeda time-delay systems, we consider the unidirectional linear direct feedback coupling of the form Eq. (9.9) given in Sect. 9.6 with the functional form (8.15) for the Mackey-Glass system and (8.16) for the Ikeda system. One can obtain the same stability condition (9.10) as in the case of piecewise linear time-delay system discussed in Sect. 9.6. For the reasons stated in Sect. 9.6, we proceed numerically by varying the coupling strength b_3 in order to identify the mechanism of synchronization transition and to demarcate the regions of approximate and perfect synchronization regions in the parameter space.

We have fixed the same values for the parameters of the coupled Mackey-Glass system as in the previous section and increased the value of the coupling strength b_3 from zero. Intermittency transition (approximate GS) is found only in the narrow range of the coupling strength $b_3 \in (0.06, 0.064)$, and for $b_3 > 0.064$ perfect GS is observed. Approximate GS is shown in Fig. 9.31a and approximate CS is shown in Fig. 9.31b for the value $b_3 = 0.062$, whereas perfect GS and perfect CS are shown in Fig. 9.31c, d, respectively, for the value of the coupling strength $b_3 = 0.07$. The time trajectory of the difference variable $\Delta = x_2(t) - x_3(t)$ corresponding to Fig. 9.31b is shown in Fig. 9.32. The intermittent bursts appear at random time intervals as seen from Fig. 9.32b. The statistical distribution of the laminar phases shows a universal asymptotic -1.5 power law behavior as seen in Fig. 9.33 when bursts of amplitude less than a certain threshold value $|\Delta| = 0.0001$ are considered. The

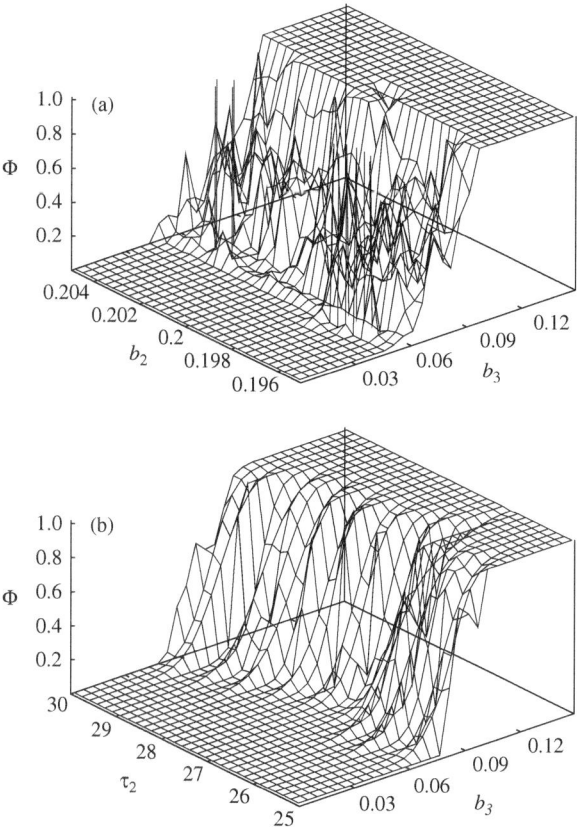

Fig. 9.25 The probability of synchronization $\Phi(b_3)$ in 3-dimensional plots for the Mackey-Glass system (**a**) as a function of the system parameter b_2 and the coupling strength b_3 and (**b**) as a function of the coupling delay τ_2 and the coupling strength b_3, exhibiting broad range intermittency transition to GS for linear error feedback coupling

narrow range intermittency transition for the linear direct feedback coupling configuration is also characterized by the probability of synchronization $\Phi(b_3)$ as shown in Fig. 9.34. It is evident from Fig. 9.34 that the value of $\Phi(b_3)$ fluctuates between zero and some finite value less than unity in the narrow range of the coupling strength $b_3 \in (0.06, 0.064)$ confirming the existence of narrow range intermittency transition to GS. For $b_3 > 0.064$, the value of $\Phi(b_3)$ attains unity indicating the existence of perfect GS. The robustness of the intermittency transition in a narrow range of the coupling strength b_3 for a range of values of the parameter $b_2 \in (0.196, 0.204)$ and the delay $\tau_1 = \tau_2 \in (25, 30)$ is shown in Fig. 9.35. The 3-dimensional plot of the probability of synchronization as a function of the system parameter b_2 and the coupling strength b_3 is shown in Fig. 9.35a, while Fig. 9.35b shows the 3-dimensional plot of $\Phi(b_3)$ as a function of the coupling delay τ_2 and the coupling strength b_3.

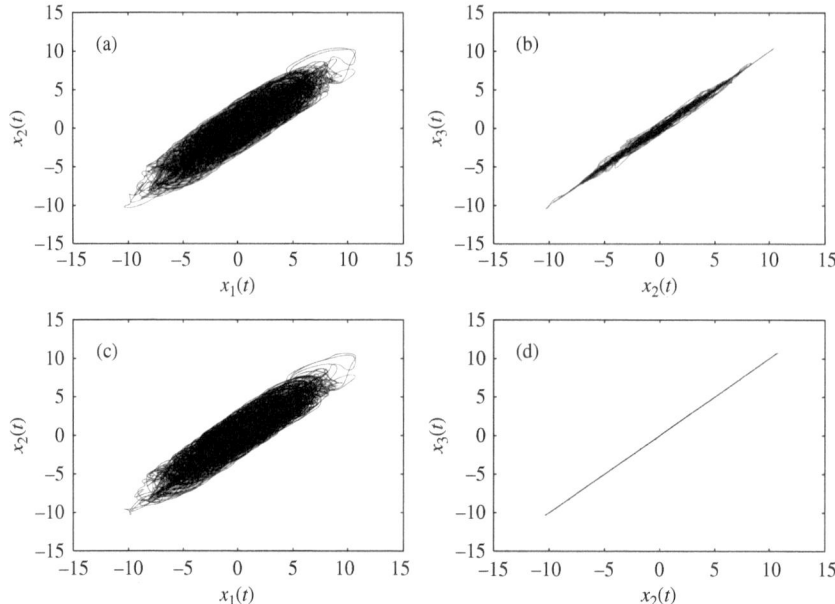

Fig. 9.26 Dynamics in the phase space of the systems (9.1) for the functional form of the Ikeda system (8.16). (**a**) and (**b**) Approximate GS and CS, respectively, for the value of the coupling strength $b_3 = 9$. (**c**) and (**d**) Perfect GS and CS, respectively, for the value of the coupling strength $b_3 = 10$

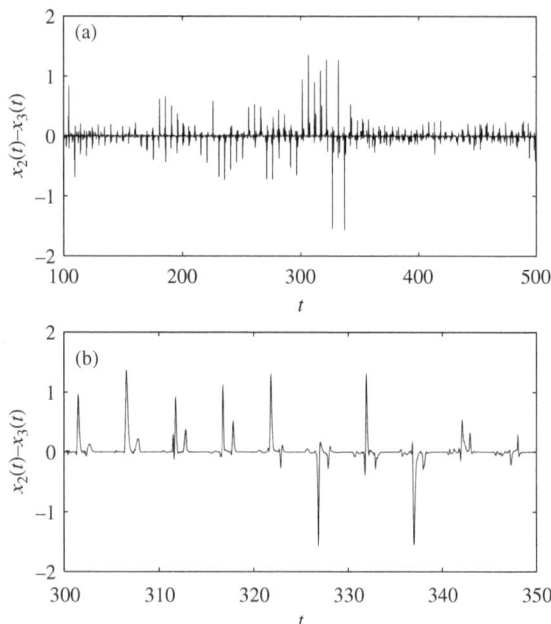

Fig. 9.27 The intermittent dynamics of the response $x_2(t)$ and auxiliary $x_3(t)$ systems for the value of the coupling strength $b_3 = 9$. (**a**) Time traces of the difference $x_2(t) - x_3(t)$ corresponding to Fig. 9.26b and (**b**) Enlarged in x scale to show bursts at periodic intervals when bursts of larger amplitudes are considered

Fig. 9.28 The statistical distribution of laminar phases for the Fig. 9.27

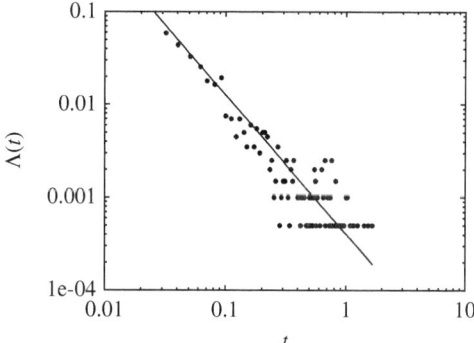

Fig. 9.29 The probability of synchronization $\Phi(b_3)$ between the response $x_2(t)$ and the auxiliary $x_3(t)$ systems for the functional form of Ikeda system (8.16), that is $f(x) = \sin(x)$

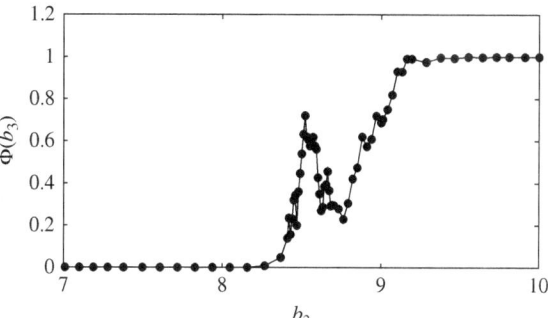

Next, we will show the existence of narrow range intermittency transition in linear direct feedback coupling configuration for the Ikeda system. The values of the parameters are now fixed as $a = 1.0, b_2 = 20, b_2 = 5, \tau_1 = 3, \tau_2 = 2$ and the coupling strength b_3 is varied to study the nature of transition in the coupled Ikeda systems (9.9) and (8.16). As the coupling strength is increased from zero, intermittency transition is observed in the range of $b_3 \in (6, 6.2)$ and for $b_3 > 6.2$ perfect GS is observed. Approximate GS and approximate CS are shown in Fig. 9.36a, b, respectively, for the value of the coupling strength $b_3 = 6.1$, while perfect GS and perfect CS are seen in Fig. 9.36c, d respectively for $b_3 = 7$. The time traces of the difference variable Δ is shown in Fig. 9.37 corresponding to Fig. 9.36b. The enlarged part of Fig. 9.37a is shown in Fig. 9.37b to show that the intermittent burst appears at random time intervals in this case of narrow range intermittency transition to GS. The corresponding intermittent dynamics is also characterized by a universal asymptotic -1.5 power law behavior as shown in Fig. 9.38 when bursts of amplitude less than a certain threshold value $\Delta = |0.0001|$ are considered. The narrow range intermittency transition is also characterized by the probability of synchronization $\Phi(b_3)$ as a function of the coupling strength as shown in Fig. 9.39. The value of $\Phi(b_3)$ is practically zero in the range $b_3 \in (0, 6)$ attributing to the desynchronized

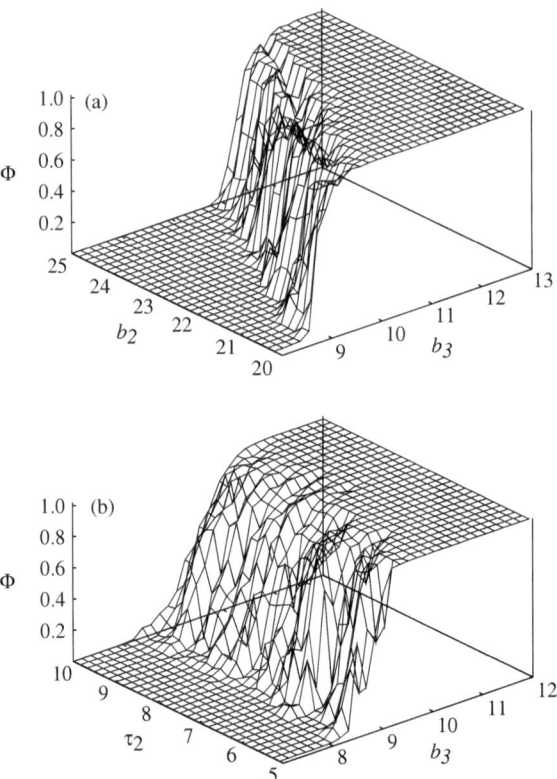

Fig. 9.30 The probability of synchronization $\Phi(b_3)$ in 3-dimensional plots for Ikeda system (**a**) as a function of the system parameter b_2 and the coupling strength b_3 and (**b**) as a function of the coupling delay τ_2 and the coupling strength b_3, exhibiting broad range intermittency transition to GS for linear error feedback coupling

state and it also oscillates between zero and finite value less than unity indicating the intermittency transition in the narrow range $b_3 \in (6, 6.2)$. For $b_3 > 6.2$, the value of probability of synchronization reaches unit value indicating the existence of perfect GS. The robustness of the intermittency transition in a narrow range of the coupling strength b_3 for a range of values of the parameter $b_2 \in (5, 10)$ and the delay $\tau_1 = \tau_2 \in (5, 10)$ is shown in Fig. 9.40. The 3-dimensional plot of the probability of synchronization as a function of the system parameter b_2 and the coupling strength b_3 is shown in Fig. 9.40a, while Fig. 9.40b shows the 3-dimensional plot of $\Phi(b_3)$ as a function of the coupling delay τ_2 and the coupling strength b_3.

  To summarize, broad range intermittency transition to GS for the case of nonlinear error feedback coupling configuration and narrow range intermittency transition to GS for the case of nonlinear direct feedback coupling configuration have also been observed for both the Mackey-Glass and Ikeda systems as in the case of piecewise linear time-delay systems. The intermittent dynamics is also characterized by bursts at periodic intervals approximately equal to the delay time of the response

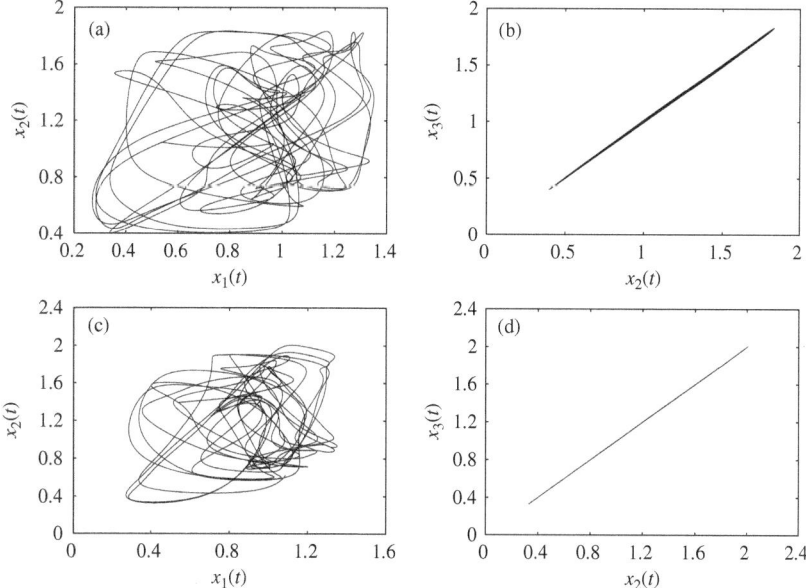

Fig. 9.31 Dynamics in the phase space of the systems (9.9) for the functional form of the Mackey-Glass system (8.15). (**a**) and (**b**) Approximate GS and CS, respectively, for the value of the coupling strength $b_3 = 0.062$. (**c**) and (**d**) Perfect GS and CS, respectively, for the value of the coupling strength $b_3 = 0.07$

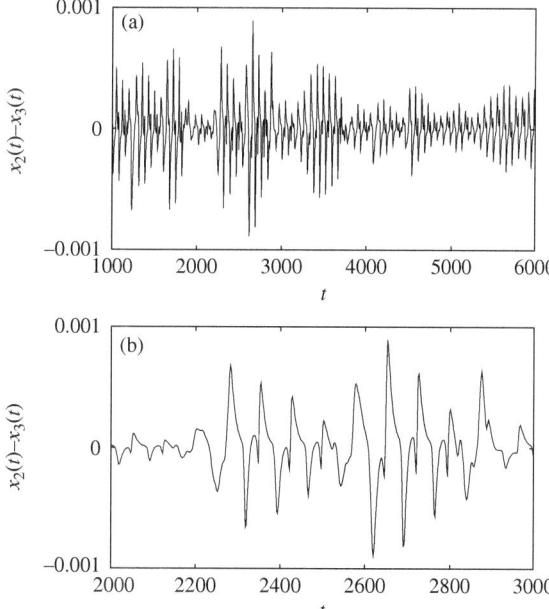

Fig. 9.32 The intermittent dynamics of the response $x_2(t)$ and auxiliary $x_3(t)$ systems for the value of the coupling strength $b_3 = 0.062$. (**a**) Time traces of the difference $x_2(t) - x_3(t)$ corresponding to Fig. 9.31b and (**b**) Enlarged in x scale to show bursts at random intervals

Fig. 9.33 The statistical
distribution of laminar phases
for the Fig. 9.32

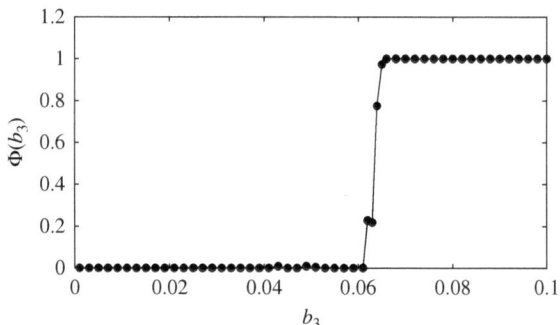

Fig. 9.34 The probability of
synchronization $\Phi(b_3)$
between the response $x_2(t)$
and the auxiliary $x_3(t)$
systems for the functional
form of Mackey-Glass
system (8.15) with the linear
direct feedback coupling as in
Eq. (9.9)

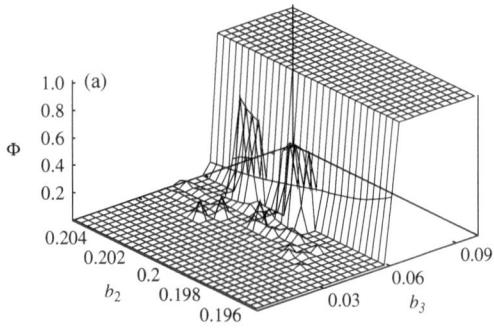

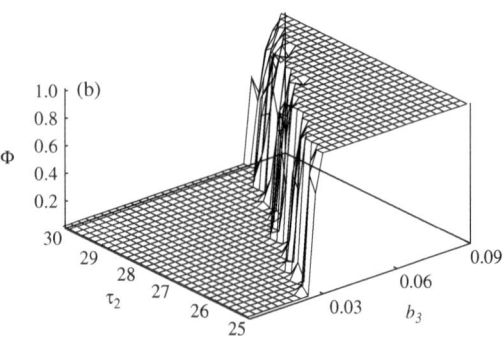

Fig. 9.35 The probability of
synchronization $\Phi(b_3)$ in
3-dimensional plots for the
coupled Mackey-Glass
system (**a**) as a function of the
system parameter b_2 and the
coupling strength b_3 and (**b**)
as a function of the coupling
delay τ_2 and the coupling
strength b_3, exhibiting narrow
range intermittency transition
to GS for linear error
feedback coupling

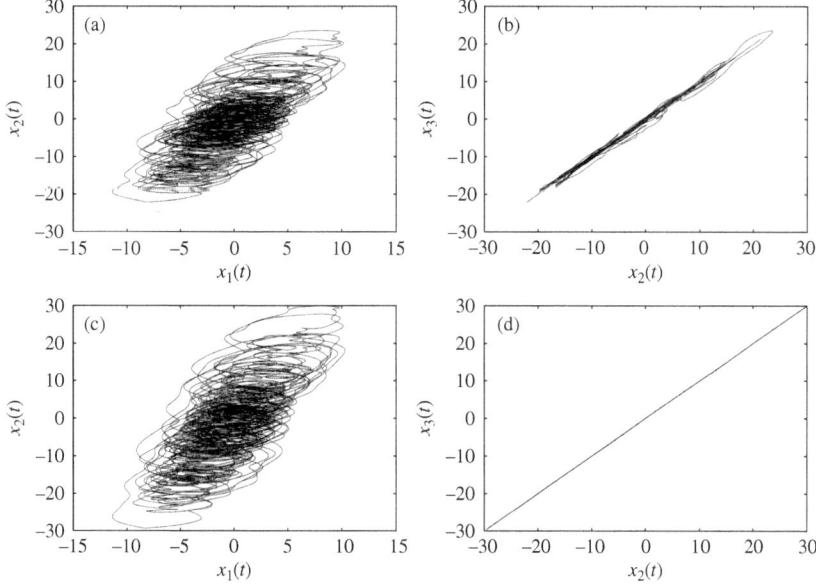

Fig. 9.36 Dynamics in the phase space of the systems (9.9) for the functional form of the Ikeda system (8.16). (**a**) and (**b**) Approximate GS and CS, respectively, for the value of the coupling strength $b_3 = 6.1$. (**c**) and (**d**) Perfect GS and CS, respectively, for the value of the coupling strength $b_3 = 7$

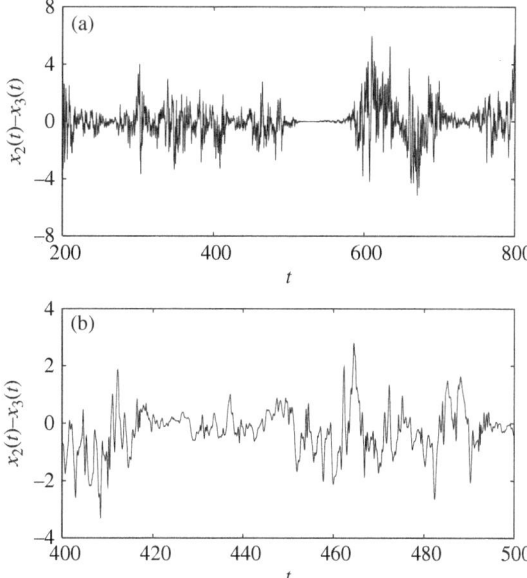

Fig. 9.37 The intermittent dynamics of the response $x_2(t)$ and auxiliary $x_3(t)$ systems for the value of the coupling strength $b_3 = 6.1$. (**a**) Time traces of the difference $x_2(t) - x_3(t)$ corresponding to Fig. 9.36b and (**b**) Enlarged in x scale to show bursts at random intervals

Fig. 9.38 The statistical distribution of laminar phases for the Fig. 9.37

Fig. 9.39 The probability of synchronization $\Phi(b_3)$ between the response $x_2(t)$ and the auxiliary $x_3(t)$ systems for the functional form of Ikeda system (8.16) with the linear direct feedback coupling as in Eq. (9.9)

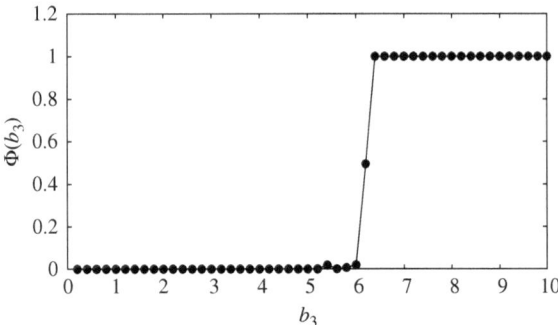

Fig. 9.40 The probability of synchronization $\Phi(b_3)$ in 3-dimensional plots for Ikeda system (**a**) as a function of the system parameter b_2 and the coupling strength b_3 and (**b**) as a function of the coupling delay τ_2 and the coupling strength b_3, exhibiting narrow range intermittency transition to GS for linear error feedback coupling

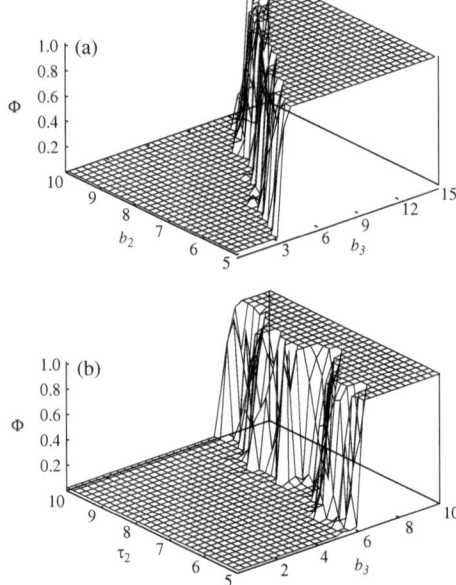

system in the case of broad range intermittency transition and by random time intervals in the case of narrow range intermittency transition for nonlinear coupling configuration.

References

1. N.F. Rulkov, M.M. Sushchik, L.S. Tsimring, H.D.I. Abarbanel, Phys. Rev. E **51**, 980 (1995)
2. H.D.I. Abarbanel, N.F. Rulkov, M.M. Sushchik, Phys. Rev. E **53**, 4528 (1996)
3. R. Brown, Phys. Rev. Lett. **81**, 4835 (1998)
4. L. Kocarev, U. Parlitz, Phys. Rev. Lett. **76**, 1816 (1996)
5. K. Pyragas, Phys. Rev. E **54**, R4508 (1996)
6. B.R. Hunt, E. Ott, J.A. Yorke, Phys. Rev. E **55**, 4029 (1997)
7. N.F. Rulkov, C.T. Lewis, Phys. Rev. E **63**, 065204(R) (2001)
8. A.E. Hramov, A.A. Koronovskii, P.V. Popov, Phys. Rev. E **72**, 037201 (2005)
9. A.E. Hramov, A.A. Koronovskii, Europhys. Lett. **79**, 169 (2005)
10. S. Boccaletti, D.L. Valladares, Phys. Rev. E **62**, 7497 (2000)
11. D.L. Valladares, S. Boccaletti, Int. J. Bifurcat. Chaos **11**, 2699 (2001)
12. A. Pikovsky, G. Osipov, M. Rosenblum, M. Zaks, J. Kurths, Phys. Rev. Lett. **79**, 47 (1997)
13. A. Pikovsky, M. Zaks, M. Rosenblum, G. Osipov, J. Kurths, Chaos **7**, 680 (1997)
14. K.J. Lee, Y. Kwak, T.K. Lim, Phys. Rev. Lett. **81**, 321 (1998)
15. L. Zhao, Y.C. Lai, C.W. Shih, Phys. Rev. E **72**, 036212 (2005)
16. M. Zhan, X. Wang, X. Gong, G.W. Wei, C.H. Lai, Phys. Rev. E **68**, 036208 (2003)
17. E.M. Shahverdiev, K.A. Shore, Phys. Rev. E **71**, 016201 (2005)
18. D.V. Senthilkumar, M. Lakshmanan, J. Kurths, Phys. Rev. E **74**, 035205(R) (2006)
19. D.V. Senthilkumar, M. Lakshmanan, Phys. Rev. E **71**, 016211 (2005)
20. N.N. Krasovskii, *Stability of Motion* (Stanford University Press, Stanford, CA, 1963)
21. K. Pyragas, Phys. Rev. E **58**, 3067 (1998)
22. N. Platt, S.M. Hammel, J. F. Heagy, Phys. Rev. Lett. **72**, 3498 (1994)
23. J.F. Heagy, N. Platt, S.M. Hammel, Phys. Rev. E **49**, 1140 (1994)

Chapter 10
Transition from Phase to Generalized Synchronization

10.1 Introduction

Chaotic phase synchronization (CPS) has become the focus of recent research as it plays a crucial role in understanding the behavior of a large class of weakly inter-acting dynamical systems in diverse natural systems including circadian rhythm, cardio-respiratory systems, neural oscillators, population dynamics, etc [1–3]. The definition of CPS is a direct extension of the classical definition of synchronization of periodic oscillations and can be referred to as entrainment between the phases of interacting chaotic systems, while the amplitudes remain chaotic and, in general, non-correlated [4] (see also Appendix B).

The notion of CPS has been investigated mainly in oscillators driven by external periodic force [5, 6], chaotic oscillators with different natural frequencies and/or with parameter mismatches [7–10], arrays of coupled chaotic oscillators [4, 11] and also in essentially different chaotic systems [12, 13]. In addition CPS has also been demonstrated experimentally in various systems, such as electrical cir-cuits [12, 14–16], lasers [17, 18], fluids [19], biological systems [20, 21], climatol-ogy [22], etc. On the other hand CPS in nonlinear time-delay systems, have been identified only very recently. A main problem here is to define even the notion of phase in time-delay systems due to the intrinsic multiple characteristic time scales in these systems. Studying CPS in such chaotic time-delay systems is of considerable importance in many fields, as in understanding the behavior of nerve cells (neuro-science), in physiological studies, in ecology, in lasers, etc [1–3, 23–27], where memory effects play a prominent role.

In this chapter we will present the progress on the identification and exis-tence of CPS in coupled piecewise-linear time-delay systems with parameter mis-matches [28]. We will show the entrainment of phases of the coupled systems from asynchronous state and its subsequent transition to generalized synchroniza-tion (GS) as a function of the coupling strength. Phase of the piecewise linear time-delay system is calculated using the Poincaré method after a newly introduced transformation of the corresponding attractors, which transforms the original non-phase-coherent attractors of both the systems into smeared limit cycle like attractors. Further, the existence of CPS and GS is characterized by recently proposed methods based on recurrence quantification analysis and in terms of Lyapunov exponents of

M. Lakshmanan, D.V. Senthilkumar, *Dynamics of Nonlinear Time-Delay Systems*, Springer Series in Synergetics, DOI 10.1007/978-3-642-14938-2_10,
© Springer-Verlag Berlin Heidelberg 2010

the coupled time-delay systems. Then, CPS and GS are demonstrated in the case of coupled Mackey-Glass and Ikeda systems.

10.2 Phase-Coherent and Non-phase-coherent Attractors

In this section, we will first make clear the distinction between the terminologies phase-coherent and non-phase-coherent chaotic attractors, which occur repeatedly in this chapter.

The distinction between *phase-coherent* and *non-phase-coherent* chaotic attractors depends on the topology of the corresponding chaotic attractor. If the projected attractor onto the phase space resembles that of a smeared limit cycle, where the phase point always rotates around a fixed origin with monotonically increasing phase, then the corresponding attractor is referred to as a phase-coherent attractor in the literature [1, 2, 4, 8, 29, 30]. Consequently, the phase of such phase-coherent chaotic attractors can be introduced straightforwardly as discussed below in the next section. On the other hand, the structure of a non-phase-coherent chaotic attractor does not look like a smeared limit cycle with a fixed center around which the phase point rotates. Most importantly such a non-phase-coherent chaotic attractor is not characterized by a monotonically increasing phase. Hence the phase of such a non-phase-coherent attractor cannot be defined straightforwardly as in the case of a phase-coherent attractor. So specialized techniques/tools have to identified to introduce phase in non-phase-coherent attractors.

As an example, let us consider the widely studied Rössler system in the context of phase synchronization with standard parameters [4, 8]. The underlying coupled dynamical equations are the following,

$$\dot{x}_{1,2} = -\omega_{1,2}y_{1,2} - z_{1,2} + C(x_{2,1} - x_{1,2}), \qquad (10.1a)$$

$$\dot{y}_{1,2} = \omega_{1,2}x_{1,2} + ay_{1,2}, \qquad (10.1b)$$

$$\dot{z}_{1,2} = 0.2 + z_{1,2}(x_{1,2} - 10). \qquad (10.1c)$$

The topology of the attractor is determined by the parameter a. For $a = 0.15$, a phase-coherent attractor (Fig. 10.1a) is observed with rather simple topological properties [31, 32] and hence the phase can be calculated straightforwardly. The topology of the Rössler attractor changes dramatically if the parameter a exceeds the value 0.21 and the phase in this case is not well defined. Funnel (non-phase-coherent) attractor for the value $a = 0.25$ is shown in Fig. 10.1b. There are large and small loops (see Fig. 10.1b) on the (x, y) plane and it is not obvious which phase gain should be attributed to these loops and hence phase cannot be calculated simply as in the case of the phase-coherent chaotic attractor (Fig. 10.1a). One way of calculating the phase for the funnel attractor (Fig. 10.1b) is to use Eq. (10.7) discussed in the next section. As a second example, one may refer to the butterfly attractor of the Lorenz system discussed in the next section and the transformation (10.6)

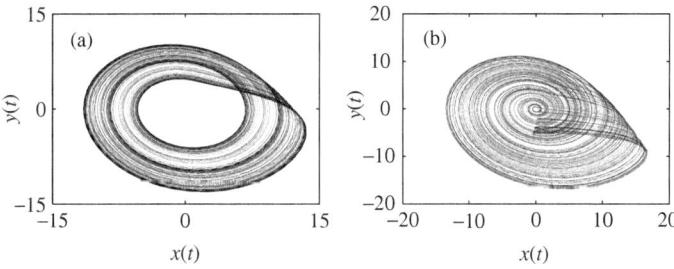

Fig. 10.1 Phase-coherent and funnel (non-phase-coherent) Rössler attractors with parameters (**a**) $a = 0.15$ and (**b**) $a = 0.25$

given below. When higher dimensional systems are considered, they will very often exhibit more complicated attractors with more than one positive Lyapunov exponent and they often fall in the class of non-phase-coherent attractors.

10.3 CPS in Chaotic Systems

The definition of CPS (chaotic phase synchronization) in coupled chaotic systems is derived from the classical definition of phase synchronization in periodic oscillators. Interacting chaotic systems are said to be in phase synchronized state when there exists entrainment between phases of the systems, while their amplitudes may remain chaotic and uncorrelated. In other words, CPS exists when their respective frequencies and phases are locked [1, 2, 4]. To study CPS, one has to identify a well defined phase variable in both the coupled systems. If the flow of the chaotic oscillators has a proper rotation around a certain reference point, the phase can be defined in a straightforward way. For example, for the Rössler system (10.1) with standard parameters the projection of the chaotic attractor on the (x, y) plane looks like a smeared limit cycle (Fig. 10.1a). In this and similar cases one can define the phase [1, 2] as

$$\phi(t) = \arctan(y(t)/x(t)). \tag{10.2}$$

A more general approach to define the phase in chaotic oscillators is the analytic signal approach [1, 2] introduced in [33]. The analytic signal $\chi(t)$ is given by

$$\chi(t) = s(t) + i\tilde{s}(t) = A(t)\exp^{i\Phi(t)}, \tag{10.3}$$

where $\tilde{s}(t)$ denotes the Hilbert transform of the observed scalar time series $s(t)$,

$$\tilde{s}(t) = \frac{1}{\pi}P.V.\int_{-\infty}^{\infty}\frac{s(t')}{t - t'}dt', \tag{10.4}$$

where P.V. stands for the Cauchy principle value of the integral and this method is especially useful for experimental applications [1, 2].

The phase of a chaotic attractor can also be defined based on an appropriate Poincaré section which the chaotic trajectory crosses once for each rotation. Each crossing of the orbit with the Poincaré section corresponds to an increment of 2π of the phase, and the phase in between two crossings is linearly interpolated [1, 2],

$$\Phi(t) = 2\pi k + 2\pi \frac{t - t_k}{t_{k+1} - t_k}, \qquad (t_k < t < t_{k+1}) \tag{10.5}$$

where t_k is the time of kth crossing of the flow with the Poincaré section. For the phase coherent chaotic oscillators, that is, for flows which have a proper rotation around a certain reference point, the phases calculated by the above three different ways are in good agreement [1, 2]. However, we often come across non-phase-coherent attractors where one cannot encounter flows with a proper rotation around a fixed reference point (with the origin coinciding with the center of rotation), in which case a single characteristic time scale does not exist in general. In such circumstances it is difficult or impossible to find a proper center of rotation and it is also intricate to find a Poincaré section that is crossed transversally by all the trajectories of the chaotic attractor. Therefore the above definitions of phase are no longer applicable for non-phase-coherent chaotic attractors.

However, it has also been demonstrated that certain non-phase-coherent chaotic attractors can be transformed into smeared limit-cycle like attractors by introducing a suitable transformation of the original variables. For example, in the case of the Lorenz system, a transformation of the form

$$u(t) = \sqrt{x^2 + y^2} \tag{10.6}$$

has been introduced in [2] to transform the butterfly attractor into a smeared limit cycle like attractor and the projected trajectory in the plane (u, z) resembles that of the Rössler attractor. The butterfly attractor of the Lorenz system for the standard parameter values is shown in Fig. 10.2a. It is difficult to use the standard approach to determine the phase of the butterfly attractor since it does not have a single fixed center of rotation. However, the transformed attractor (Fig. 10.2b) projected in the plane (u, z) using the above transformation (Eq. (10.6)) resembles that of a smeared limit cycle like attractor and now the phase of the attractor can be introduced using the above approaches.

Recently, another definition of the phase based on the general idea of curvature has been proposed by Osipov et al. [30]. For any two-dimensional curve $\mathbf{r} = (u, v)$, the angle velocity at each point is

$$\nu = (ds/dt)/R,$$

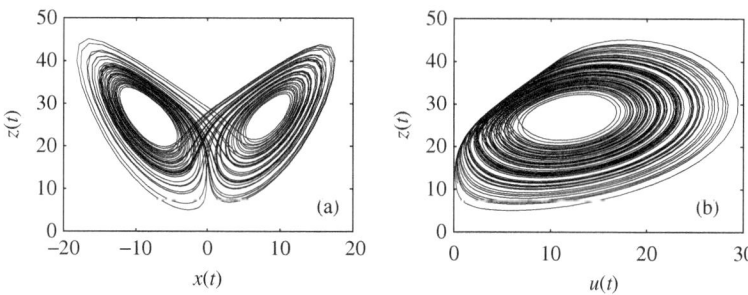

Fig. 10.2 (**a**) Butterfly attractor of the Lorenz system for standard parameter values, and (**b**) Transformed attractor projected in the plane (u, z) resembling that of smeared limit cycle like attractor

where $ds/dt = \sqrt{\dot{u}^2 + \dot{v}^2}$ is the speed along the curve and $R = (\dot{u}^2 + \dot{v}^2)^{3/2}/(\dot{v}\ddot{u} - \ddot{v}\dot{u})$ is the radius of the curvature. If $R > 0$ at each point, then

$$\nu = \frac{d\Phi}{dt} = \frac{\dot{v}\ddot{u} - \ddot{v}\dot{u}}{\dot{u}^2 + \dot{v}^2}$$

is always positive and hence the variable

$$\Phi = \int \nu dt = \arctan \frac{\dot{v}}{\dot{u}} \tag{10.7}$$

is a monotonically increasing function of time and can be considered as the phase of the oscillator. These definitions of frequency and phase are general for any dynamical system if the projection of the phase trajectory on some plane is a curve with a positive curvature. For example, for the non-phase-coherent Rössler attractor in the funnel regime (Fig. 10.1b), the projections of chaotic trajectories on the plane $(\dot{x}, \dot{y})$ always rotate around the origin, and the phase can be defined as $\Phi = \arctan(\dot{y}/\dot{x})$ [30]. However, it is not clear whether an appropriate plane can always be found, on which the projected trajectories rotate around the origin for higher dimensional chaotic systems.

10.4 CPS and Time-Delay Systems

CPS has been studied extensively during the last decade in various nonlinear dynamical systems as discussed in the introduction. However, only a few methods have been available in the literature   [1, 2]   (discussed in the previous section) to calculate the phase of chaotic attractors but unfortunately they are valid only in the case of phase-coherent chaotic attractors of low-dimensional systems. On the other hand even the definition of phase itself is not so clear in non-phase-coherent chaotic attractors and in particular in high-dimensional systems such as time-delay systems. Correspondingly methods to calculate the phase of non-phase-coherent hyperchaotic attractors of time-delay systems are not readily available. The most promising approach available in the literature to calculate the phase of non-phase-coherent attractors is based on the concept of curvature [30], discussed above but this is often restricted to low-dimensional systems. However, one finds that even this procedure does not work in the case of nonlinear time-delay systems, in general, where very often the attractor is non-phase-coherent and high-dimensional. In the following we will show the existence of CPS in a system of two coupled non-identical scalar piecewise linear time-delay systems possessing highly non-phase-coherent hyperchaotic attractors.

We first consider the following unidirectionally coupled drive $x_1(t)$ and response $x_2(t)$ systems, which we have discussed in Chap. 3 [34–36],

$$\dot{x}_1(t) = -ax_1(t) + b_1 f(x_1(t - \tau)), \tag{10.8a}$$
$$\dot{x}_2(t) = -ax_2(t) + b_2 f(x_2(t - \tau)) + b_3 f(x_1(t - \tau)), \tag{10.8b}$$

where b_1, b_2 and b_3 are constants, $a > 0$, τ is the delay time and $f(x)$ is the piece-wise linear function of the form (3.2).

We have chosen the values of parameters (same values as studied in [28]) as $a = 1.0, b_1 = 1.2, b_2 = 1.1$ and $\tau = 15$. For this parametric choice, in the absence of coupling, the drive $x_1(t)$ and the response $x_2(t)$ systems evolve independently. Further in this case, both the drive $x_1(t)$ and the response $x_2(t)$ systems exhibit hyperchaotic attractors with five and four positive Lyapunov exponents, respectively, i.e. both subsystems are qualitatively different (due to $b_1 \neq b_2$). The corresponding attractors are shown in Fig. 10.3a, b, respectively, which clearly show the non-phase-coherent nature. The Kaplan and Yorke [37, 38] dimension for the above attractors turn out to be 8.4085 and 7.007, respectively, obtained by using the formula

$$D_L = j + \frac{\sum_{i=1}^{j} \lambda_i}{|\lambda_{j+1}|}, \tag{10.9}$$

where j is the largest integer for which $\lambda_1 + \ldots + \lambda_j \geq 0$. The parameter b_3 is the coupling strength of the unidirectional nonlinear coupling (10.8b), while the parameters b_1 and b_2 play the role of parameter mismatch resulting in nonidentical

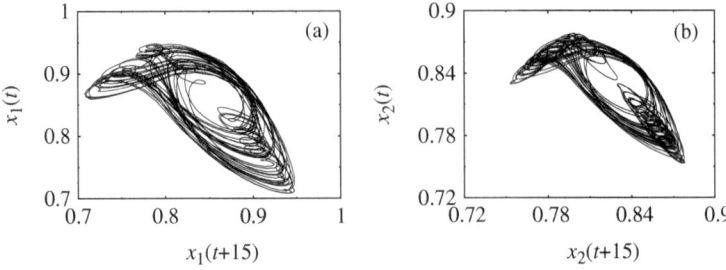

Fig. 10.3 (**a**) The non-phase coherent hyperchaotic attractor of the drive (10.8a) and (**b**) The non-phase coherent hyperchaotic attractor of the uncoupled response (10.8b)

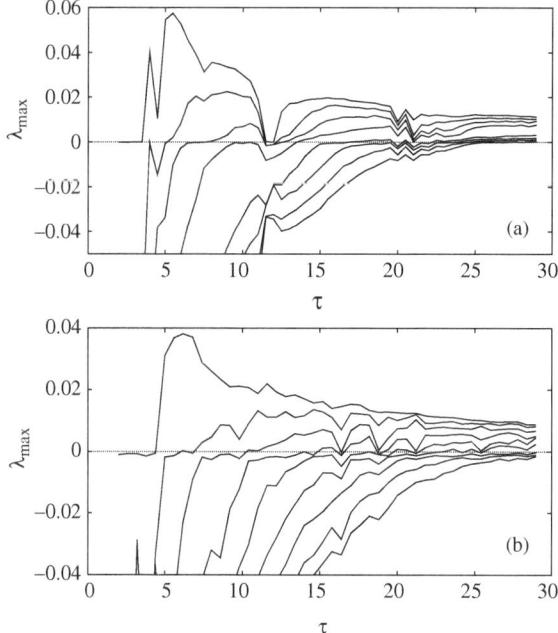

Fig. 10.4 The first ten maximal Lyapunov exponents λ_{max} of (**a**) the scalar time-delay system (10.8a) for the parameter values $a = 1.0, b_1 = 1.2, \tau \in (2, 29)$ and (**b**) the scalar time-delay system (10.8b) for the parameter values $a = 1.0, b_1 = 1.1$ in the same range of delay time in the absence of the coupling b_3

coupled time-delay systems. The spectrum of the first ten largest Lyapunov exponents of the uncoupled system (10.8a) for the values of the parameters $a = 1.0$ and $b_1 = 1.2$ in the range of time-delay $\tau \in (2, 29)$ is shown again in Fig. 10.4a for comparison and that of the system (10.8b) for the parameter value $b_2 = 1.1$ and $b_3 = 0$ in the same range of delay time is also shown in Fig. 10.4b.

Now the task is to identify and to characterize the existence of CPS in the coupled time-delay systems (10.8), possessing highly non-phase-coherent hyperchaotic attractors, when the coupling is introduced ($b_3 > 0$). In the following we present three different approaches to study CPS in coupled piecewise linear time-delay systems (10.8).

10.5 CPS from Poincaré Surface of Section of the Transformed Attractor

We now introduce a transformation to successfully capture the phase in the present problem. It transforms the non-phase coherent attractor (Fig. 10.5a) into a smeared limit cycle-like form with well-defined rotations around one center (Fig. 10.5b). This transformation is performed by introducing the new state variable

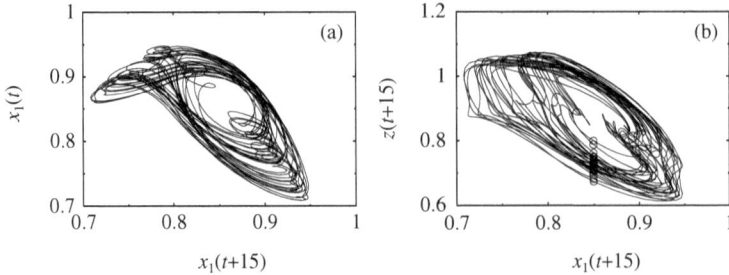

Fig. 10.5 (**a**) The non-phase coherent hyperchaotic attractor of the uncoupled drive (10.8a) and (**b**) Transformed attractor in the $x_1(t+\tau)$ and $z(t+\tau)$ space along with the Poincaré points represented as *open circles*

$$z(t + \tau) = z(t + \tau, \hat{\tau}) = x_1(t)x_1(t + \hat{\tau})/x_1(t + \tau), \qquad (10.10)$$

where $\hat{\tau}$ is an optimal value of delay time to be chosen (so as to rescale the original non-phase coherent attractor into a smeared limit cycle-like form), and then we plot the above attractor (Fig. 10.5a) in the $(x_1(t + \tau), z(t + \tau))$ phase space. The functional form of this transformation (along with a delay time $\hat{\tau}$) has been identified by generalizing the transformation used in the case of chaotic attractors in the Lorenz system [1], so as to unfold the original non-phase-coherent attractor (Fig. 10.5a) into a phase-coherent attractor. We find optimal value of $\hat{\tau}$ to be 1.6. It is to be noted that on closer examination of the transformed attractor (Fig. 10.5b) in the vicinity of the common center that it does not have any closed loop (unlike the case of the original attractor (Fig. 10.5a)) even though the trajectories show sharp turns in some regimes of the phase space. If it is so, such closed loops will lead to phase mismatch, and one cannot obtain exact matching of phases of both the drive and response systems as shown in Fig. 10.6 and discussed below. Now the attractor (Fig. 10.5b) looks indeed like a smeared limit cycle with nearly well defined rotations around a fixed center.

It is to be noted that the above type of transformation (10.10) can be applied to the non-phase-coherent attractors of any time-delay system in general, except for the fact that the optimal value of $\hat{\tau}$ should be chosen for each system appropriately. We have adopted here a geometric approach for the selection of $\hat{\tau}$ and look for an optimum transform which leads to a phase-coherent structure. This is indeed demonstrated for the attractor of Mackey-Glass system in the next section. The main point that we want to stress here is that even for highly non-phase-coherent hyperchaotic attractors of time-delay systems, one can identify suitable transformations to unfold the attractor to identify phase.

Therefore, the phase of the transformed attractor can be now defined based on an appropriate Poincaré section which is transversally crossed by all trajectories using Eq. (10.5). Open circles in Fig. 10.5b correspond to the Poincaré points of the smeared limit-cycle-like attractor. Phases, $\phi_1^z(t)$ and $\phi_2^z(t)$, of the drive $x_1(t)$ and the response $x_2(t)$ systems, respectively, are calculated from the state variables $z_1(t + \tau)$ and $z_2(t + \tau)$ according to Eq. (10.10). The existence of 1:1 CPS between the systems (10.8) is characterized by the phase locking condition

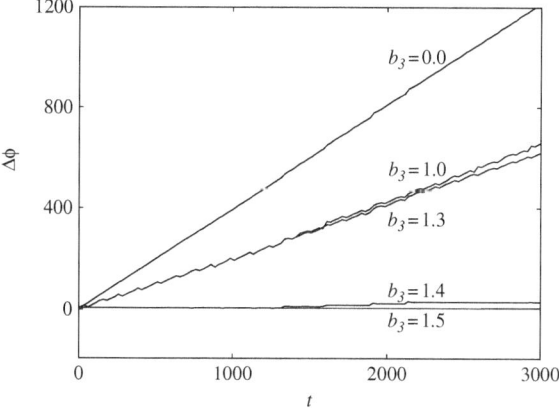

Fig. 10.6 Phase differences $\left(\Delta\phi = \phi_1^z(t) - \phi_2^z(t)\right)$ between the systems (10.8a) and (10.8b) for different values of the coupling strength $b_3 = 0.0, 1.0, 1.3, 1.4$ and 1.5

$$\left|\phi_1^z(t) - \phi_2^z(t)\right| < const. \tag{10.11}$$

The phase difference $\left(\Delta\phi = \phi_1^z(t) - \phi_2^z(t)\right)$ between the systems (10.8a) and (10.8b) is shown in Fig. 10.6 for different values of the coupling strength b_3. For $b_3 = 0.0$ (uncoupled), $\Delta\phi$ increases monotonically as a function of time confirming that both systems are in an asynchronous state (and that they are also nonidentical) in the absence of coupling between them. For the values of $b_3 = 1.0$ and 1.3, the phase slips in the corresponding phase difference $\Delta\phi$ show that the systems are in a transition state. The strong boundedness of the phase difference specified by Eq. (10.11) is obtained for $b_3 > 1.382$ and $\Delta\phi$ becomes zero for the value of the coupling strength $b_3 = 1.5$, showing a high quality CPS.

The mean frequency of the chaotic oscillations is defined as [4, 13]

$$\Omega_{1,2} = \left\langle d\phi_{1,2}^z(t)/dt \right\rangle = \lim_{T\to\infty} \frac{1}{T}\int_0^T \dot{\phi}_{1,2}(t)dt, \tag{10.12}$$

and the 1:1 CPS between the drive $x_1(t)$ and the response $x_2(t)$ systems can also be characterized by a weaker condition of frequency locking, that is, the equality of their mean frequencies $\Omega_1 = \Omega_2$. The mean frequency ratio Ω_2/Ω_1 and its difference $\Delta\Omega = \Omega_2 - \Omega_1$ are shown in Fig. 10.7 as a function of the coupling strength $b_3 \in (0, 3)$. It is also evident from this figure that the mean frequency locking criterion (10.12) is satisfied for $b_3 > 1.382$ and that both the frequency ratio Ω_2/Ω_1 and frequency difference $\Delta\Omega$ show substantial saturation in their values confirming the strong boundedness in the phases of both the systems.

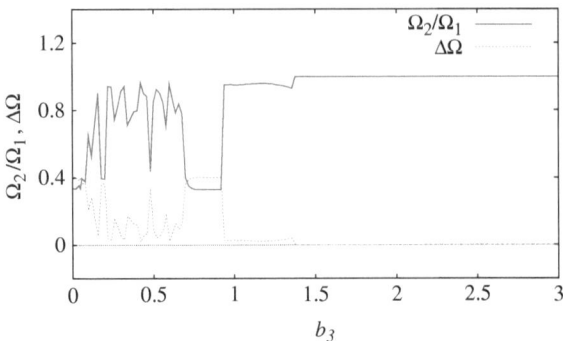

Fig. 10.7 Mean frequency ratio Ω_2/Ω_1 and the frequency difference $\Delta\Omega = \Omega_2 - \Omega_1$ as a function of the coupling strength $b_3 \in (0, 3)$

10.6 CPS from Recurrence Quantification Analysis

The complex synchronization phenomena in the coupled time-delay systems (10.8) can also be analyzed by means of the recently proposed methods based on recurrence plots [39, 40], see Appendix C also. These methods help to identify and quantify CPS (particularly in non-phase coherent attractors) and GS. For this purpose, the generalized autocorrelation function $P(t)$ has been introduced in [39, 40] as

$$P(t) = \frac{1}{N-t} \sum_{i=1}^{N-t} \Theta(\varepsilon - ||X_i - X_{i+t}||), \tag{10.13}$$

where Θ is the Heaviside function, X_i is the ith data corresponding to either the drive variable x_1 or the response variable x_2 specified by Eqs. (10.8) and ε is a predefined threshold. $||.||$ is the Euclidean norm and N is the number of data points. $P(t)$ can be considered as a statistical measure about how often ϕ has increased by 2π or multiples of 2π within the time t in the original space. If two systems are in CPS, their phases increase on the average by $K.2\pi$, where K is a natural number, within the same time interval t. The value of K corresponds to the number of cycles when $||X(t+T) - X(t)|| \sim 0$, or equivalently when $||X(t+T) - X(t)|| < \varepsilon$, where T is the period of the system. Hence, looking at the coincidence of the positions of the maxima of $P(t)$ for both the systems, one can qualitatively identify CPS.

A criterion to quantify CPS is the cross correlation coefficient between the drive, $P_1(t)$, and the response, $P_2(t)$, which can be defined as Correlation of Probability of Recurrence (CPR),

$$CPR = \langle \bar{P}_1(t) \bar{P}_2(t) \rangle / \sigma_1 \sigma_2, \tag{10.14}$$

where $\bar{P}_{1,2}$ means that the mean value has been subtracted and $\sigma_{1,2}$ are the standard deviations of $P_1(t)$ and $P_2(t)$, respectively. If both the systems are in CPS, the probability of recurrence is maximal at the same time t and CPR $\approx$ 1. If they are not in CPS, the maxima do not occur simultaneously and hence one can expect a drift in both the probability of recurrences and low values of CPR.

When the systems (10.8) are in generalized synchronization, two close states in the phase space of the drive variable correspond to that of the response. Hence the neighborhood identity is preserved in phase space. Since the recurrence plots are nothing but a record of the neighborhood of each point in the phase space, one can expect that their respective recurrence plots are almost identical. Based on these facts two indices are defined to quantify GS.

First, the authors of [39] proposed the Joint Probability of Recurrences (JPR),

$$JPR = \frac{\frac{1}{N^2} \sum_{i,j}^{N} \Theta(\varepsilon_x - ||X_i - X_j||)\Theta(\varepsilon_y - ||Y_i - Y_j||) - RR}{1 - RR} \qquad (10.15)$$

where RR is the rate of recurrence, ε_x and ε_y are thresholds corresponding to the drive and response systems, respectively, and X_i is the ith data corresponding to the drive variable x_1 and Y_i is the ith data corresponding to the response variable x_2, specified by Eqs. (10.8). RR measures the density of recurrence points and it is fixed as 0.02 [39]. JPR is close to 1 for systems in GS and is small when they are not in GS. The second index depends on the coincidence of the probability of recurrence, which is defined as Similarity of Probability of Recurrence (SPR),

$$SPR = 1 - \langle (\bar{P}_1(t) - \bar{P}_2(t))^2 \rangle / \sigma_1 \sigma_2. \qquad (10.16)$$

SPR is of order 1 if both systems are in GS and approximately zero or negative if they evolve independently.

Now, we will apply these concepts to the original (non-transformed) attractor (Fig. 10.5a). We estimate these recurrence based measures from 5,000 data points after sufficient transients with the integration step $h = 0.01$ and sampling rate $\Delta t = 100$. The generalized autocorrelation functions $P_1(t)$ and $P_2(t)$ (Fig. 10.8a) for the coupling strength $b_3 = 0.6$ show that the maxima of both systems do not occur simultaneously and there exists a drift between them, so there is no synchronization at all. This is also reflected in the rather low value of CPR $= 0.381$. For $b_3 \in (0.78, 1.381)$, from Fig. 10.9, we observe the first substantial increase of recurrence reaching CPR $\approx 0.5 - 0.6$. Looking into the details of the generalized correlation function $P(t)$, we find that now the main oscillatory dynamics becomes locked, i.e. the main maxima of P_1 and P_2 coincide. For $b_3 \in (1.382, 2.2)$, CPR reaches almost 1 as seen in Fig. 10.9, while now the positions of all the maxima of P_1 and P_2 are also in coincidence, and this is in accordance with the strongly bounded nature of phase differences. This is a strong indication for CPS. Note, however that the heights of the peaks are clearly different (Fig. 10.8b). The differences in the peak heights indicate that there is no strong interrelation in the amplitudes. Further increase of the coupling strength (here $b_3 = 2.21$) leads to the coincidence of both the positions and the heights of the peaks (Fig. 10.8c), confirming the onset of GS in the systems (10.8). This is also further confirmed by the maximal values of the indices JPR $= 1$

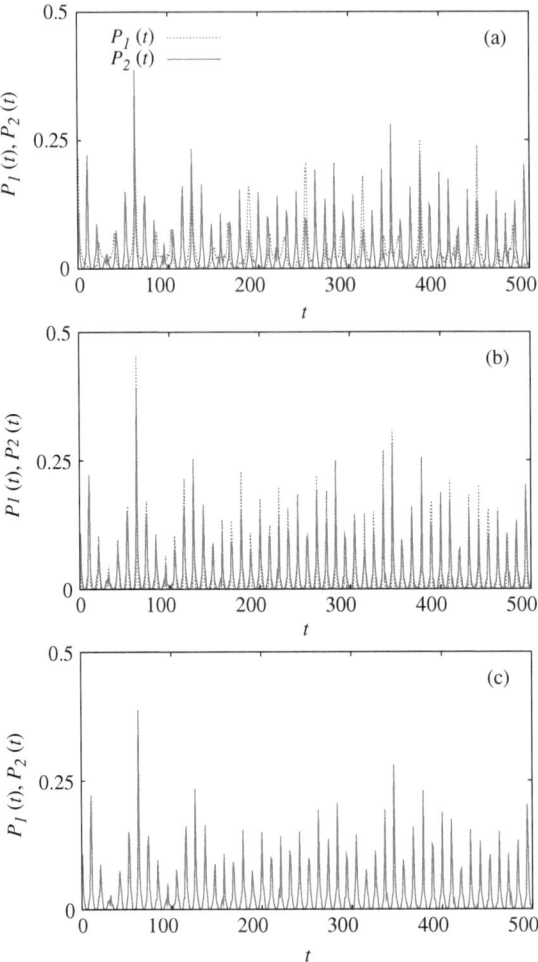

Fig. 10.8 Generalized autocorrelation functions of both the drive $P_1(t)$ and the response $P_2(t)$ systems. (**a**) Non-phase synchronization for $b_3 = 0.6$, (**b**) Phase synchronization for $b_3 = 1.5$ and (**c**) Generalized synchronization for $b_3 = 2.3$

and SPR $= 1$, which is due to the strong correlation in the amplitudes of both systems. It is clear from the construction of SPR that it measures the similarity between the generalized autocorrelation functions $P_1(t)$ and $P_2(t)$. In the regimes of CPS, as the generalized autocorrelation functions coincide in almost all the regimes except for the heights of their maxima, it is also quantified by larger values of SPR. The index SPR in Fig. 10.9 also shows the onset of CPS and it fluctuates around the value 1 in the regime of CPS ($b_3 \in (1.382, 2.2)$) before reaching saturation. This confirms the strong correlation in the amplitudes of both the systems, thereby quantifying the existence of GS. The transition from non-synchronized state via CPS to GS is characterized by the maximal values of CPR, SPR and JPR (Fig. 10.9). As

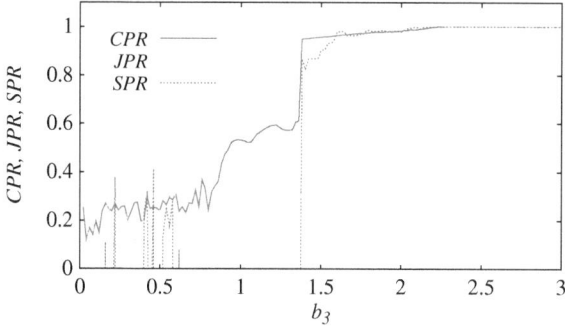

Fig. 10.9 Indices CPR, JPR and SPR as a function of coupling strength $b_3 \in (0, 3)$

expected from the construction of these functions, CPR refers mainly to the onset of CPS, whereas JPR quantifies clearly the onset of GS and SPR indicates both the onset of CPS and GS. In this connection, we have also confirmed the onset and existence of GS by using the auxiliary system approach [41] introduced by Abarbanel et al for the range of the coupling strength $b_3 > 2.2$.

It is to be noted that a variety of nonlinear techniques [42] such as mutual information, methods of predictability, etc. to identify the basic types of complex synchronization, particularly phase and generalized synchronization, are discussed in [40]. It is clearly shown here that recurrence based approach is one of the most efficient tools to identify the existence of different kinds of synchronization phenomena, while all the other techniques are in general not or only partly appropriate for the identification problem.

10.7 CPS from the Lyapunov Exponents

The transition from non-synchronization to CPS can also be characterized by the changes in the Lyapunov exponents of the coupled time-delay system (10.8). The eight largest Lyapunov exponents of the coupled systems are shown in Fig. 10.10. From this figure one can find that all the positive Lyapunov exponents, except the largest one $\left(\lambda_{max}^{(2)}\right)$, corresponding to the response system suddenly become negative at the value of the coupling strength $b_3 = 0.78$ which is an indication of the onset of the transition regime. One may also note that at this value of b_3 already one of the Lyapunov exponents of the response system attains negative saturation while another one reaches negative saturation slightly above $b_3 = 0.78$. This is a strong indication that in this rather complex attractor the amplitudes become somewhat interrelated already at the transition to CPS (as in the funnel attractor [30] of the Rössler system). Also the third positive Lyapunov exponent of the response system gradually becomes more negative from $b_3 = 0.78$ and reaches its saturation value at $b_3 = 1.381$, and thereby confirming the onset of CPS (which is also indicated by the transition of the indices of CPR and SPR in Fig. 10.9 in the range

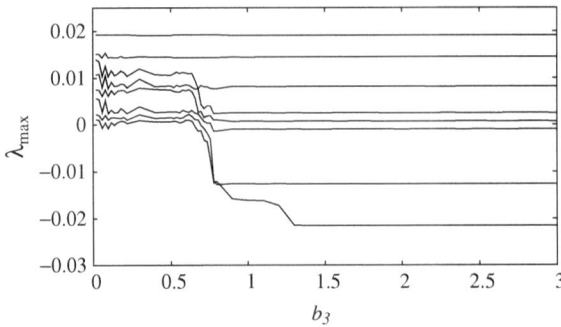

Fig. 10.10 Spectrum of first eight largest Lyapunov exponents of the coupled systems (10.8) as a function of coupling strength $b_3 \in (0, 3)$

of $b_3 \in (0.78, 1.381)$). It is interesting to note that the Lyapunov exponents of the response system $\lambda_i^{(2)}$ (other than $\lambda_{max}^{(2)}$) are changing already at the early stage of CPS ($b_3 \in (0.78, 1.381)$), where the complete CPS is not yet attained. This has been also observed for the onset of CPS in phase-coherent oscillators [7].

10.8 Concept of Localized Sets

Recently, an interesting framework for identifying phase synchronization without having explicitly the measure of the phase, namely the concept of localized sets, has been introduced [43]. The basic idea of this concept is that one has to define a typical event in one of the coupled oscillators and then observe the other oscillator whenever this event occurs. These observations give rise to a set D. Depending upon the property of this set D one can state whether there PS exists or not. The coupled oscillators evolve independently if the sets obtained by observing the corresponding events in both the oscillators spread over the attractors of the oscillators. On the other hand, if the sets are localized on the attractors then PS exist between the interacting oscillators.

We have also confirmed the existence of CPS in coupled piecewise linear time-delay systems using the concept of localized set. We have defined the event in the attractors of both the drive and response system as Poincaré sections, which are shown as open squares in Figs. 10.11. The sets obtained by observing the response system whenever the defined event occurs in the drive system and that corresponds to drive system whenever the defined event occurs in the response system are shown as filled circles. The observed sets are distributed over both the drive and response attractors as shown in Fig. 10.11a, b, respectively, for the value of the coupling strength $b_3 = 0.1$, at which both the systems evolve independently. For the value of $b_3 = 1.5$, the observed sets are localized on both the attractors as shown in Fig. 10.11c, d confirming the existence of CPS between both the drive and response systems.

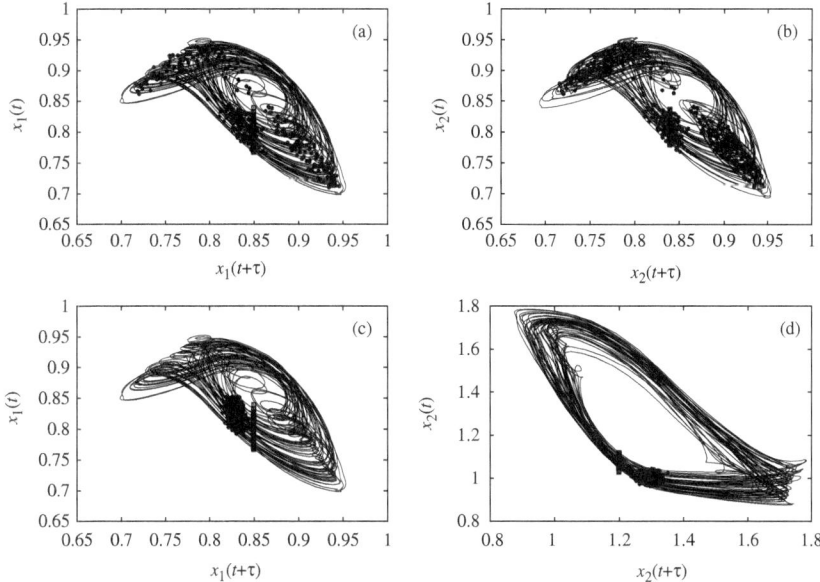

Fig. 10.11 Localized sets and CPS: (**a**) and (**c**) attractor of the drive system, (**b**) and (**d**) attractor of the response system. The *open squares* indicate the events in the corresponding attractors. In (**a**) and (**b**) the sets spread over the attractor and hence there is no CPS for the value of the coupling strength $b_3 = 0.1$ and, in (**c**) and (**d**) the sets are localized confirming the existence of CPS for $b_3 = 1.5$

10.9 Transition from Phase to Generalized Synchronization: Mackey-Glass & Ikeda Systems

Following the above analysis, in this section, we present the results on the identification and existence of CPS in coupled Mackey-Glass time-delay systems with parameter mismatches. We will demonstrate the entrainment of phases of the coupled systems starting from an asynchronous state and its subsequent transition to generalized synchronization (GS) as a function of the coupling strength. The phases of these time-delay systems are calculated using the Poincaré method after a newly introduced transformation of the corresponding attractors, which transforms the original non-phase-coherent attractors of both the systems into smeared limit cycle like attractors. Further, the existence of CPS and GS are characterized by the methods based on recurrence quantification analysis and in terms of Lyapunov exponents of the coupled time-delay systems. Finally, we briefly present the results of CPS in coupled Ikeda systems.

We will first consider the coupled Mackey-Glass systems of the form represented in Eq. (10.8) with the functional form (8.15) for the nonlinear function $f(x)$.

We have chosen the parameter values (cf. [26, 37]) as $a = 0.1$, $b_1 = 0.2$, $b_2 = 0.205$, $\tau_1 = \tau_2 = 20$ and varied the coupling strength b_3. The non-phase-coherent chaotic attractor of the system $x_1(t)$, Eq. (10.8a), for the above choice of

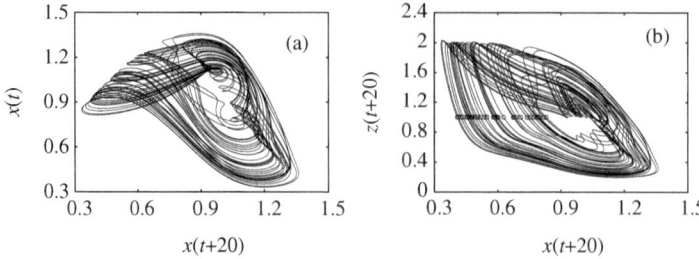

Fig. 10.12 (**a**) The non-phase coherent chaotic attractor of the uncoupled drive (10.8a) and (**b**) Transformed attractor in the $x_1(t+\tau)$ and $z(t+\tau)$ space along with the Poincaré points represented as *open circles* for the Mackey-Glass system (8.15)

parameters is shown in Fig. 10.12a and it possesses one positive and one zero Lyapunov exponents, besides negative ones. Similarly, the second system $x_2(t)$, Eq. (10.8b), also exhibits a non-phase-coherent chaotic attractor with one positive and one zero Lyapunov exponents for the chosen parametric values in the absence of the coupling strength b_3. The parameters b_1 and b_2 contribute to the parameter mismatch between the systems $x_1(t)$ and $x_2(t)$. The first four maximal Lyapunov exponents of both the systems (10.8a) and (10.8b) are shown in Fig. 10.13a, b,

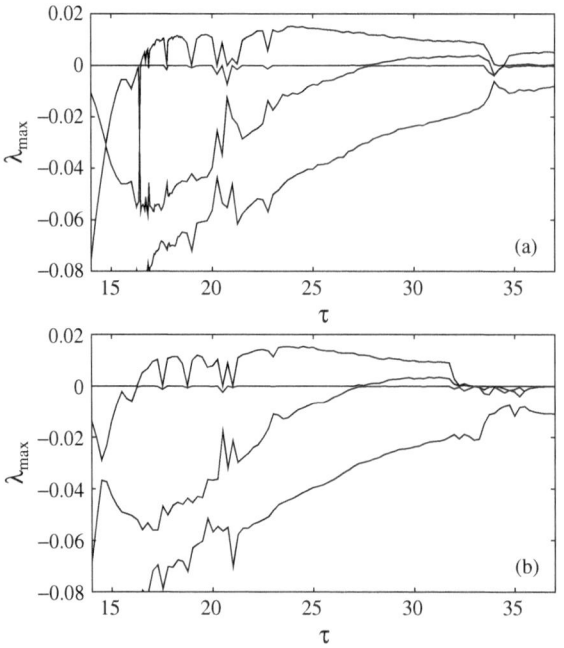

Fig. 10.13 The first four maximal Lyapunov exponents λ_{max} of (**a**) the Mackey-Glass time-delay system (10.8a) for the parameter values $a = 0.1$, $b_1 = 0.2$, $\tau \in (14, 37)$ and (**b**) time-delay system (10.8b) for the parameter values $a = 0.1$, $b_1 = 0.205$ in the same range of delay time in the absence of the coupling b_3

respectively, as a function of the delay time $\tau \in (14, 37)$ when $b_3 = 0$. The Kaplan and Yorke [37, 38] dimension calculated using (10.9) for the present systems ((10.8a) and (10.8b)) work out to be 2.27969 and 2.21096, respectively. Now, the existence of CPS as a function of the coupling strength in the coupled Mackey-Glass systems (10.8) will be discussed using the three approaches used for identifying CPS in coupled piecewise-linear time-delay systems.

10.9.1 CPS from Poincaré Section of the Transformed Attractor

The non-phase-coherent chaotic attractor (Fig. 10.12a) of the Mackey-Glass system is transformed into a smeared limit cycle-like attractor (Fig. 10.12b) using the same transformation (10.10) as used for the piecewise linear time-delay systems. For the attractor (Fig. 10.12a) of the Mackey-Glass system, the optimal value of the delay time $\hat{\tau}$ in Eq. (10.10) is found to be 8.0. The Poincaré points are shown as open circles in Fig. 10.12b from which the instantaneous phase $\phi_1^z(t)$ is calculated using (10.5). The existence of CPS in the coupled Mackey-Glass systems (10.8) is also characterized by the phase locking condition (10.11) as shown in Fig. 10.14. The phase differences $\Delta\phi = \phi_1^z(t) - \phi_2^z(t)$ between the systems (10.8a) and (10.8b) for the values of the coupling strength $b_3 = 0.04, 0.08, 0.11, 0.12$ and 0.3 are shown in Fig. 10.14. For the value of the coupling strength $b_3 = 0.3$, there exists a strong boundedness in the phase difference showing high quality CPS. The mean frequency ratio Ω_2/Ω_1 calculated from (10.12) along with the mean frequency difference $\Delta\Omega$ is shown in Fig. 10.15. The value of the mean frequency ratio is $\Omega_2/\Omega_1 \approx 1$ in the range of $b_3 \in (0.12, 0.23)$ corresponding to the transition regime (which is also to be confirmed from the indices CPR and JPR in the next subsection), see the inset of Fig. 10.15. Similarly the mean frequency difference is also $\Delta\Omega \approx 0$ confirming the

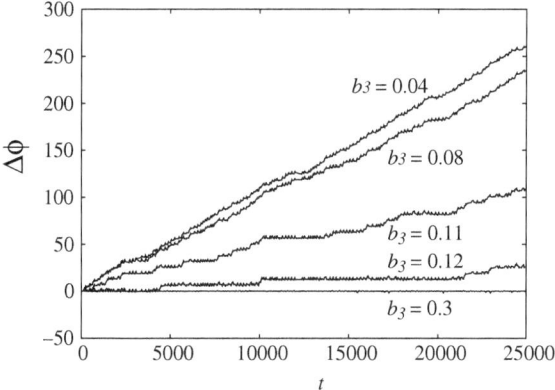

Fig. 10.14 Phase differences $\left(\Delta\phi = \phi_1^z(t) - \phi_2^z(t)\right)$ between the systems (10.8a) and (10.8b) Mackey-Glass system for different values of the coupling strength $b_3 = 0.04, 0.08, 0.11, 0.12$ and 0.3

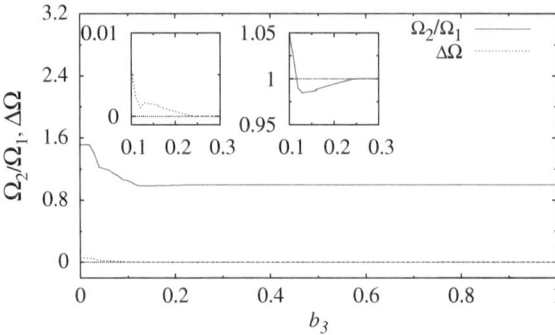

Fig. 10.15 Mean frequency ratio Ω_2/Ω_1 and their difference $\Delta\Omega = \Omega_2 - \Omega_1$ as a function of the coupling strength $b_3 \in (0, 1)$

transition regime. For the value of $b_3 > 0.23$, both the quantities Ω_2/Ω_1 and $\Delta\Omega$ acquire complete saturation in their values confirming existence of CPS.

10.9.2 CPS from Recurrence Quantification Analysis

The existence of CPS from the original non-phase-coherent chaotic attractors of the systems (10.8) for the Mackey-Glass system is analyzed in this section using the recurrence quantification measures defined earlier in Sect. 10.6. We have estimated these measures again using a set of 5,000 data points, and the same integration step and the sampling rate as used in the case of coupled piecewise linear time-delay systems. The maxima of generalized autocorrelations of both the drive $P_1(t)$ and the response $P_2(t)$ systems (Fig. 10.16a) do not occur simultaneously for $b_3 = 0.1$, which indicates the independent evolution of both the systems without any correlation and this is also reflected in the rather low value of CPR = 0.4. For $b_3 = 0.3$, the maxima of both $P_1(t)$ and $P_2(t)$ are in good agreement (Fig. 10.16b) and this shows the strongly bounded phase difference. It is to be noted that even though both the maxima coincide, the heights of the peaks are clearly of different magnitudes contributing to the fact that there is no strong correlation in the amplitudes of both the systems in spite of the existence of CPS. Both the positions and the peaks are in coincidence (Fig. 10.16c) for the value of coupling strength $b_3 = 0.9$ in accordance with the strong correlation in the amplitudes of both the systems (10.8) corresponding to GS. This is also reflected in the maximal values of both JPR=1 and SPR=1. The indices CPR, JPR and SPR are shown in Fig. 10.17. The onset of CPS is shown by the first substantial increase of the index CPR at $b_3 = 0.11$ and the transition regime is shown by the successive plateaus of CPR in the range $b_3 \in (0.12, 0.23)$. The maximal values of CPR for $b_3 > 0.23$ indeed confirm the existence of high quality CPS. The existence of GS is also confirmed from both the indices JPR and SPR.

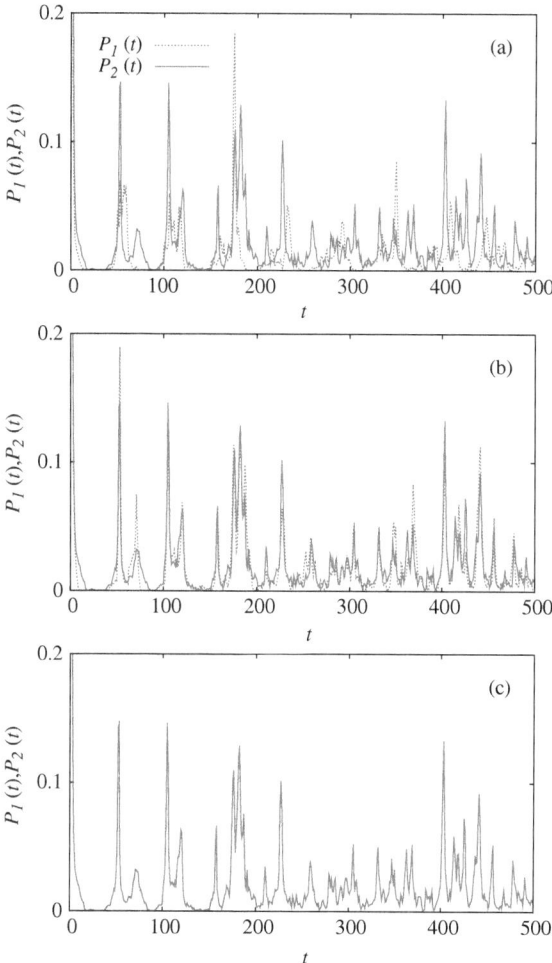

Fig. 10.16 Generalized autocorrelation functions of both the drive system (10.8a), $P_1(t)$, and the response system (10.8b), $P_2(t)$, for coupled Mackey-Glass systems. (**a**) Non-phase synchronization for $b_3 = 0.1$, (**b**) Phase synchronization for $b_3 = 0.3$ and (**c**) Generalized synchronization for $b_3 = 0.9$

10.9.3 CPS from the Lyapunov Exponents

The onset of CPS is also characterized by the changes in the Lyapunov exponents of the coupled Mackey-Glass systems (10.8). The first four largest Lyapunov exponents of the corresponding coupled systems (10.8) are shown in Fig. 10.18. The zero Lyapunov exponent of the response system $x_2(t)$ already becomes negative as soon as the coupling is introduced and the onset of CPS is indicated by the negative saturation of the zero Lyapunov exponent at $b_3 = 0.11$. The positive Lyapunov exponent of the response system becomes gradually negative in the transition regime

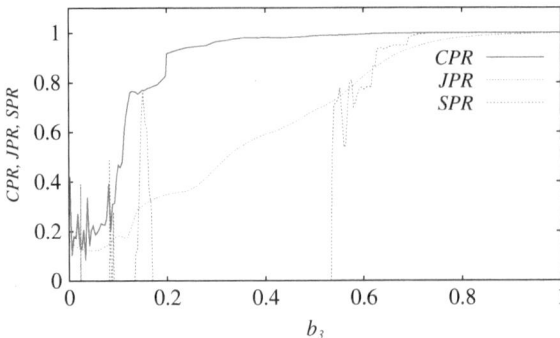

Fig. 10.17 Indices CPR, JPR and SPR as a function of the coupling strength $b_3 \in (0, 1)$

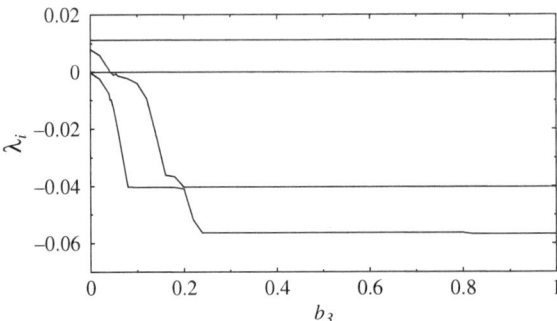

Fig. 10.18 Spectrum of the first four largest Lyapunov exponents of the coupled Mackey-Glass systems (10.8) as a function of coupling strength $b_3 \in (0, 1)$

($b_3 \in (0.12, 0.23)$) and it reaches its negative saturation at $b_3 = 0.23$ at which high quality CPS exists. The transition of the positive Lyapunov exponent to negativity in this rather complex attractor is a firm indication of the rather strong correlation in the amplitudes of both the systems even before the onset of CPS. This behavior of negative transition of the positive Lyapunov exponent of the response system before CPS has also been observed in [13, 44].

As discussed in the case of piecewise linear time-delay systems, we have also confirmed the existence of CPS using the concept of localized sets. The event is chosen as Poincaré sections, which are plotted as open squares in Fig. 10.19. Observed sets from drive/response attractors collected during the Poincaré sections in response/drive attractors are depicted as closed circles. When both the drive and the response systems evolve independently, the observed sets are spread all over the attractors of both the systems as shown in Fig. 10.19a, b for the value of $b_3 = 0.04$. The observed sets are localized on the attractor during the phase synchronized state as shown in Fig. 10.19c, d for the value of coupling strength $b_3 = 1.5$, confirming the existence of CPS.

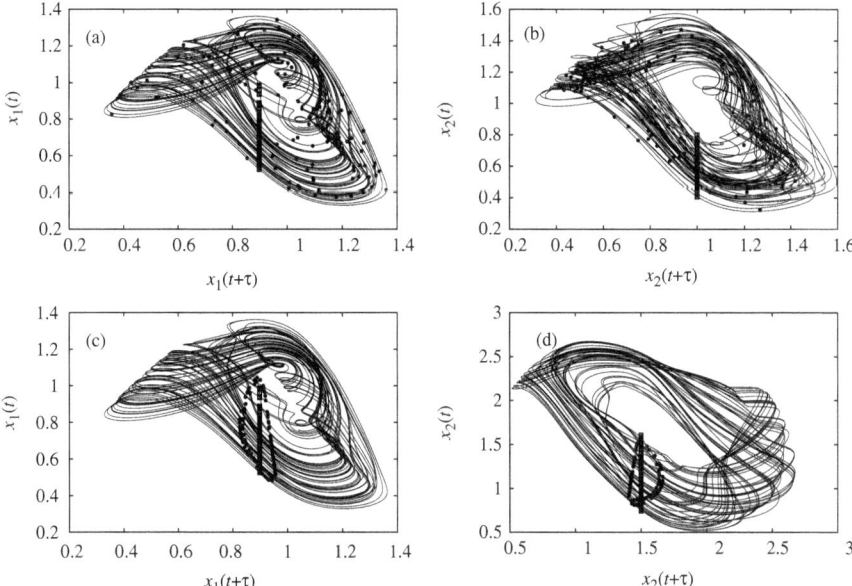

Fig. 10.19 Localized sets and CPS: (**a**) and (**c**) attractor of the drive system, (**b**) and (**d**) attractor of the response system of the coupled Mackey-Glass systems. The *open squares* indicate the events in the corresponding attractors. In (**a**) and (**b**) the sets spread over the attractor and hence there is no CPS for the value of the coupling strength $b_3 = 0.04$ and, in (**c**) and (**d**) the sets are localized confirming the existence of CPS for $b_3 = 0.3$

10.9.4 CPS in Coupled Ikeda Systems

Finally, we consider the coupled Ikeda systems of the form (10.8) with the functional form (8.16) for $f(x)$. Now we have chosen the parameter values as $a = 1.0$, $b_1 = b_2 = 5$, $\tau_1 = 2$ and $\tau_2 = 3$ for the Ikeda systems. It is to be noted that here we have chosen parameter mismatch in the value of the feedback delay time of the drive τ_1 and the response system τ_2 whereas we have considered parameter mismatch in b_1 and b_2 alone in the case of piecewise linear time-delay system and in Mackey-Glass system. One may also obtain the same results with parameter mismatch either in $b_1 \neq b_2$ or $\tau_1 \neq \tau_2$, or in both.

As the topology of the hyperchaotic attractor of the Ikeda system is more complicated with large number of highly non-uniform irregular loops, one is not able to unfold the original attractor with the transformation (10.10) used for the piecewise linear time-delay system and for the Mackey-Glass system. It appears that it is more difficult to identify an appropriate transformation to unloop the attractor so that the projected attractor in the new phase space looks like a phase-coherent attractor. However, one may try to find an optimal transformation to unfold the original attractor of the Ikeda system. Nevertheless, we have identified the existence of CPS in the coupled Ikeda systems also, using the recurrence indices we have used for both

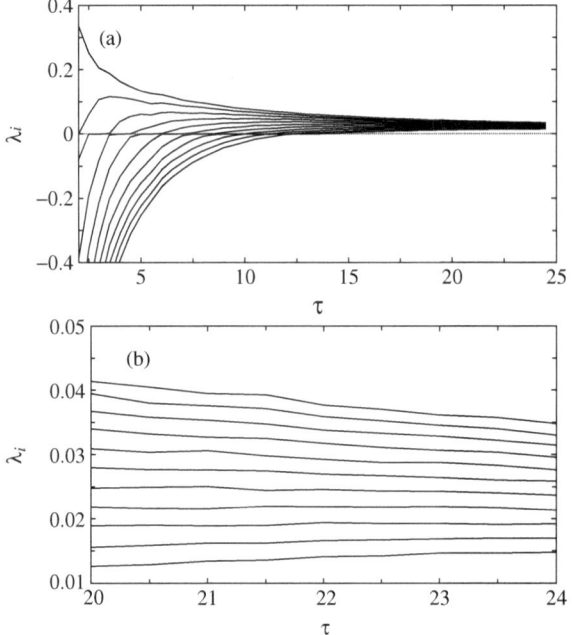

Fig. 10.20 First eleven largest Lyapunov exponents of the Ikeda system as a function of delay time $\tau \in (2, 25)$

the piecewise linear time-delay system and Mackey-Glass system. We have also characterized the existence of CPS using the Lyapunov exponents of the coupled Ikeda systems.

The set of first eleven largest Lyapunov exponents of the Ikeda systems for the above values of the parameters a and b is shown in Fig. 10.20 as a function of delay time τ. Figure 10.20b shows an enlarged part of Fig. 10.20 in the x-axis to view clearly the eleven largest positive Lyapunov exponents. The existence of CPS in the coupled Ikeda system is calculated from the original non-phase-coherent chaotic attractor using the recurrence based indices. Generalized autocorrelation function of the drive $P_1(t)$ and the response $P_2(t)$ are plotted in Fig. 10.21 for three different values of the coupling strength b_3. Maxima of both the quantities, $P_1(t)$ and $P_2(t)$, do not occur simultaneously (Fig. 10.21a) for the value of $b_3 = 4$, which corresponds to asynchronous state of the drive and response systems, whereas for the value of the coupling strength $b_3 = 6$, the coincidence of maxima of generalized autocorrelation functions, $P_1(t)$ and $P_2(t)$, contributes to the existence of CPS between the coupled Ikeda systems. The existence of GS is again indicated by the coincidence of both the positions and peaks of the generalized autocorrelation functions, $P_1(t)$ and $P_2(t)$, of both the systems for the value of $b_3 = 20$. The onset of CPS and GS is also characterized both by the recurrence based indices and by the Lypunov exponents of the coupled systems. The onset of CPS is shown by a

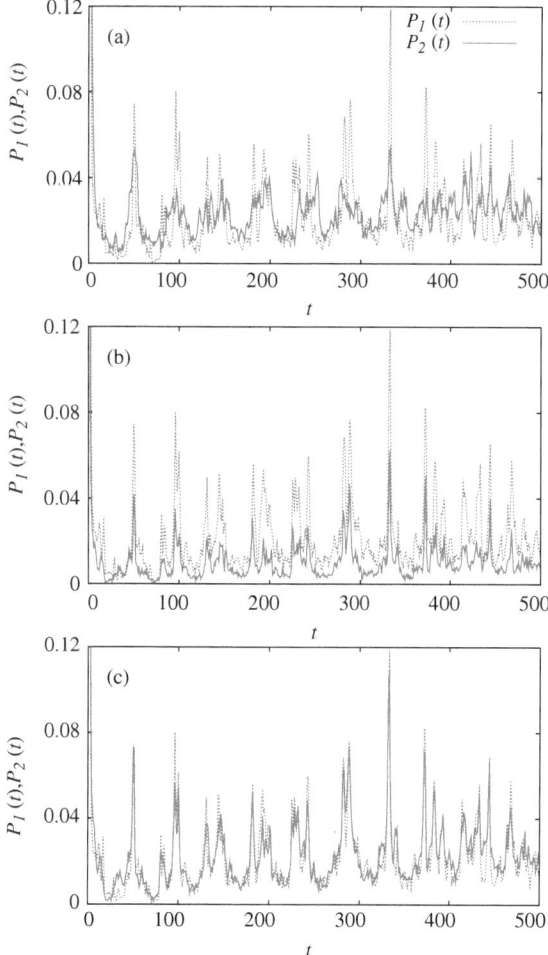

Fig. 10.21 Generalized autocorrelation functions of the coupled Ikeda systems (drive system, $P_1(t)$, and the response system, $P_2(t)$). (**a**) Non-phase synchronization for $b_3 = 4$, (**b**) Phase synchronization for $b_3 = 6$ and (**c**) Generalized synchronization for $b_3 = 20$

gradual increase in the value of the index CPR (Fig. 10.22) and it shows substantial saturation for $b_3 > 19$ where the GS appears as indicated by the maximal value of the index SPR. JPR also shows substantial saturation above $b_3 > 19$. The largest Lyapunov exponents of the coupled Ikeda system shown in Fig. 10.23 clearly indicates the onset of CPS and GS in their values. The onset of CPS is indicated by the transition of the least positive (almost zero) Lyapunov exponent (Fig. 10.23) of the response system to negative value at $b_3 = 5.52$ and the existence of CPS is indicated by the transition of one of the positive Lyapunov exponents of the response system to negative value at $b_3 = 13.56$ as in the case of the piecewise linear time-delay

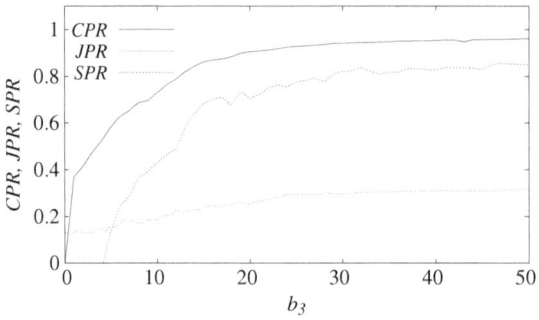

Fig. 10.22 Indices CPR, JPR and SPR as a function of coupling strength $b_3 \in (0, 50)$

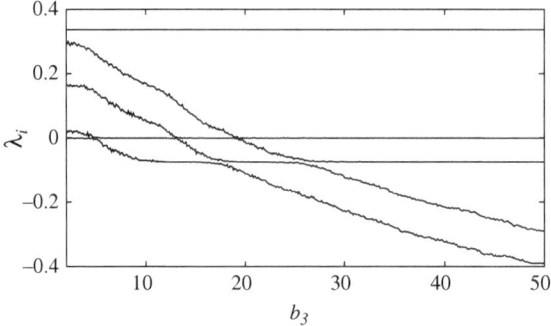

Fig. 10.23 Spectrum of first five largest Lyapunov exponents of the coupled Ikeda systems as a function of the coupling strength $b_3 \in (2, 50)$

system and the Mackey-Glass system. The onset of GS is shown by the transition of the largest positive Lyapunov exponent of the response system to negative value at $b_3 = 19.3$.

We have confirmed the existence of CPS in the coupled Ikeda time-delay systems also by using the concept of localized set. We have defined the event in the attractor of the drive system as a segment characterized by $x_1(t+\tau) = 0$ and $x_1(t) > 2.0$ and another event in the response system as a segment characterized by $x_2(t+\tau) = 0$ and $x_2(t) < -3.0$, which are shown as black lines in Fig. 10.24. The sets obtained by observing the response Ikeda system whenever the defined event occurs in the drive system and vice versa are shown as dots in Fig. 10.24a, b, respectively, for the value of the coupling strength $b_3 = 4.0$, for which there is no CPS as discussed earlier and hence the sets are spread over the attractors. On the other hand for the value of the coupling strength $b_3 = 20$ for which CPS exists as seen from Figs. 10.21, 10.22 and 10.23, the sets are localized as shown in Fig. 10.24c, d confirming the existence of CPS in the coupled Ikeda systems.

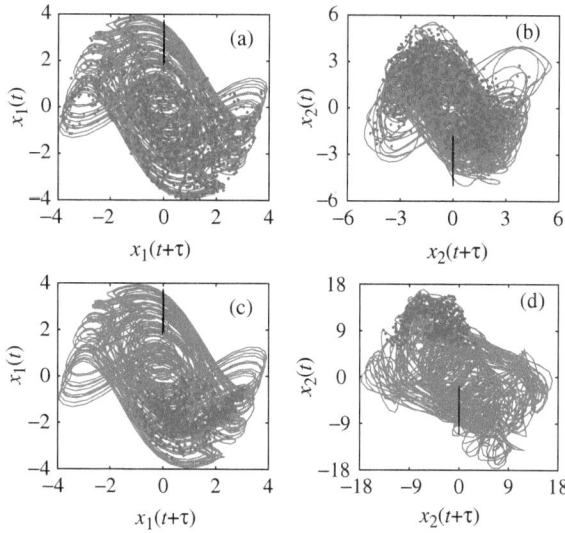

Fig. 10.24 Coupled Ikeda system: (**a**) and (**c**) attractor of the drive system, (**b**) and (**d**) attractor of the response system. The *bars* indicate the events in the corresponding attractors. In (**a**) and (**b**) the sets spread over the attractor and hence there is no CPS for the value of the coupling strength $b_3 = 4.0$ and, in (**c**) and (**d**) the sets are localized confirming the existence of CPS for $b_3 = 20.0$

10.10 Summary

It is clear from the above analysis that CPS in coupled time delay systems exhibiting highly nonphase coherent hyperchaotic attractors can be realized by various methods. These include unfolding transformation of the hyperchaotic attractor, recurrence based indices characterization and changes in the Lyapunov exponents with typical examples, including piecewise linear, Mackey-Glass and Ikeda systems. We have also pointed out that the onset of CPS is also characterized based on recurrence quantification analysis. Furthermore, existence of CPS in these systems is also confirmed using the concept of localized sets. To conclude, CPS is a typical dynamical phenomenon associated with coupled nonlinear time-delay systems.

References

1. A.S. Pikovsky, M.G. Rosenblum, J. Kurths, *Synchronization – A Unified Approach to Nonlinear Science* (Cambridge University Press, Cambridge, 2001)
2. S. Boccaletti, J. Kurths, G. Osipov, D.L. Valladares, C.S. Zhou, Phys. Rep. **366**, 1 (2002)
3. J. Kurths (ed.), Special issue on phase synchronization. Int. J. Bifurcat. Chaos **10**, 2289 (2000)
4. G.V. Osipov, A.S. Pikovsky, M.G. Rosenblum, J. Kurths, Phys. Rev. E **55**, 2353 (1997)
5. A.S. Pikovsky, G.V. Osipov, M.G. Rosenblum, M. Zaks, J. Kurths, Phys. Rev. Lett. **79**, 47 (1997)
6. A.S. Pikovsky, M.G. Rosenblum, G.V. Osipov, J. Kurths, Physica D **219**, 104 (1997)
7. M.G. Rosenblum, A.S. Pikovsky, J. Kurths, Phys. Rev. Lett. **76**, 1804 (1996)

8. M.G. Rosenblum, A.S. Pikovsky, J. Kurths, Phys. Rev. Lett. **78**, 4193 (1997)
9. M. Zhan, G.W. Wei, C.H. Lai, Phys. Rev. E **65**, 036202 (2002)
10. U. Parlitz, L. Junge, W. Lauternorn, L. Kocarev, Phys. Rev. E **54**, 2115 (1996)
11. M. Zhan, Z. Zheng, G. Hu, X. Peng, Phys. Rev. E **62**, 3552 (2000)
12. E. Rosa Jr., C.M. Ticos, W.B. Pardo, J.A. Walkenstein, M. Monti, J. Kurths, Phys. Rev. E **68**, 025202(R) (2003)
13. S. Guan, C.H. Lai, G.W. Wei, Phys. Rev. E **72**, 016205 (2005)
14. A. Pujol-Peré, O. Calvo, M.A. Matias, J. Kurths, Chaos **13**, 319 (2003)
15. M.S. Baptista, T.P. Silva, J.C. Sartorelli, I.L. Caldas, E. Rosa Jr., Phys. Rev. E **67**, 056212 (2003)
16. S.K. Dana, B. Blasius, J. Kurths, Chaos **16**, 023111 (2006)
17. K.V. Volodehenko, V.N. Ivanov, S.H. Gong, M. Choi, Y.J. Park, C.M. Kim, Phys. Rev. Lett. **26**, 1406 (2001)
18. D.J. DeShazer, R. Breban, E. Ott, R. Roy, Phys. Rev. Lett. **87**, 044101 (2001)
19. D. Maza, A. Vallone, H. Macini, S. Boccaletti, Phys. Rev. Lett. **85**, 5567 (2000)
20. P. Tass, M.G. Rosenblum, J. Weule, J. Kurths, A. Pikovsky, J. Volkmann, A. Schnitzler, H.J. Freund, Phys. Rev. Lett. **81**, 3291 (1998)
21. R.C. Elson, A.I. Selvertson, R. Huerta, N.F. Rulkov, M.I. Rabinovich, H.D.I. Abarbanel, Phys. Rev. Lett. **81**, 5692 (1998)
22. D. Maraun, J. Kurths, Geophys. Res. Lett. **32**, L15709 (2005)
23. T. Heil, I. Fischer, W. Elsaber, B. Krauskopf, K. Green, A. Gavrielides, Phys. Rev. E **67**, 066214 (2003)
24. N. Kopell, G.B. Ermentrout, M.A. Whittington, R.D. Traub, PNAS **97**, 1867 (2000)
25. M. Kostur, P. Hanggi, P. Talkner, J.L. Mateos, Phys. Rev. E **72**, 036210 (2005)
26. M.C. Mackey, L. Glass, Science **197**, 287 (1977)
27. L.B. Shaw, I.B. Schwartz, E.A. Rogers, R. Roy, Chaos **16**, 015111 (2006)
28. D.V. Senthilkumar, M. Lakshmanan, J. Kurths, Phys. Rev. E **74**, 035205(R) (2006)
29. M. Lakshmanan, R. Sahadevan (eds.), *Proceedings of the 3rd National Conference on Nonlinear Systems and Dynamics* (Allied Publishers, Chennai, 2006), p. 202
30. G.V. Osipov, B. Hu, C. Zhou, M.V. Ivanchenko, J. Kurths, Phys. Rev. Lett. **91**, 024101 (2003)
31. J.D. Farmer, Ann. N.Y. Acad. Sci. **357**, 453 (1980)
32. E.F. Stone, Phys. Lett. A **163**, 367 (1992)
33. D. Gabor, J. IEE Lond. **93**, 429 (1946)
34. D.V. Senthilkumar, M. Lakshmanan, Int. J. Bifurcat. Chaos **15**, 2985 (2005)
35. D.V. Senthilkumar, M. Lakshmanan, Phys. Rev. E **71**, 016211 (2005)
36. D.V. Senthilkumar, M. Lakshmanan, J. Phys.: Conf. Ser. **23**, 300 (2005)
37. J.D. Farmer, Physica D **4**, 366 (1982)
38. J. Kaplan, J. Yorke, in *Functional Differential Equations and Approximation of Fixed Points*, ed. by H.O. Peitgen, H.O. Walther. Lecture Notes in Mathematics, vol 730 (Springer, Berlin, 1979)
39. M.C. Romano, M. Thiel, J. Kurths, I.Z. Kiss, J.L. Hudson, Europhys. Lett. **71**, 466 (2005)
40. N. Marwan, M.C. Romano, M. Thiel, J. Kurths, Phys. Rep. **438**, 237 (2007)
41. H.D.I. Abarbanel, N.F. Rulkov, M.M. Sushchik, Phys. Rev. E **53**, 4528 (1996)
42. H. Kantz, T. Schriber, *Nonlinear Time Series Analysis* (Cambridge University Press, New York, 1997)
43. T. Pereira, M.S. Baptista, J. Kurths, Phys. Rev. E **75**, 026216 (2007)
44. B. Hu, G.V. Osipov, H.Y. Yang, J. Kurths, Phys. Rev. E **67**, 066216 (2003)

Chapter 11
DTM Induced Oscillating Synchronization

11.1 Introduction

So far we have considered nonlinear dynamical systems with constant time delay. However, it is also possible that the delay could also vary as a function of time. The notion of time dependent delay (TDD) with stochastic or chaotic modulation in time-delay systems was introduced by Kye et al. [1] to understand the behaviour of dynamical systems with time dependent topology. They have reported that in a time-delay system with TDD, the reconstructed phase trajectory does not collapse to a simple manifold, a property different from that of delayed systems with fixed delay time (which is considered to be a serious drawback of the latter type of systems). It has been shown very recently that a distributed delay enriches the characteristic features of the delayed system over that of the fixed delay systems [2]. Based on these considerations, current studies on chaotic synchronization in time-delay systems are also focused towards time-delay systems with time dependent delay [3–6]. In this connection it is also of considerable interest to study the effect of simple modulations such as periodic modulation [6] on the nature of the chaotic attractor.

Earlier, in Chap. 8, we have studied chaotic synchronization in a system of two unidirectionally coupled odd piecewise linear time-delay systems [7] with two different constant delay times: one in the coupling term and the other in the individual systems, namely, feedback delay. We have shown that there exists a transition from anticipatory to lag synchronization through complete synchronization as a function of a system parameter. Suitable stability criteria were also obtained. In this chapter, we wish to investigate whether there arises any new phenomenon due to the introduction of periodic delay time modulation in the coupled time-delay systems which we have studied earlier and its effects on the various synchronization scenario. Interestingly, one finds that even with simple periodic modulation, the time-delay system cannot be collapsed into a simple manifold and that the delay time cannot be extracted using standard methods. More interestingly, one finds that the fully rectified sinusoidal modulation of delay time introduces a new type of oscillating synchronization that oscillates between anticipatory, complete and lag synchronizations for the case of constant coupling delay. This is further corroborated by suitable stability condition based on Krasovskii-Lyapunov theory. Intermittent

M. Lakshmanan, D.V. Senthilkumar, *Dynamics of Nonlinear Time-Delay Systems*,
Springer Series in Synergetics, DOI 10.1007/978-3-642-14938-2_11,
© Springer-Verlag Berlin Heidelberg 2010

anticipatory and lag synchronizations are also found to exist in the present system for the case of identical modulation in both the coupling and feedback delays, for a range of modulational frequencies. In addition, one also finds that there exist regions of exact anticipatory and lag synchronizations for lower values of modulational frequencies. The results can be corroborated by the nature of similarity functions and the intermittent behavior by the probability distribution of the laminar phase, satisfying universal $-\frac{3}{2}$ power law behavior of on-off intermittency [8, 9]. These studies are also extended to coupled Mackey-Glass and coupled Ikeda systems.

11.2 Estimation of the Effect of Delay Time Modulation

The concept of time dependent delay with stochastic or chaotic modulation was introduced by Kye et al. [1] in time-delay systems, and they have shown in the case of Mackey-Glass system that the delay time carved out of the time series of the time-delay system is undetectable by the conventional measures and hence any reconstruction of phase space of the delayed system is hardly possible. This fact has motivated some authors [3, 5, 6] to look for delay systems with delay time modulation as an ideal candidate for secure communication.

Interestingly we find here that even with a fully rectified sinusoidal delay time modulation of the form

$$\tau(t) = \tau_0 + \tau_a \left| \sin(\omega t) \right|, \tag{11.1}$$

where τ_0 is the zero frequency component, τ_a is the amplitude and ω/π is the frequency of the modulation, we can realize the effects of modulation. Note that in the delay term, we have introduced the fully rectified sinusoidal modulational form (absolute of the sine term) so as to keep the delay time positive even for values of $\tau_a > \tau_0$ so as to avoid acausality problem in Eq. (3.1) for negative values of τ when $\tau_a > \tau_0$. However, for values of τ_0 sufficiently greater than τ_a the rectification in the modulation (11.1) is not required.

The scalar piecewise linear system (3.1) also exhibits the properties studied by Kye et al. with stochastic or chaotic modulation. In order to demonstrate the effect of fully rectified sinusoidal delay time modulation of the form (11.1) on the time series of the piecewise linear time-delay system (3.1), we will now calculate (1) *filling factor* [10], (2) *length of polygon line* [11] and (3) *average mutual information* [1, 12, 13], both in the presence and in the absence of delay time modulation, and show how periodic modulation removes any imprints of the time-delay.

11.2.1 Filling Factor

Now we will compute the filling factor [10] for the chaotic trajectory $x(t)$ of the time-delay system (3.1) by projecting it onto the pseudospace $(x, x_{\hat{\tau}}, \dot{x})$ with P^{3N}

equally sized hypercubes, where the delayed time series $x_{\hat{\tau}} = x(t - \hat{\tau})$ is constructed from $x(t)$ for various values of $\hat{\tau}$. The filling factor is the number of hypercubes which are visited by the projected trajectory, normalized to the total number of hypercubes, P^{3N}. Figure 11.1a shows the filling factor for constant delay $\tau_0 = 10$ when $\tau_a = 0$ in Eq. (11.1), where one can identify the existence of an underlying time-delay induced instability [10] which induces local minima in the filling factor at $\hat{\tau} \approx n\tau_0$, $n = 1, 2, 3....$ From the latter, one can identify the value of the time-delay parameter τ of the system (3.1) under consideration. Figure 11.1b shows the

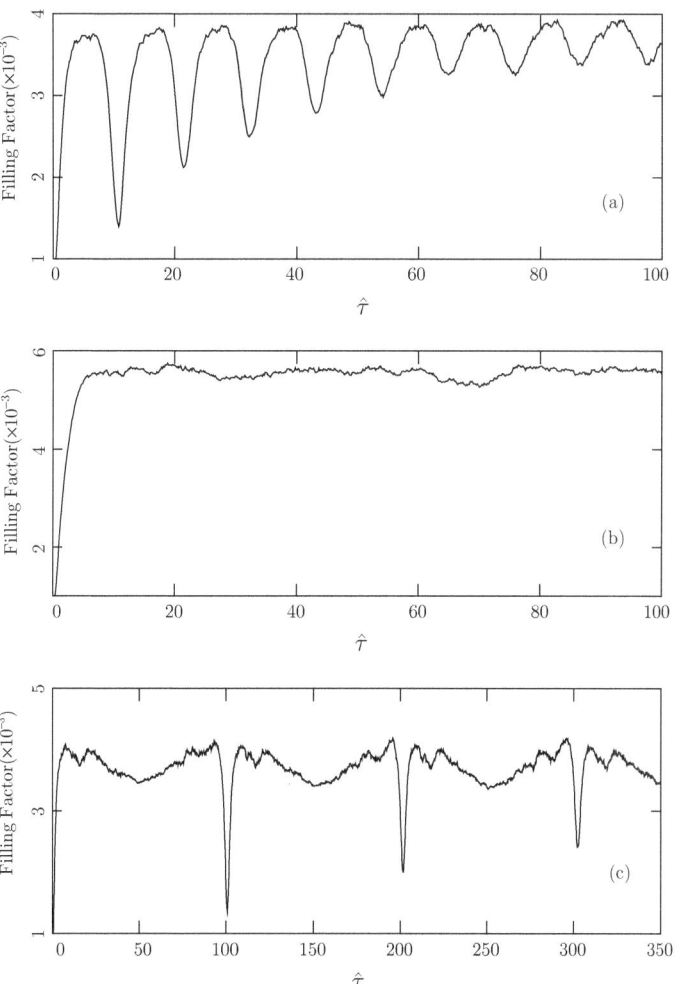

Fig. 11.1 Filling factor as a function of delay time $\hat{\tau}$ (of delayed time series $x_{\hat{\tau}}$). (**a**) with constant delay $\tau_0 = 10$ when $\tau_a = 0$, (**b**) with delay time modulation of the form (11.1) with $\tau_0 = 10$, $\tau_a = 90$ and $\omega = 10^{-4}$ and (**c**) with large constant delay $\tau_0 = 100 (\tau_a = 0)$

filling factor with delay time modulation of the form (11.1) with $\tau_0 = 10$, $\tau_a = 90$ and $\omega = 0.0001$, where no local minima occurs. Figure 11.1c is plotted for $\tau_0 = 100$ and $\tau_a = 0$ to show that the disappearance of local minima in Fig. 11.1b is not due to large delays but only because of the delay time modulation. From the figures one can realize that the imprints of the delay time embedded in the projected trajectory is completely ironed out due to the presence of delay time modulation.

11.2.2 Length of Polygon Line

Next, to calculate the length of polygon line [11], the trajectory in $(x, x_{\hat{\tau}}, \dot{x})$ space is restricted to a two dimensional surface. The restriction in dimension is effected by intersecting the projected trajectory with a surface $k(x, x_{\hat{\tau}}, \dot{x}) = 0$. Consequently the number of times the trajectory traverses the surface and the corresponding intersection points can be calculated. One then orders the points with respect to the values of $x_{\hat{\tau}}$, and a simple measure for the alignment of the points is the length L of polygon line connecting all the ordered points. Figure 11.2a shows the length of polygon line L with constant delay $\tau_0 = 10$, where the local minima correspond to the delay time of the system we have considered. Figure 11.2b shows length of polygon line L with delay time modulation where there is no remanance of information about delay time from the trajectory, whereas Fig. 11.2c is plotted for $\tau_0 = 100$, $\tau_a = 0$, to show that the imprints of delay time carved out in the trajectory vanishes in Fig. 11.2b only due to the delay time modulation and not because of large delay.

11.2.3 Average Mutual Information

As a final example, we will calculate average mutual information defined by (see for example, [1, 12, 13] and references therein)

$$I(\hat{\tau}) = \sum_{x(n),x(n+\hat{\tau})} P(x(n), x(n + \hat{\tau})) \times \log_2 \left[\frac{P(x(n), x(n + \hat{\tau}))}{P(x(n))P(x(n + \hat{\tau}))} \right], \quad (11.2)$$

where $P(x(n), x(n + \hat{\tau}))$ is the joint probability density for measurements in the chaotic time series $X = (x(1), x(2), ..., x(m))$ and in the constructed delay time series $X_{\hat{\tau}} = (x(1 + \hat{\tau}), x(2 + \hat{\tau}), ..., x(m + \hat{\tau}))$ by varying $\hat{\tau}$, resulting in values $x(n)$ and $x(n + \hat{\tau})$. $P(x(n))$ and $P(x(n + \hat{\tau}))$ are the individual probability densities for the measurements of X and $X_{\hat{\tau}}$. Figure 11.3 shows the average mutual information for the cases of constant delay time with $\tau_0 = 10$, $\tau_a = 0$ (Fig. 11.3a) and with delay time modulation (Fig. 11.3b). Figure 11.3c is plotted for $\tau_0 = 100$ and $\tau_a = 0$ to show that the absence of local peaks in Fig. 11.3b is due to delay time modulation and not because of large delay. For fixed delay time the average

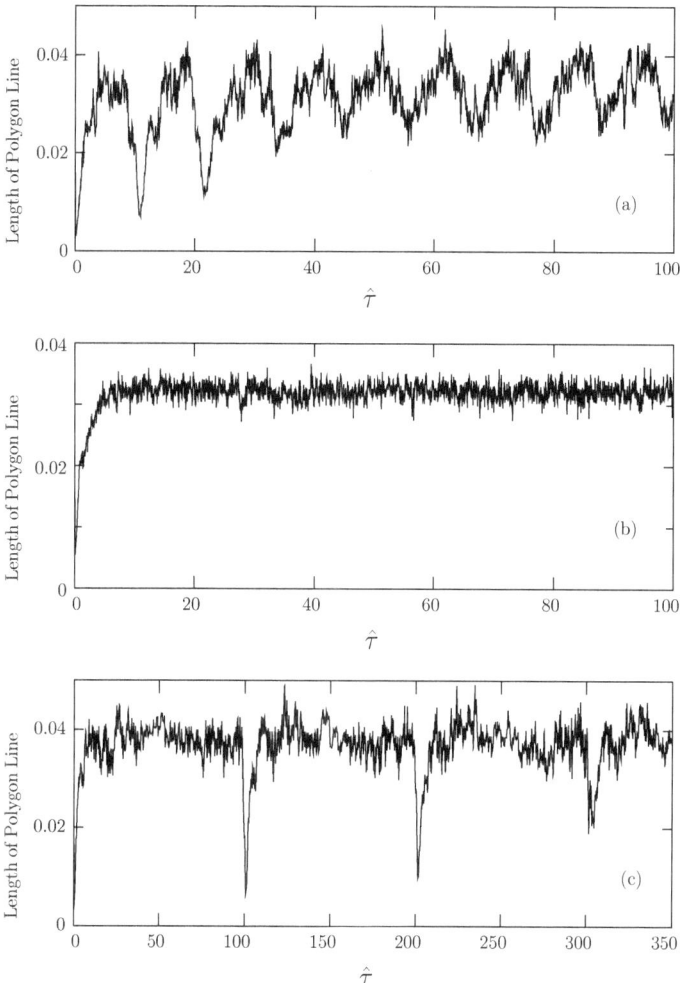

Fig. 11.2 Length of polygon line as a function of delay time $\hat{\tau}$ (of the delayed time series $x_{\hat{\tau}}$). (**a**) with constant delay $\tau_0 = 10(\tau_a = 0)$, (**b**) with $\tau_0 = 10$ and $\tau_a = 100$ and (**c**) with large constant delay $\tau_0 = 100(\tau_a = 0)$

mutual information shows local peaks at the time-delay $\hat{\tau} = \tau_0$ (or at multiples of it, $\hat{\tau} = n\tau_0$) of the system, whereas for the case of delay time modulation the average mutual information has no such peaks to identify the delay time of the delayed system.

One can also obtain similar results with other measures such as autocorrelation function, onestep prediction error and average fitting error [10, 11, 14, 15]. However, we are not presenting these results here for lack of space. In order to perform the phase space reconstruction, the first step is to find out the delay time for the projected

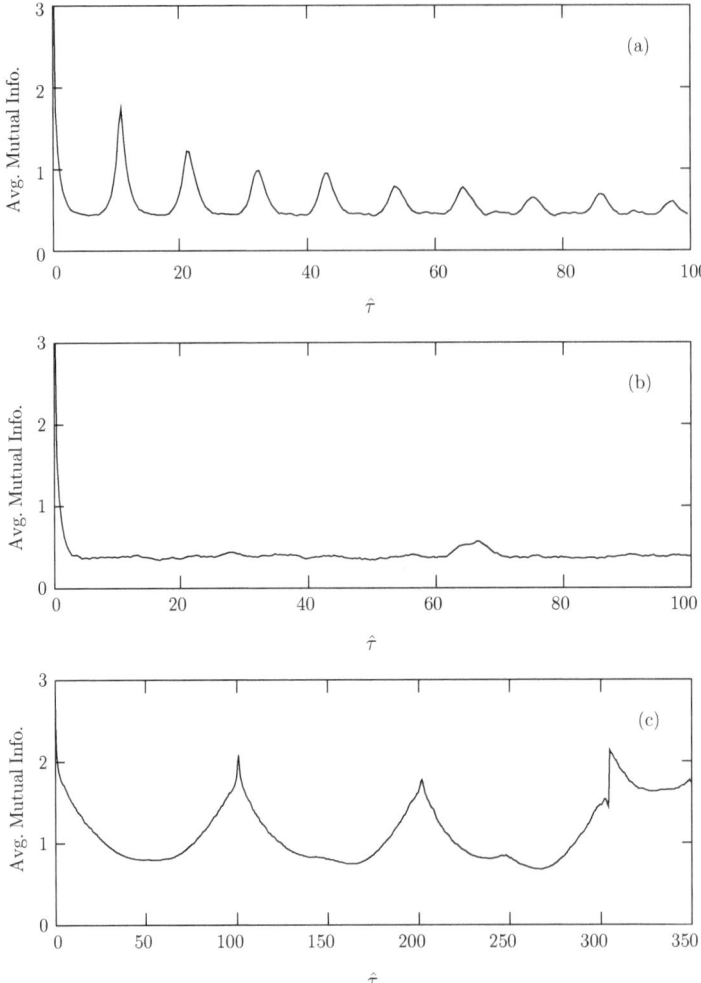

Fig. 11.3 Average mutual information as a function of delay time $\hat{\tau}$ (of delayed time series $x_{\hat{\tau}}$). (**a**) with constant delay $\tau_0 = 10$, $\tau_a = 0$, (**b**) with delay time modulation of the form (11.1) with $\tau_0 = 10$, $\tau_a = 100$ and (**c**) with large constant delay $\tau_0 = 100$, $\tau_a = 0$

trajectories. By introducing the delay time modulation the imprints of delay time in the projected trajectory is completely removed as seen above for the present system, inhibiting any possibility of phase space reconstruction. This is essentially a consequence of the fact that when the delay time is modulated by the fully rectified sine term, the delay time effectively gets increased in which case the number of positive Lyapunov exponents also increases (as noted in Fig. 11.2 in [16]). Consequently, the study of chaos synchronization in a system of such coupled delay time modulated oscillators will be of considerable interest.

11.3 Coupled System and Stability Condition in the Presence of Delay Time Modulation

Now let us consider the following unidirectionally coupled drive $x_1(t)$ and response $x_2(t)$ systems with two different modulated time-delays $\tau_1(t)$ and $\tau_2(t)$ as feedback and coupling time-delays, respectively (hereafter we write $\tau_1(t)$ and $\tau_2(t)$ simply as τ_1 and τ_2, respectively),

$$\dot{x}_1(t) = -ax_1(t) + b_1 f(x_1(t - \tau_1)), \tag{11.3a}$$

$$\dot{x}_2(t) = -ax_2(t) + b_2 f(x_2(t - \tau_1)) + b_3 f(x_1(t - \tau_2)), \tag{11.3b}$$

where b_1, b_2 and b_3 are constants, $a > 0$, and $f(x)$ is of the same form as in Eq. (3.2) with

$$\tau_1 = \tau_{10} + \tau_{1a} |\sin(\omega_1 t)|, \tag{11.4a}$$

$$\tau_2 = \tau_{20} + \tau_{2a} |\sin(\omega_2 t)|, \tag{11.4b}$$

where τ_{10} and τ_{20} are the zero frequency components of feedback delay and coupling delay, τ_{1a} and τ_{2a} are the amplitudes of the time dependent components of τ_1 and τ_2, respectively, and ω_1/π and ω_2/π are the corresponding frequencies of their modulations.

Now we can deduce the stability condition for synchronization of the two coupled time-delay systems, Eqs. (11.3a) and (11.3b), in the presence of the delay coupling $b_3 f(x_1(t - \tau_2))$ with time delay modulation in both the feedback delay and coupling delay. The time evolution of the difference system with the state variable $\Delta = x_{1\tau_2 - \tau_1} - x_2$, where $x_{1\tau_2 - \tau_1} = x_1(t - (\tau_2 - \tau_1))$, can be written for small values of Δ by using the evolution equations (11.3) as

$$\dot{\Delta} = -a\Delta + (b_2 + b_3 - b_1) f(x_1(t - \tau_2)) + b_2 f'(x_1(t - \tau_2))\Delta_{\tau_1}, \quad \Delta_\tau = \Delta(t - \tau). \tag{11.5}$$

Then $\Delta = 0$ corresponds to anticipatory synchronization when $\tau_2 < \tau_1$, identical or complete synchronization for $\tau_2 = \tau_1$ and lag synchronization when $\tau_2 > \tau_1$. In order to study the stability of the synchronization manifold as in the case of constant time delay [7], we choose the parametric condition,

$$b_1 = b_2 + b_3, \tag{11.6}$$

so that the evolution equation for the difference system Δ becomes

$$\dot{\Delta} = -a\Delta + b_2 f'(x_1(t - \tau_2))\Delta_{\tau_1}. \tag{11.7}$$

Note that here the coefficient in front of the Δ_{τ_1} term is a function of time t. In any case, the synchronization manifold is locally attracting if the origin of this equation is stable. Following again the Krasovskii-Lyapunov functional approach [17, 18], we define a positive definite Lyapunov functional of the form

$$V(t) = \frac{1}{2}\Delta^2 + \mu \int_{-\tau_1(t)}^{0} \Delta^2(t+\theta)d\theta, \tag{11.8}$$

where μ is an arbitrary positive parameter, $\mu > 0$. Note that $V(t)$ approaches zero as $\Delta \to 0$.

To estimate a sufficient condition for the stability of the solution $\Delta = 0$, we require the derivative of the functional $V(t)$ along the trajectory of Eq. (11.7),

$$\frac{dV}{dt} = -a\Delta^2 + b_2 f'(x_1(t-\tau_2))\Delta\Delta_{\tau_1} + \mu\left[\Delta_{\tau_1}^2 \tau_1' + \Delta^2 - \Delta_{\tau_1}^2\right], \tag{11.9}$$

to be negative. Note that in the case of constant modulation $\tau_1' = \frac{d\tau_1}{dt}$ vanishes. The above equation can be rewritten as

$$\frac{dV}{dt} = -\mu\Delta^2 \Gamma(X,\mu), \tag{11.10}$$

where $\Gamma = \left[((a-\mu)/\mu) - (b_2 f'(x_1(t-\tau_2))/\mu)X + X^2/(1-\tau_1')\right]$, $X = \Delta_{\tau_1}/\Delta$. In order to show that $\frac{dV}{dt} < 0$ for all Δ and Δ_τ and so for all X, it is sufficient to show that $\Gamma_{min} > 0$. One can easily check that the absolute minimum of Γ occurs at $X = b_2 f'(x_1(t-\tau_2))/2\mu(1-\tau_1')$ with $\Gamma_{min} = [4\mu(a-\mu) (1-\tau_1') - b_2^2 f'^2(x_1(t-\tau_2))]/4\mu^2(1-\tau_1')$. Consequently, we have the condition for stability as

$$a > \frac{b_2^2 f'^2(x_1(t-\tau_2))}{4\mu(1-\tau_1')} + \mu = \Phi(\mu). \tag{11.11}$$

Again $\Phi(\mu)$ as a function of μ for a given $f'(x)$ has an absolute minimum at $\mu = \left(\left|\frac{b_2 f'(x_1(t-\tau_2))}{2\sqrt{(1-\tau_1')}}\right|\right)$ with $\Phi_{min} = \left|\frac{b_2 f'(x_1(t-\tau_2))}{\sqrt{(1-\tau_1')}}\right|$. Since $\Phi \geq \Phi_{min} = \left|\frac{b_2 f'(x_1(t-\tau_2))}{\sqrt{(1-\tau_1')}}\right|$, from the inequality (11.11), it turns out that a sufficient condition for asymptotic stability is

$$a > \left|\frac{b_2 f'(x_1(t-\tau_2))}{\sqrt{(1-\tau_1')}}\right| \tag{11.12}$$

along with the condition (11.6) on the parameters b_1, b_2 and b_3.

Now from the form of the piecewise linear function $f(x)$ given by Eq. (3.2), we have,

$$\left| f'(x_1(t - \tau_2)) \right| = \begin{cases} 1.5, & 0.8 \le |x_1| \le \frac{4}{3} \\ 1.0, & |x_1| < 0.8 \end{cases} \qquad (11.13)$$

Consequently the stability condition (11.12) becomes $a > 1.5 \left| \dfrac{b_2}{\sqrt{(1-\tau_1')}} \right| >$ $\left| \dfrac{b_2}{\sqrt{(1-\tau_1')}} \right|$ along with the parametric restriction $b_1 = b_2 + b_3$.

Thus one can take $a > \left| \dfrac{b_2}{\sqrt{(1-\tau_1')}} \right|$ as a less stringent condition for (11.12) to be valid, while

$$a > 1.5 \left| \frac{b_2}{\sqrt{\left(1 - \tau_1'\right)}} \right| \qquad (11.14)$$

can be considered as the most general condition specified by (11.12) for asymptotic stability of the synchronized state $\Delta = 0$. The condition (11.14) indeed corresponds to the stability condition for exact anticipatory/lag as well as exact complete synchronizations for a given value of the coupling delay τ_2 in a general sense. It may be noted that the stability condition (11.14) is valid irrespective of the nature of the coupling delay, that is whether it is constant or modulated. However, when the feed back delay τ_1 is constant the condition (11.14) reduces to $a > 1.5|b_2|$ as discussed in Chap. 8, see [7]. In the following, we will consider both the possibilities of constant ($\tau_2 = \tau_{20}$) and periodically modulated ($\tau_2 = \tau_{20} + \tau_{2a} |\sin(\omega_2 t)|$) coupling delays with a periodically modulated feedback delay ($\tau_1 = \tau_{10} + \tau_{1a} |\sin(\omega_1 t)|$). We demonstrate through detailed numerical analysis that there exists oscillating synchronization that oscillates between anticipatory, complete and lag synchronizations for the case of constant coupling delay $\tau_2 = \tau_{20}$. Intermittent anticipatory/lag and complete synchronizations are shown to exist for the case of coupling delay with delay time modulation $\tau_2 = \tau_{20} + \tau_{2a} |\sin(\omega_2 t)|$, when $\tau_{2a} = \tau_{1a}$ and $\omega_1 = \omega_2$. For $\tau_{2a} \neq \tau_{1a}$ and $\omega_1 \neq \omega_2$, more complicated oscillating type synchronizations occur.

11.4 Oscillating Synchronization

To start with, we consider the case of constant coupling delay, $\tau_2 = \tau_{20}$, and show that there exists oscillating synchronization that oscillates between anticipatory, complete and lag synchronizations as a function of time for suitable range of parameters.

Now we will choose the delay time modulation in the form (11.4a) for the feedback delay $\tau_1 (= \tau_{10} + \tau_{1a} |\sin(\omega_1 t)|)$ with $\tau_{10} = 10$, $\tau_{1a} = 90$ and $\omega_1 = 10^{-4}$. We have fixed the value of $\tau_{2a} = 0$ in (11.4b), so that the coupling delay becomes

constant $\tau_2 = \tau_{20} = 45$ with the parameters $a = 1, b_1 = 1.2$ in Eq. (11.3) and
the values of b_2 and b_3 are chosen according to the parametric restriction (11.6)
depending upon the stability condition to be satisfied. For the chosen values of τ_{10}
and τ_{1a}, one can find that τ_1 oscillates between $(\tau_1(t) = \tau_{10} + \tau_{1a}|\sin(\omega_1 t)| =
10 + 90|\sin(\omega_1 t)|)$ the values 10 and 100. With the chosen value of constant cou-
pling delay $\tau_2 = 45$ and time dependent feedback delay τ_1, as time evolves one finds
that the feedback delay $\tau_1(t)$ is lesser than the value of constant coupling delay τ_2
initially for some time (in which case $\tau(t) = \tau_2 - \tau_1(t) > 0$, so that there exists lag
synchronization $x_1(t - \tau(t)) = x_2(t)$ with varying lag time $\tau(t) = \tau_2 - \tau_1(t)$). As
time evolves, $\tau_1(t)$ increases eventually and it approaches $\tau_1 = 45$ at a certain later
time $(T = \pi/\omega_1)$, where $\tau(t) = \tau_2 - \tau_1(t) = 0$, so that $x_1(t) = x_2(t)$ and a complete
synchronization occurs at a specific value of time. As $\tau_1(t)$ increases further above
the value of $\tau_2 = 45$, the delay time $\tau(t)$ becomes negative, $\tau(t) = \tau_2 - \tau_1(t) < 0$
with $x_1(t - \tau(t)) = x_2(t))$ and there exists anticipatory synchronization with varying
anticipating time $\tau(t) = \tau_2 - \tau_1(t)$. This anticipatory synchronization continues till

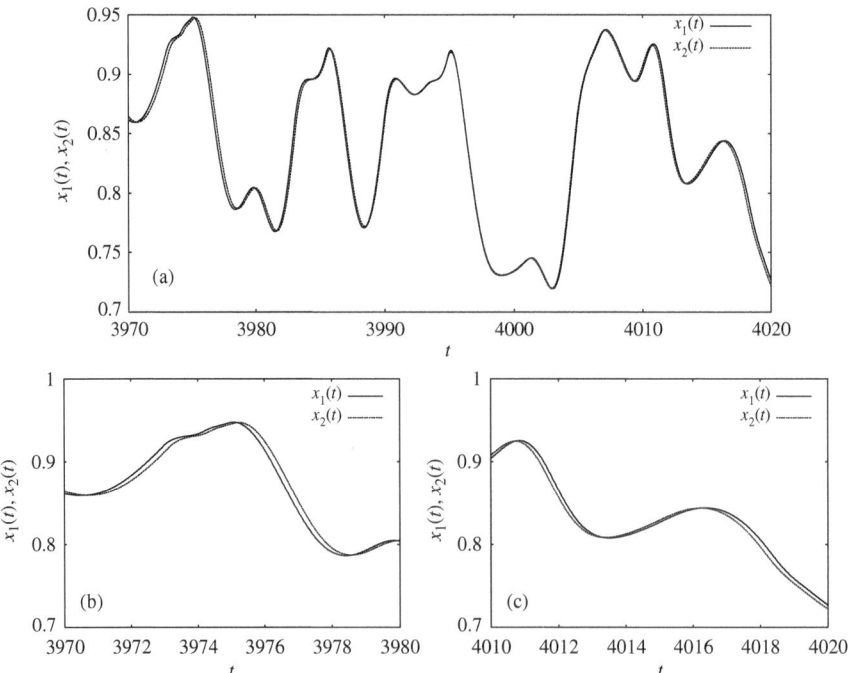

Fig. 11.4 Oscillating synchronization for the constant coupling delay $\tau_2 = 45$ with time dependent
feedback delay of the form (11.4a) with $\tau_{10} = 10$, $\tau_{1a} = 90$ and $\omega = 10^{-4}$. (**a**) Oscillating from
lag to anticipatory synchronization via complete synchronization in the region $t \in (3970, 4020)$,
(**b**) enlarged figure showing lag synchronization in the region $t \in (3970, 3980)$ and (**c**) enlarged
figure showing anticipatory synchronization in the region $t \in (4010, 4020)$

the value of time dependent feed back delay $\tau_1(t)$ decreases to approach the value of the constant coupling delay $\tau_2 = 10$ after reaching its maximum value of 100. Therefore as time evolves there is oscillation between lag, complete and anticipatory synchronizations with time dependent anticipating and lag times.

Figure 11.4a shows the evolution of the drive $x_1(t)$ and the response $x_2(t)$ at the transition between lag to anticipatory synchronization via complete synchronization for the value of $b_2 = 0.1$, where the general stability condition (11.14) is satisfied, whereas Figs. 11.4b, c are the enlarged part of lag and anticipatory synchronization regimes in Fig. 11.4a, respectively. Figure 11.5a shows the evolution of the drive $x_1(t)$ and the response $x_2(t)$ at the next transition between anticipatory to lag via complete synchronization and the enlarged part of Fig. 11.5a in the anticipatory and lag synchronization regimes are shown in Figs. 11.5b, c, respectively. In Figs. 11.6 a, b, the difference signals $x_1(t - \tau) - x_2(t), \tau > 0$ and $x_1(t - \tau) - x_2(t), \tau < 0$ are plotted respectively for the value of parameters satisfying the general stability condition corresponding to the Fig. 11.4, confirming the transition between lag to anticipatory synchronization. Thus as a consequence of delay time modulation there

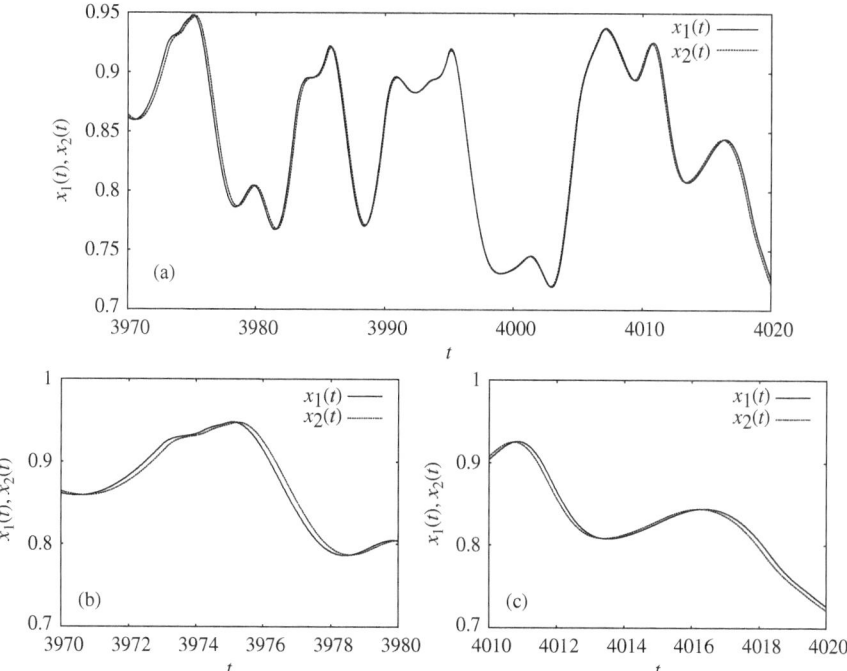

Fig. 11.5 Oscillating synchronization for the constant coupling delay $\tau_2 = 45$ with time dependent feedback delay of the form (11.4a) with $\tau_{10} = 10$, $\tau_{1a} = 90$ and $\omega = 10^{-4}$. (**a**) Oscillating from anticipatory to lag synchronization at the next transition in the region $t \in (27400, 27450)$, (**b**) enlarged figure showing anticipatory synchronization in the region $t \in (27400, 27410)$ and (**c**) enlarged figure showing lag synchronization in the region $t \in (27440, 27450)$

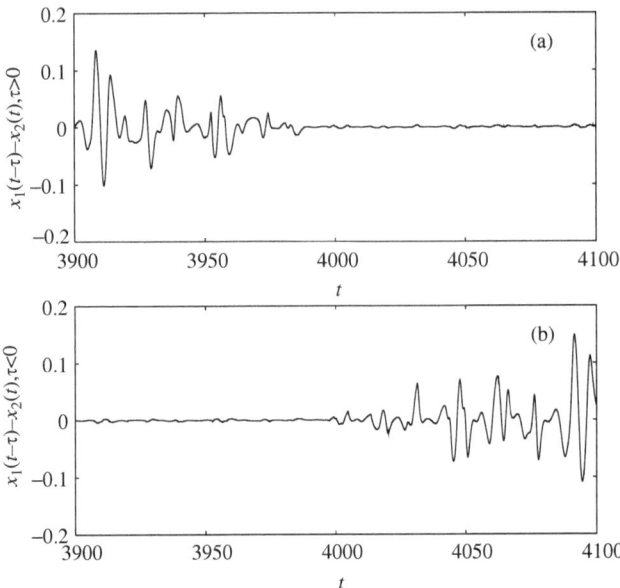

Fig. 11.6 (**a**) Difference between $x_1(t - \tau), \tau > 0$ and $x_2(t)$, showing lag synchronization for certain time and (**b**) difference between $x_1(t - \tau), \tau < 0$ and $x_2(t)$, showing anticipatory synchronization for the following period of time for $b_2 = 0.1$ satisfying the general stability condition (11.14). Note that complete synchronization occurs in the transition regime

exists a new type of oscillating synchronization that oscillates between anticipatory, complete and lag synchronizations with varying anticipating and lag times.

11.5 Intermittent Anticipatory Synchronization

Now we consider the coupled time-delay system (11.3) with delay time modulation of the form (11.4) in both the feedback and coupling delays for further studies. We have fixed the values of the parameters as $a = 1, b_1 = 1.2, \tau_{1a} = \tau_{2a} = 90, \omega_1 = \omega_2 = 10^{-5}$ (identical modulations) and the values of b_2 and b_3 are chosen according to the parametric restriction $b_1 = b_2 + b_3$, depending upon the stability condition to be satisfied. For τ_1, the zero frequency component of amplitude is fixed as $\tau_{10} = 10$ and for τ_2, it is fixed as $\tau_{20} = 5$, so that a constant difference is maintained between the feedback and the coupling time delays throughout the time evolution. With the coupling delay $\tau_2(= 5 + 90 \left|\sin(10^{-5}t)\right|)$ being less than the feedback delay $\tau_1(= 10 + 90 \left|\sin(10^{-5}t)\right|)$, that is $\tau_2(t) < \tau_1(t)$, the value of the anticipating time $\tau = \tau_2 - \tau_1$ turns out to be negative such that the relation between drive $x_1(t)$ and the response $x_2(t)$ now becomes $x_1(t - \tau) = x_2(t), \tau < 0$, demonstrating anticipatory synchronization, provided the stability condition (11.14) is satisfied with the parametric restriction specified by Eq. (11.6).

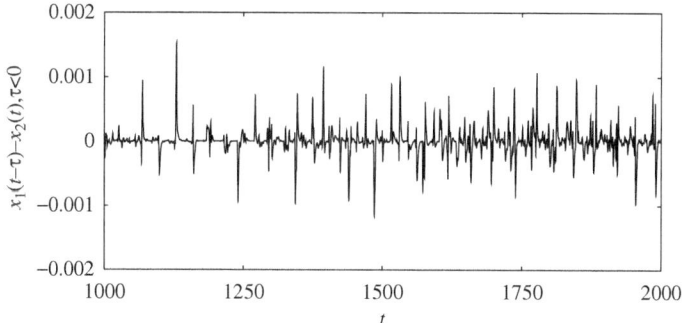

Fig. 11.7 The time series $x_1(t - \tau) - x_2(t)$, $\tau < 0$, for $b_2 = 0.7$ and $b_3 = 0.5$ (so that the less stringent condition $a > \left| b_2 / \sqrt{1 - \tau_1'} \right|$ is satisfied while (11.14) is violated) corresponding to intermittent anticipatory synchronization with the amplitude of the laminar phase approximately zero

Now let us choose the parameter b_2 as the control parameter, whose value determines the stability condition given by Eq. (11.12).

1. For $b_2 = 0.7$, $1.5 \left| \dfrac{b_2}{\sqrt{1 - \tau_1'}} \right| > a > \left| \dfrac{b_2}{\sqrt{1 - \tau_1'}} \right|$, the less stringent condition is satisfied with $\sqrt{1 - \tau_1'} \approx 1$ for the chosen values of ω and τ_a. One can observe intermittent anticipatory synchronization as shown in Fig. 11.7, exhibiting typical features of *on-off* intermittency [8, 9] with the *off* state near the laminar phase and the *on* state showing a random burst. For this value of b_2 the amplitude of the laminar phase corresponding to the synchronized state is approximately zero (of the order 10^{-5}).

2. For $b_2 = 0.1$, $a > 1.5 \left| \dfrac{b_2}{\sqrt{1 - \tau_1'}} \right| > \left| \dfrac{b_2}{\sqrt{1 - \tau_1'}} \right|$, the general stability condition (11.14) is satisfied and correspondingly the numerical analysis reveals that here the intermittent anticipatory synchronization is such that the amplitude of the laminar phase corresponding to the synchronized state is exactly zero (in the sense that the difference $\Delta = x_1(t - \tau) - x_2(t)$, $\tau < 0$ is of the order 10^{-16} in the laminar phases) as shown in Fig. 11.8.

To analyze the statistical features associated with the intermittent nature in Fig. 11.8 for the value of $b_2 = 0.1$, we have calculated the distribution of laminar phases $\Lambda(t)$ with the amplitude less than a threshold value $\Delta < 10^{-10}$ and we have observed a universal asymptotic $-\frac{3}{2}$ power law distribution as shown in Fig. 11.9, which is quite typical for on-off intermittency [8, 9]. One can also find a similar power law distribution for the value of $b_2 = 0.7$ discussed above but now with a *larger* threshold value ($\Delta < 10^{-4}$) of the laminar region.

Now we use the similarity function $S_a(\tau)$ (8.13) to characterize the existence of anticipatory synchronization. Figure 11.10 shows the similarity function $S_a(\tau)$ as a function of the difference between the feedback and the coupling delays, $\tau = \tau_2 - \tau_1$

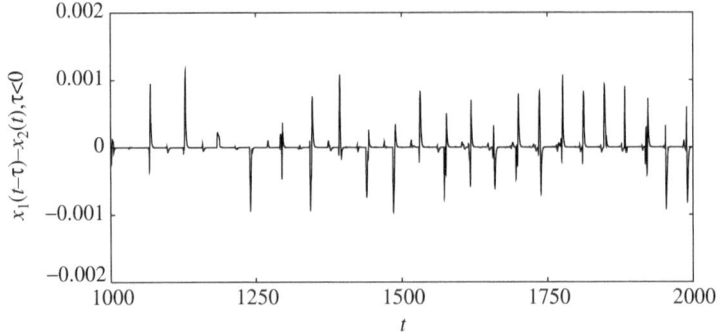

Fig. 11.8 The time series $x_1(t-\tau) - x_2(t)$, $\tau < 0$, for $b_2 = 0.1$ and $b_3 = 1.1$. Here the general stability criterion (11.14) is satisfied corresponding to intermittent anticipatory synchronization with the amplitude of the laminar phase exactly zero

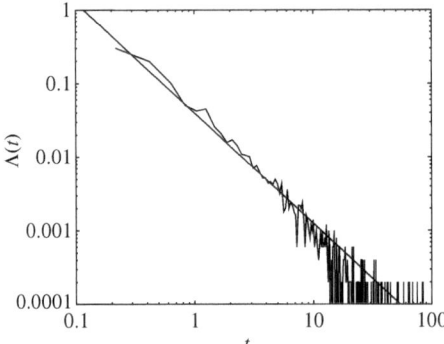

Fig. 11.9 The statistical distribution of laminar phase satisfying $-\frac{3}{2}$ power law scaling for $b_2 = 0.1$ and $b_3 = 1.1$, where the general stability criterion (11.14) is satisfied

for three different values of b_2, the parameter whose value determines the stability condition (11.12). In Fig. 11.10, the Curve 3 is plotted for the value of $b_2 = 1.1$, $(1.5 \left| \frac{b_2}{\sqrt{1-\tau_1'}} \right| > \left| \frac{b_2}{\sqrt{1-\tau_1'}} \right| > a)$, where both the less stringent condition and the most general condition are violated. From the curve 3 one can find that the minimum value of $S_a(\tau)$ is greater than zero for all values of τ, resulting in the lack of exact time shift (anticipating time) between the drive and the response signals. On the other hand the curve 2 corresponds to the value of $b_2 = 0.7$ such that the less stringent condition is satisfied while the general stability criterion (11.14) is violated as seen above. Curve 2 shows that the minimum of similarity function $S_a(\tau)$ is approximately zero (of the order 10^{-4}) for $\tau < 0$, as may be seen in the inset of Fig. 11.10, indicating the existence of intermittent anticipatory synchronization with the amplitude of the laminar phases of the difference signal $\Delta = x_1(t-\tau) - x_2(t)$, $\tau < 0$, being approximately zero ($< 10^{-5}$). On the other hand, the curve 1 is plotted for $b_2 = 0.1$, satisfying the general stability criterion Eq. (11.14), which shows that the

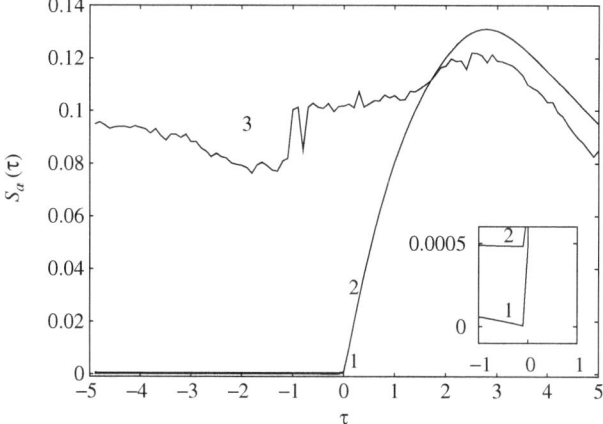

Fig. 11.10 Similarity function for intermittent anticipatory synchronization $S_a(\tau)$ for different values of b_2, the other system parameters are $a = 1.0, b_1 = 1.2$ and $\omega = 10^{-5}$. (*Curve 1:* $b_2 = 0.1, b_3 = 1.1$, *Curve 2*: $b_2 = 0.7, b_3 = 0.5$ and *Curve 3*: $b_2 = 1.1, b_3 = 0.1$)

minimum of similarity function is much closer to zero (of order 10^{-8}), $\tau < 0$, indicating that there exists an intermittent anticipatory synchronization with the amplitude of the laminar phase of the difference signal becoming exactly zero with the anticipating time equal to the difference between the two time delays $\tau = \tau_2 - \tau_1$.

Next, by reducing the value of the modulational frequencies $\omega = \omega_1 = \omega_2$ further, we find that the lengths of the laminar phases increase gradually with a corresponding decrease in the number of turbulent phases. Finally at an appropriate value of the modulational frequency all the turbulent phases disappear and there exists only exact anticipatory synchronization without any intermittent bursts provided the exact stability condition is satisfied. Correspondingly the similarity function $S_a(\tau)$ becomes zero exactly (which is of the order 10^{-16}) for $\tau < 0$ in this case, as shown in Chap. 8, also see [7].

11.6 Complete Synchronization

Complete synchronization follows the anticipatory synchronization when the value of the coupling time-delay τ_2 equals the feedback time-delay τ_1, that is $\tau_2 = \tau_1$, where the anticipating time becomes $\tau = \tau_2 - \tau_1 = 0$. Here also, the same stability criterion Eq. (11.14) holds good with the same parametric restriction specified by (11.6). In this case of complete synchronization ($\tau_2 = \tau_1$), the delay time modulation does not induce any intermittent nature in the dynamical behavior of the coupled systems for any value of the modulational frequency ($\omega_1 = \omega_2$) as inferred from Eq. (11.4). Figure 11.11a shows as an illustration the plot of $x_1(t)$ Vs $x_2(t)$ for the values of $b_2 = 0.7$ and $\omega_1 = \omega_2 = 10^{-5}$, such that the less stringent condition is satisfied and the general stability criterion (11.14) is violated. The plot shows

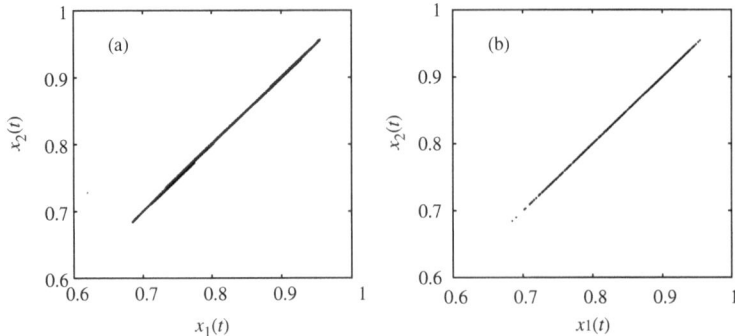

Fig. 11.11 Complete synchronization between $x_1(t)$ Vs $x_2(t)$ when $\tau_{20} = \tau_{10}$. (**a**) Approximate complete synchronization for $b_2 = 0.7$ and (**b**) Exact complete synchronization for $b_2 = 0.1$

small deviations from the localized diagonal line implying an approximate synchronization, whereas Fig. 11.11b shows an entirely localized sharp diagonal line for the value of $b_2 = 0.1$, where the general stability condition (11.14) is satisfied, indicating the complete synchronization.

11.7 Intermittent Lag Synchronization

When the value of the coupling delay τ_2 is increased above the value of the feedback delay $\tau_1 (\tau_2 > \tau_1)$, then the value of the retarded time $\tau = \tau_2 - \tau_1$ turns out to be positive such that the relation between the drive $x_1(t)$ and the response $x_2(t)$ now becomes $x_1(t - \tau) = x_2(t)$, $\tau > 0$, depicting the existence of lag synchronization, provided the general stability condition (11.14) is satisfied along with the parametric condition (11.6).

We have fixed the same values for all the parameters as in the case of intermittent anticipatory synchronization except for the zero frequency component τ_{20} of coupling delay τ_2 which is now fixed at $\tau_{20} = 15$. Figure 11.12 shows the intermittent lag synchronization for the value of $b_2 = 0.7$, in which case only the less stringent stability condition is satisfied, where the laminar phase has an amplitude which is nearly zero (of the order 10^{-5}). Figure 11.13 shows intermittent lag synchronization for the value of $b_2 = 0.1$, where the amplitude of the laminar phase vanishes exactly. In the later case the most general stability criterion (11.14) is satisfied. The statistical behavior associated with the intermittent nature in this case of intermittent lag synchronization is also characterized by the probability distribution of laminar phases having amplitudes less than a threshold value $\Delta < 10^{-10}$ corresponding to a universal asymptotic $-\frac{3}{2}$ power law distribution as shown in the Fig. 11.14.

The figure shows the probability distribution $\Lambda(t)$ of intermittent lag synchronization for the value of $b_2 = 0.1$. One can also verify that the intermittent lag

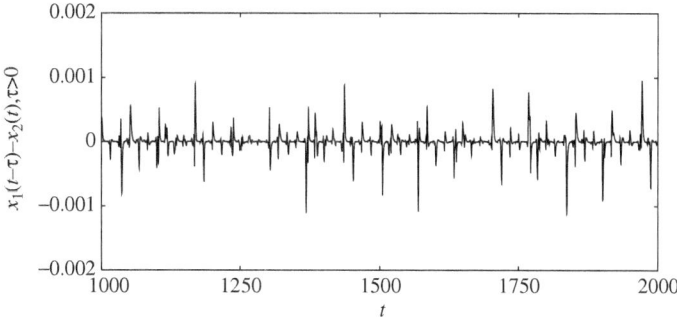

Fig. 11.12 The time series $x_1(t - \tau) - x_2(t)$, $\tau > 0$, for $b_2 = 0.7$ and $b_3 = 0.5$ (so that the less stringent condition $a > \left| b_2/\sqrt{1 - \tau_1'} \right|$ is satisfied while (11.14) is violated) corresponding to intermittent lag synchronization with the amplitude of the laminar phase approximately zero

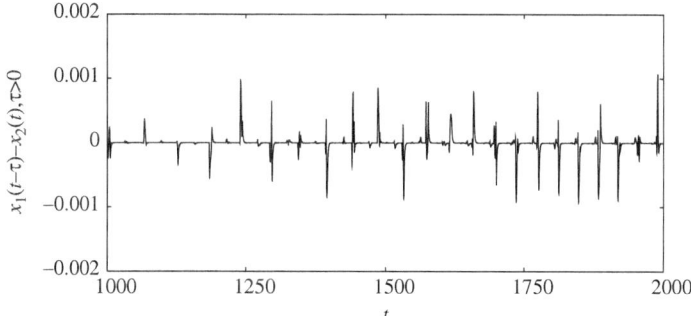

Fig. 11.13 The time series $x_1(t - \tau) - x_2(t)$, $\tau > 0$, for $b_2 = 0.1$ and $b_3 = 1.1$. Here the general stability criterion (11.14) is satisfied corresponding to intermittent lag synchronization with the amplitude of the laminar phase exactly zero

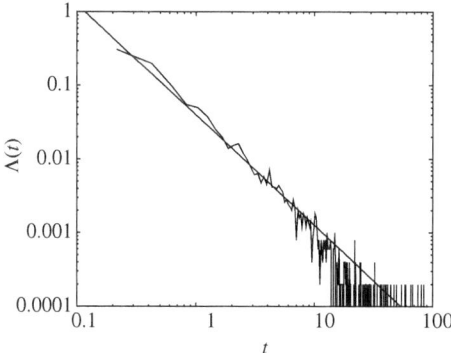

Fig. 11.14 The statistical distribution of laminar phase satisfying $-\frac{3}{2}$ power law scaling for $b_2 = 0.1$ and $b_3 = 1.1$, where the general stability criterion (11.14) is satisfied

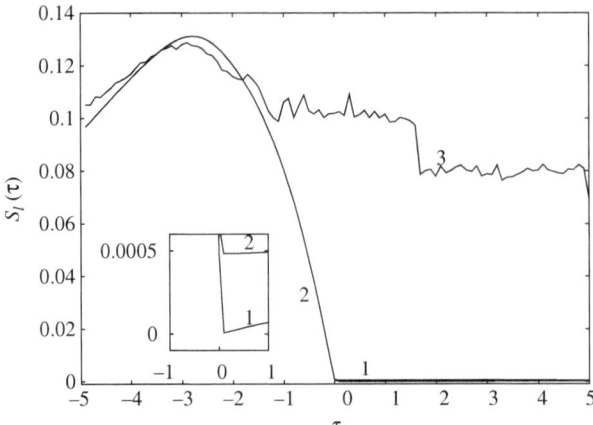

Fig. 11.15 Similarity function for intermittent lag synchronization $S_l(\tau)$ for different values of b_2, the other system parameters are $a = 1.0$, $b_1 = 1.2$ and $\omega = 10^{-5}$ (*Curve 1*: $b_2 = 0.1, b_3 = 1.1$, *Curve 2*: $b_2 = 0.7, b_3 = 0.5$ and *Curve 3*: $b_2 = 1.1, b_3 = 1.0$)

synchronization for the value of $b_2 = 0.7$ has also similar power law distribution for *larger* threshold value ($\Delta < 10^{-4}$)of amplitude of the laminar phases.

The existence of intermittent lag synchronization is also characterized by a similarity function $S_l(\tau)$ defined in Eq. (8.12). Figure 11.15 shows the similarity function $S_l(\tau)$ for intermittent lag synchronization as a function of the retarded time $\tau = \tau_2 - \tau_1$. Curve 3 is plotted for the value of $b_2 = 1.1$ (which is greater than both $a\sqrt{1 - \tau_1'}$ and $a\sqrt{1 - \tau_1'}/1.5$), where the minimum of similarity function $S_l(\tau)$ occurs at a finite value of $S_l(\tau) > 0$ and hence there is a lack of lag synchronization between the drive and the response signals indicating asynchronization. Curve 2 corresponds to the value of $b_2 = 0.7$, (which is less than $a\sqrt{1 - \tau_1'}$ but greater than $a\sqrt{1 - \tau_1'}/1.5$), where the minimum of similarity function $S_l(\tau)$ is approximately zero (of the order of 10^{-4}, as may be seen in the inset of Fig. 11.15) indicating the existence of intermittent lag synchronization with the amplitude of the laminar phase being approximately zero. However, for the value of $b_2 = 0.1$, for which the general condition (11.14) is obeyed, the minimum of similarity function for Curve 1 becomes much closer to zero (of the order 10^{-8}) which corresponds to intermittent lag synchronization with exact time shift between the two signals during the laminar phase.

Next, as in the case of intermittent anticipatory synchronization, by reducing the value of modulational frequency one can find that the lengths of the laminar phases increase with vanishing turbulent phases and finally at an appropriate value of the modulational frequency there exists exact lag synchronization without any intermittent bursts provided the exact stability condition is satisfied. Correspondingly the similarity function $S_l(\tau)$ becomes zero exactly (which is of the order 10^{-16}) for $\tau > 0$ in this case.

11.8 Complex Oscillating Synchronization

Finally, when $\tau_{1a} \neq \tau_{2a}$ or/and $\omega_1 \neq \omega_2$ the frequencies as well as amplitudes of the modulated feedback delay $\tau_1(t)(= \tau_{10} + \tau_{1a} |\sin(\omega_1 t)|)$ and the modulated coupling delay $\tau_2(t)(= \tau_{20} + \tau_{2a} |\sin(\omega_2 t)|)$ differ from each other resulting in a more complicated variation of the anticipating/lag time $\tau(t) = \tau_2(t) - \tau_1(t)$. This in turn results in the existence of more complex oscillating synchronization than the one presented in Sect. 11.4. It is clear that one can also introduce other kinds of modulations instead of periodic modulation to obtain varying forms of oscillating synchronizations.

11.9 DTM Induced Oscillating Synchronization: Mackey-Glass & Ikeda Systems

In this section, we will present brief details of the delay time induced oscillating synchronization in both the Mackey-Glass and Ikeda time-delay systems. Now, let us consider again the unidirectionally coupled systems of the form Eq. (11.3) with two different modulated time-delays $\tau_1(t)$ and $\tau_2(t)$ as feedback and coupling delays, respectively (Hereafter we will write $\tau_1(t)$ and $\tau_2(t)$ simply as τ_1 and τ_2, respectively.), We will use the same functional form (8.15) and (8.16) for the Mackey-Glass and Ikeda systems, respectively, with the modulations given by Eqs. (11.4). One can show that again the stability condition obtained in Sect. 11.3 for the existence of oscillating synchronization in the presence of delay time modulation holds good here also with appropriate function $f(x)$.

For our further discussion, we now consider a constant coupling delay, $\tau_2 = \tau_{20}$, with a modulated feedback delay τ_1, given by (11.4a), and show that there exists an oscillating synchronization that oscillates between anticipatory, complete and lag synchronizations as a function of time for suitable range of parameters. The results of simulation for the coupled Mackey-Glass and Ikeda systems are presented briefly in the following.

11.9.1 Coupled Mackey-Glass Systems

Oscillating synchronization of the Mackey-Glass system is shown in Fig. 11.16 for the value of the constant coupling delay $\tau_2 = 45$ with time dependent feedback delay of the form (11.4a) with $\tau_{10} = 10, \tau_{1a} = 90$, and $\omega = 10^{-4}$, and for the value of the coupling strength $b_3 = 0.06$ consistent with the parametric condition $b_1 = b_2 + b_3$. The other parameters are fixed at $a = 0.1$ and $b = 0.2$. Figure 11.16a shows oscillating synchronization which oscillates between lag and anticipatory synchronization via complete synchronization for $t \in (4000, 4150)$. An enlarged part of Fig. 11.16a in the range $t \in (4000, 4020)$ is shown in Fig. 11.16b, which clearly shows the existence of lag synchronization at one end. Similarly,

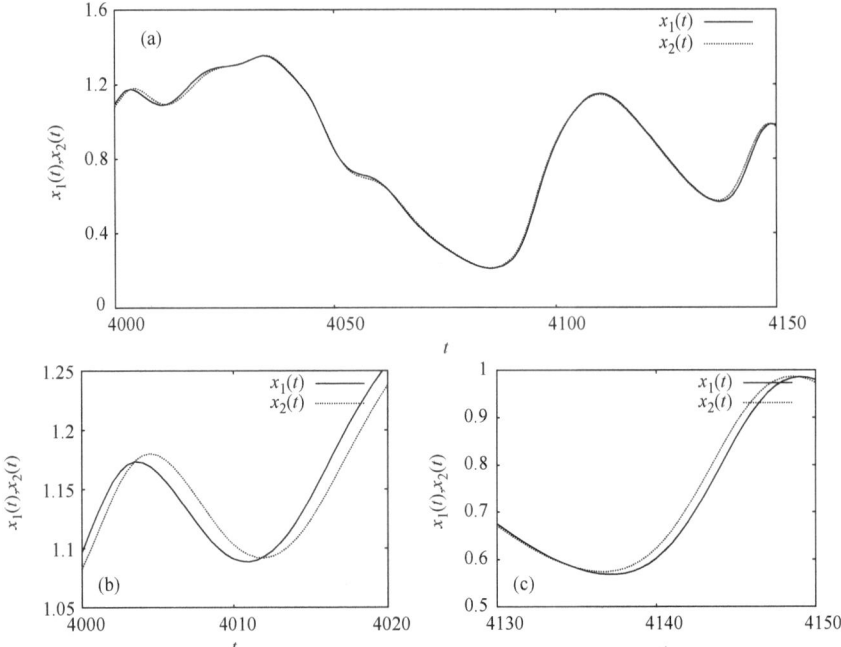

Fig. 11.16 Oscillating synchronization exhibited by the coupled Mackey-Glass system for the constant coupling delay $\tau_2 = 45$ with time dependent feedback delay of the form (11.4a) with $\tau_{10} = 10$, $\tau_{1a} = 90$ and $\omega = 10^{-4}$. The other parameters are chosen as $a = 1.0$, $b_1 = 1.2$, $b_2 = 0.1$ and $b_3 = 1.1$ (**a**) Oscillating from lag to anticipatory synchronization via complete synchronization in the region $t \in (4000, 4150)$, (**b**) lag synchronization in the range $t \in (4000, 4020)$ and (**c**) anticipatory synchronization in the range $t \in (4130, 4250)$

Fig. 11.16c in the range $t \in (4130, 4150)$ clearly depicts the existence of anticipatory synchronization at the other end of Fig. 11.16a. Oscillating synchronization which oscillates between anticipatory and lag via complete synchronization at the next transition in the range $t \in (27450, 27600)$ is shown in Fig. 11.17a. Figures 11.17 b, c are the enlarged parts of Fig. 11.17a in the ranges $t \in (27450, 27470)$ and $t \in (27580, 27600)$, respectively, which show clearly the regions of anticipatory and lag synchronizations.

11.9.2 Coupled Ikeda Systems

Next, we will point out the existence of oscillating synchronization in the coupled Ikeda system with the parameters chosen as $a = 1$, $b_1 = 20$, $b_2 = 6.5$, $b_3 = 13.5$, $\tau_2 = 2$, $\tau_{10} = 2$, $\tau_{1a} = 0.5$ and $\omega = 10^{-3}$. As the characteristic time scale of the Ikeda system is very small, the values of both the constant coupling delay and that of the time dependent feedback delay are chosen to be small when com-

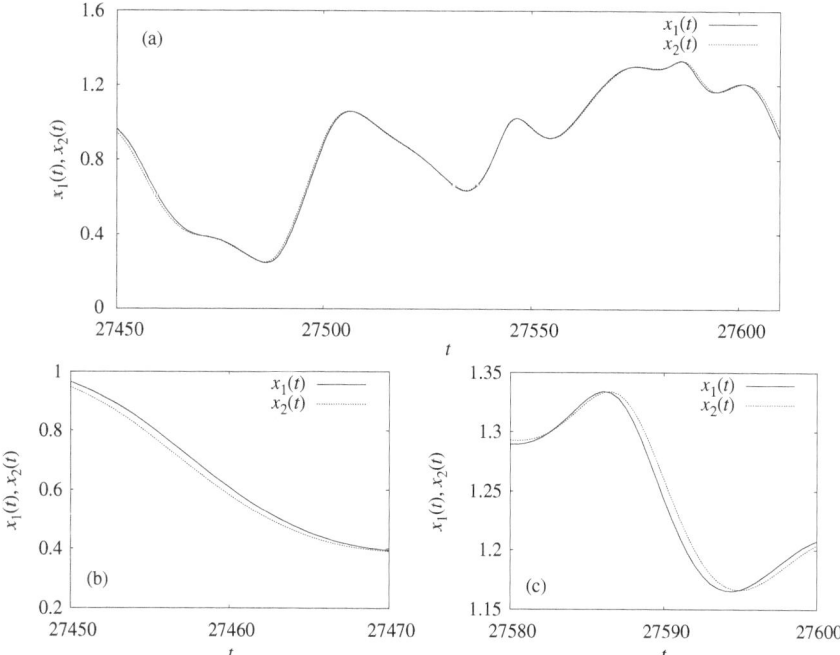

Fig. 11.17 Oscillating synchronization exhibited by the coupled Mackey-Glass system for the constant coupling delay $\tau_2 = 45$ with time dependent feedback delay of the form (11.4a) with $\tau_{10} = 10$, $\tau_{1a} = 90$ and $\omega = 10^{-4}$. (**a**) Oscillating from anticipatory to lag synchronization at the next transition in the range $t \in (27450, 27600)$, (**b**) anticipatory synchronization in the range $t \in (27450, 27470)$ and (**c**) lag synchronization in the range $t \in (27580, 27600)$

pared with the piecewise linear and Mackey-Glass systems. Also the value of the frequency of modulation is now fixed at $\omega = 10^{-3}$ for identifying the oscillating synchronization in the Ikeda system.

Oscillating synchronization for the above values of the parameters of the Ikeda system is shown in Fig. 11.18a which oscillates in the range $t \in (10240, 10300)$ between lag (Fig. 11.18b) and anticipatory (Fig. 11.18c) synchronizations via complete synchronization. Similar oscillating synchronization at the next transition is shown in Fig. 11.19a in the range $t \in (11690, 11750)$. Anticipatory and lag synchronization regimes of Fig. 11.19a are shown in the enlarged Figs. 11.19b, c, respectively.

It is also to be added that both the Mackey-Glass and Ikeda systems exhibit intermittent anticipatory synchronization for $\tau_2 < \tau_1(t)$ and intermittent lag synchronization for $\tau_2 > \tau_1(t)$ as in the case of piecewise linear time-delay system discussed in Sects. 11.5 and 11.7. This is also an extension of the results for anticipatory and lag synchronizations presented earlier for constant feedback and coupling delays as in the case of the coupled piecewise linear systems. Further characterizations can be also carried out as in Sects. 11.5 and 11.7.

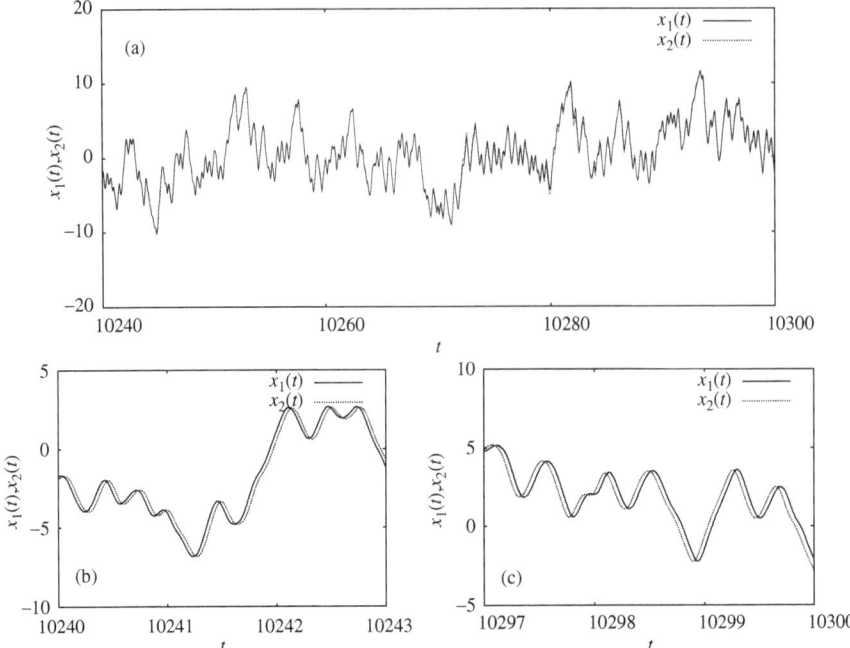

Fig. 11.18 Oscillating synchronization in the coupled Ikeda system for the constant coupling delay $\tau_2 = 2$ with time dependent feedback delay of the form (11.4a) with $\tau_{10} = 2$, $\tau_{1a} = 0.5$ and $\omega = 10^{-3}$. Other parameters are chosen as $a = 1.0$, $b_1 = 20.0$, $b_2 = 6.5$ and $b_3 = 13.4$ **(a)** Oscillating from lag to anticipatory synchronization via complete synchronization in the region $t \in (10240, 10300)$, **(b)** lag synchronization in the range $t \in (10240, 10243)$ and (c) anticipatory synchronization in the range $t \in (10297, 10300)$

Recently, it has also been shown that the simple sinusoidal modulation of delay time in coupled semiconductor lasers with two optoelectronic feedback (with delays) results in loss of signatures of time-delays [19], which is characterized using autocorrelation function. Existence of complete synchronization is also shown to occur in the coupled semiconductor lasers with both unidirectional and bidirectional delay coupling.

11.10 Summary

To conclude, delay time modulation can introduce new features in synchronizations and their transitions in coupled time delay systems which are desirable from secure communication point of view. In particular, modulation in delay time can remove any imprints of delay time in the system trajectory, thereby inhibiting/reducing the possibility of phase-space reconstruction. As illustrative examples, coupled piecewise linear, Mackey-Glass and Ikeda systems have been used.

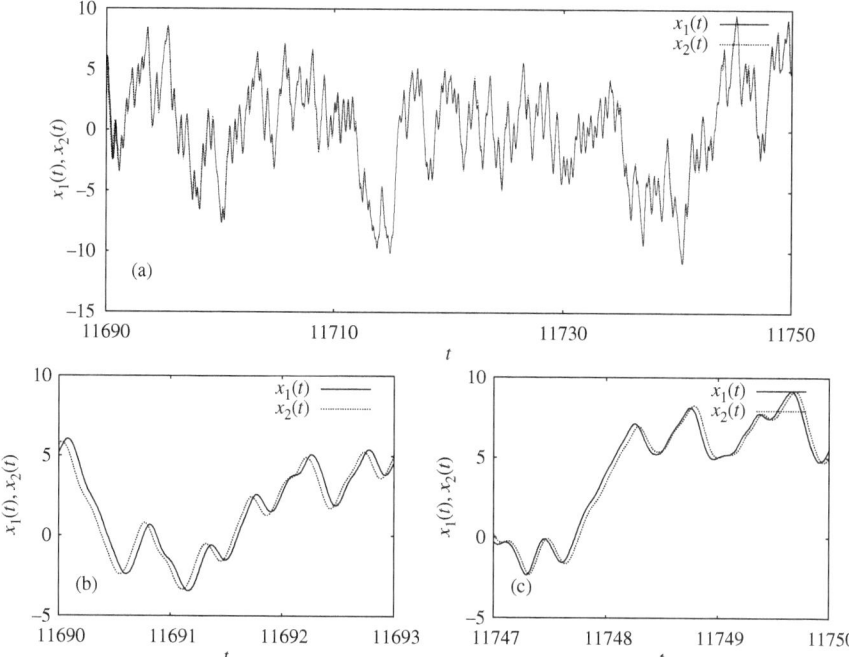

Fig. 11.19 Oscillating synchronization in the coupled Ikeda system for the constant coupling delay $\tau_2 = 2$ with time dependent feedback delay of the form (11.4a) with $\tau_{10} = 2, \tau_{1a} = 0.5$ and $\omega = 10^{-3}$. (**a**) Oscillating from lag to anticipatory synchronization via complete synchronization in the region $t \in (11690, 11750)$, (**b**) anticipatory synchronization in the range $t \in (11690, 11693)$ and (**c**) lag synchronization in the range $t \in (11747, 11750)$

References

1. W.-H. Kye, M. Choi, S. Rim, M.S. Kurdoglyan, C.-M. Kim, Y.-J. Park, Phys. Rev. E **69**, 055202(R) (2004)
2. F.M. Atay, Phys. Rev. Lett. **91**, 094101 (2003)
3. W.-H. Kye, M. Choi, M.-W. Kim, S. -Y. Lee, S. Rim, C.-M. Kim, Y.-J. Park, Phys. Lett. A **322**, 338 (2004)
4. D.V. Senthilkumar, M. Lakshmanan, Chaos **17**, 013112 (2007)
5. W.-H. Kye, M. Choi, C.-M. Kim, Y.-J. Park, Phys. Rev. E **71**, 045202(R) (2005)
6. E.M. Shahverdiev, K.A. Shore, Phys. Rev. E **71**, 016201 (2005)
7. D.V. Senthilkumar, M. Lakshmanan, Phys. Rev. E **71**, 016211 (2005)
8. N. Platt, S.M. Hammel, J.F. Heagy, Phys. Rev. Lett. **72**, 3498 (1994)
9. J.F. Heagy, N. Platt, S.M. Hammel, Phys. Rev. E **49**, 1140 (1994)
10. M.J. Bunner, Th. Meyer, A. Kittel, J. Parisi, Phys. Rev. E **56**, 5083 (1997)
11. M.J. Bunner, M. Popp, Th. Meyer, A. Kittel, J. Parisi, Phys. Rev. E **54**, 3082(R) (1996)
12. A.M. Fraser, H.L. Swinney, Phys. Rev. A **33**, 1134 (1986)
13. H.D.I. Abarbanel, *Analysis of Observed Chaotic Data* (Springer, New York, 1996)
14. C. Zhou, C.H. Lai, Phys. Rev. E **60**, 320 (1999)

15. V.I. Ponomarenko, M.D. Prokhorov, Phys. Rev. E **66**, 026215 (2002)
16. D.V. Senthilkumar, M. Lakshmanan, Int. J. Bifurcat. Chaos **15**, 2985 (2005)
17. N.N. Krasovskii, *Stability of Motion* (Stanford University Press, Stanford, CA, 1963)
18. K. Pyragas, Phys. Rev. E **58**, 3067 (1998)
19. E.M. Shahverdiev, K.A. Shore, IET Optoelectronics **3**, 326 (2009)

Chapter 12
Exact Solutions of Certain Time Delay Systems: The Car-Following Models

12.1 Introduction

In spite of the complex dynamics exhibited by even the simplest of nonlinear time delay systems, there exists a host of coupled nonlinear time delay systems which admit exact solutions. Particularly, certain coupled systems of nonlinear delay differential equations modelling traffic flow [1–3], called the car following models, possess exact analytic solutions in terms of Jacobian elliptic functions under periodic boundary conditions. However, under open boundary conditions, they admit shock-like solutions, representing the stationary propagation of a traffic jam [2, 3]. We will closely follow here the approach of Tutiya and Kanai [4] in the following discussion just to illustrate how exact solutions can arise in delay systems.

12.2 The Car-Following Models

It is interesting to note that one can treat a traffic flow, including pedestrian flow, as a compressible fluid from a macroscopic point of view, or as a many-body problem of driven particles in a microscopic sense. Consequently, models can be developed based either on hydrodynamic equations or in terms of coupled ordinary differential equations, and even in terms of cellular automata.

Consider for example, the highway traffic. One can model it as a system of particles moving in one dimension in a definite direction interacting with each other asymptotically, see Fig.12.1.

In this one-dimensional picture one essentially considers contrasting density patterns, which change quite irregularly as the density of particles increases, and which finally take the form of a stable traffic jam propagating backwards with constant speed. One can introduce a set of coupled delay- differential equations of the following form to represent the traffic flow:

$$\dot{x}_n(t) = F(\Delta x_n(t - \tau)), \qquad \Delta x_n(t) = x_{n-1}(t) - x_n(t), \qquad n = 1, 2, \cdots \quad (12.1)$$

M. Lakshmanan, D.V. Senthilkumar, *Dynamics of Nonlinear Time-Delay Systems*,
Springer Series in Synergetics, DOI 10.1007/978-3-642-14938-2_12,
© Springer-Verlag Berlin Heidelberg 2010

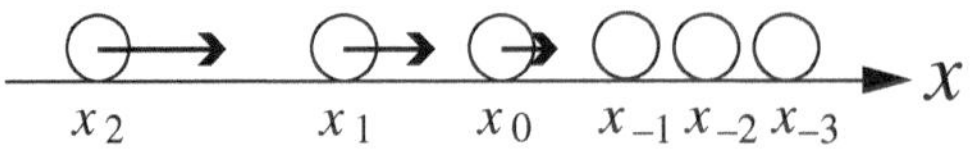

Fig. 12.1 An illustration of one-lane traffic with the assumption that cars overtaking and colliding are prohibited [4]

where $x_n(t), n = 1, 2, \cdots$, denotes the position of the nth car at time t and $\Delta x_n(t)$ is the distance between the nth car and the one in front of it (that is the $(n-1)$th). In the above car following model described by (12.1), one can observe that it represents the solution where the velocity of each car, $\dot{x}_n(t)$, is determined in terms of the distance that separates it from its predecessor with a delay τ, that is in terms of $\Delta x_n(t - \tau)$. The function $F(x)$ in (12.1), often called optimal velocity function, is usually determined from real traffic data.

Two typical models correspond to the following forms:

(i) Newell model: $F = V \left[1 - exp\left(-\frac{\gamma}{V}\left(\Delta x_n(t - \tau) - L\right)\right)\right]$, where r, V, L are parameters.

(ii) Tanh model: $F = \xi + \eta tanh\left(\frac{\Delta x_n(t-\tau)-\rho}{2A}\right)$, where ξ, η, ρ and A are parameters.

Exact solutions to these two models can be deduced using the so called Hirota bilinearization method [5], well known in the theory of soliton systems (see for example [6]).

12.3 The Newell Model

The Newell equation reads as

$$\dot{x}_n(t) = V \left[1 - exp\left(-\frac{\gamma}{V}\left(\Delta x_n(t - \tau) - L\right)\right)\right], \qquad (12.2)$$
$$\Delta x_n(t) = x_{n-1}(t) - x_n(t), \qquad n = 1, 2, \cdots$$

Here V is the maximum allowed velocity of the car, γ is the slope of the optimal velocity of function at $\Delta x_n = L$ corresponding to the sensitivity of the driver to changes in the traffic situation, and L is the minimum headway. In Fig.12.2, the optimal velocity function F is shown as a function of the headway Δx_n, where the parameters have been deduced from empirical data [4].

Now, in order to eliminate the background uniform flow, one can change the dependent variable $x_n(t)$ to $y_n(t)$ as

$$y_n(t) = x_n(t) - (V_0 t - L_0 n), \qquad (12.3)$$

where the velocity V_0 and headway L_0 satisfy the condition

$$V_0 = V \left[1 - exp\left(-\frac{\gamma}{V}(L_0 - L)\right)\right], \qquad (12.4)$$

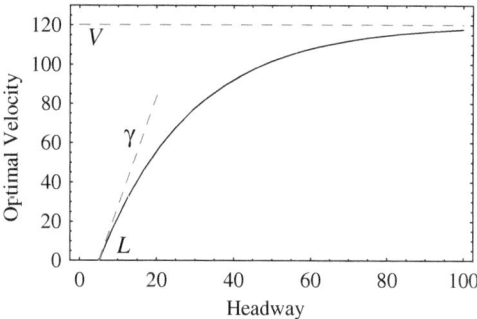

Fig. 12.2 The optimal velocity function for the Newell model. The values of the parameters are as follows: $V = 120$, $\gamma = 6$ and $L = 5$, where V indicates the maximum allowed velocity, γ is the derivative of F at $\Delta x_n = L$, and L is the minimum distance between cars (adapted from [4])

which is required of a uniform solution. Substituting (12.3) and (12.4) into (12.1), the Newell equation (12.3) can be rewritten as

$$\dot{y}_n(t) = (V - V_0)[1 - exp(-s_n(t - \tau)], \tag{12.5a}$$

where

$$s_n(t) = \frac{\gamma}{V}(y_{n-1}(t) - y_n(t)). \tag{12.5b}$$

Now differentiating (12.5b) once and using (12.5a), we can rewrite (12.5) as

$$\frac{1}{\alpha_0}\dot{s}_n(t) = -exp\left(-s_{n-1}(t - \tau)\right) + exp\left(-s_n(t - \tau)\right), \tag{12.6a}$$

where

$$\alpha_0 = \gamma\left(1 - \frac{V_0}{V}\right) = \gamma exp\left[-\frac{\gamma}{V}(L_0 - L)\right]. \tag{12.6b}$$

One can now reexpress the above Newell equation in Hirota's bilinear form. For this purpose, let us define

$$\psi_n(t) = exp(-s_n(t)). \tag{12.7}$$

Then from (12.6) one has

$$\frac{1}{\alpha_0}\frac{\dot{\psi}_n(t + \tau)}{\psi_n(t + \tau)} = \psi_{n-1}(t) - \psi_n(t). \tag{12.8}$$

Following the standard bilinearization procedure (for example for the nonlinear Schrödinger equation see for instance [6]), one can introduce the bilinearizing transformation

$$\psi_n(t) = \frac{g_n(t)}{f_n(t)}. \tag{12.9}$$

Then (12.5) can be rewritten as

$$\frac{1}{\alpha_0} \frac{\dot{g}_n(t+\tau)f_n(t+\tau) - g_n(t+\tau)\dot{f}_n(t+\tau)}{f_n(t+\tau)g_n(t+\tau)} = \frac{g_{n-1}(t)f_n(t) - g_n(t)f_{n-1}(t)}{f_{n-1}(t)f_n(t)} \tag{12.10}$$

Consequently (12.10) can be decoupled as

$$\dot{g}_n(t+\tau)f_n(t+\tau) - g_n(t+\tau)\dot{f}_n(t+\tau)$$
$$= \lambda(g_{n-1}(t)f_n(t) - g_n(t)f_{n-1}(t)), \tag{12.11a}$$

$$f_{n-1}(t)f_n(t) = \frac{\alpha_0}{\lambda}f_n(t+\tau)g_n(t+\tau). \tag{12.11b}$$

Here λ is a constant. Equations (12.11) are now in the required bilinear form.
Following the standard procedure of Hirota bilinearizition method [5], one can assume that

$$f_n(t) = 1 + exp(an + 2bt), \tag{12.12a}$$

$$g_n(t) = u + vexp(an + 2bt), \tag{12.12b}$$

where a, b, u and v are constants. Substituting (12.12) back into (12.11), one obtains a shock-like solution of the form

$$f_n(t) = 1 + exp[2b(t - n\tau)], \tag{12.13a}$$

$$g_n(t) = \frac{b}{\alpha_0(1 - e^{-2b\tau})} \{1 + exp(2b[t - \tau(n+1)])\}, \tag{12.13b}$$

where b is a free parameter. Using (12.13) and (12.9) in (12.7) one can finally obtain the solution

$$s_n(t) = log \frac{\alpha_0 sinh(b\tau)}{b} \frac{cosh[b(t - n\tau)]}{cosh[b(t - (n+1)\tau)]}. \tag{12.14}$$

Solution (12.14) represents a shock-like structure moving with velocity $U = \frac{1}{\tau}$ representing a traffic jam backwards. Here open boundary conditions have been assumed. A plot of the function (12.14) is shown in Fig.12.3 to show the solitary nature and shock-like structure. In addition one can also obtain an elliptic function wave solution of the form

Fig. 12.3 Solution $s_n(t)$ as a function of t for various values of n. The other parameters are fixed at $\alpha_0 = 5, b = 1, \tau = 2$

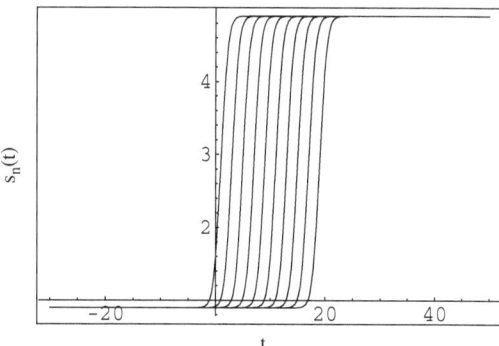

$$s_n(t) = log \frac{2\alpha_0 sn(\Omega\tau)cn(\Omega\tau)dn(\Omega\tau)}{\Omega[1 - k^2 sn^2(\Omega\tau)sn^2(\phi + \Omega\tau)](1 - k^2 sn^2(\Omega\tau)sn^2\phi)}, \quad (12.15)$$

where $\phi = \Omega(t - 2\tau n)$, and sn, cn, and dn are Jacobian elliptic functions with modulus k, while the parameter Ω satisfies a certain transcendental equation [4].

12.4 The tanh Car-Following Model

Consider the car following model introduced in [2, 3],

$$\dot{x}_n = \xi + \eta \, tanh \left(\frac{\Delta x_n(t - \tau) - \rho}{2A} \right), \quad (12.16)$$

where ξ, η, ρ and A are constant parameters. Defining the distance variable

$$h_n(t) = \frac{(\Delta x_n(t) - \rho)}{2A}, \quad (12.17)$$

Equation (12.16) can be rewritten as

$$\dot{h}_n(t + \tau) = \frac{\eta}{2A} \left[tanh \, h_{n-1}(t) - tanh \, h_n(t) \right]. \quad (12.18)$$

Several specific elliptic function solutions to (12.18) can be given [2, 3]:

$$(i) \, tanh \, h_n(t) = a \, sn \, \Omega(t - 2n\tau) + b, \quad (12.19)$$

$$(ii) \, tanh \, h_n(t) = \frac{b}{sn \, \Omega(t - 2n\tau) + a} + c, \quad (12.20)$$

$$(iii) \, tanh \, h_n(t) = \frac{b}{sn^2 \, \Omega(t - 2n\tau) + a} + c. \quad (12.21)$$

In the above Ω is a free parameter, while the parameters a, b, and c can be fixed in terms of Ω, A, η and τ.

Other interesting solutions can be given again by bilinearizing the system (12.16). Defining

$$\psi_n = \tanh h_n, \tag{12.22}$$

Equation (12.17) can be rewritten as

$$\dot{\psi}_n(t + \tau) = \frac{\eta}{2A} \left[1 - (\psi_n(t + \tau))^2 \right] (\psi_{n-1}(t) - \psi_n(t)) \tag{12.23}$$

Again defining

$$\psi_n(t) = \frac{g_n(t)}{f_n(t)}, \tag{12.24}$$

one can rewrite (12.23) into a system of bilinear equations,

$$\dot{g}_n(t + \tau) f_n(t + \tau) - g_n(t + \tau) \dot{f}_n(t + \tau)$$
$$= \lambda \left[(g_{n-1}(t) f_n(t) - g_n(t) f_{n-1}(t)) \right], \tag{12.25a}$$

$$f_{n-1}(t) f_n(t) = \frac{\eta}{2A\lambda} \left[f_n^2(t + \tau) - g_n^2(t + \tau) \right], \tag{12.25b}$$

where λ is a constant.

Making now the substitution

$$f_n(t) = 1 + exp(2bt - an), \quad g_n(t) = u + v \, exp(bt - an) \tag{12.26}$$

into (12.25), and finding consistent forms of u and v, one can obtain the solution

$$f_n(t) = 1 + exp(2bt - an),$$
$$g_n(t) = \left[1 - \frac{2bA}{\eta(1 - e^{-2b\tau})} \right] + \left[1 + e^{-2b\tau} exp(2bt - an) \right], \tag{12.27}$$

where the constant parameters a and b are related by

$$e^a = \frac{bA/\eta + 1 - e^{2b\tau}}{bA/\eta - 1 + e^{-2b\tau}}. \tag{12.28}$$

Using the above forms of $f_n(t)$ and $g_n(t)$, then one obtains the exact solution to the tanh car following model (12.16) as

$$\Delta x_n(t) = \rho + A \, log \left(\frac{2\eta \, sinh \, (b\tau) \, cosh \, \left(bt - \frac{a}{2}n\right)}{bA \, cosh \left[b(t - \tau) - \frac{a}{2}n\right]} - 1 \right) \tag{12.29}$$

Fig. 12.4 The plot of the function $\Delta x_n(t)$ Vs t for different values of n, while the other parameters are held fixed

$\Delta x_n(t)$

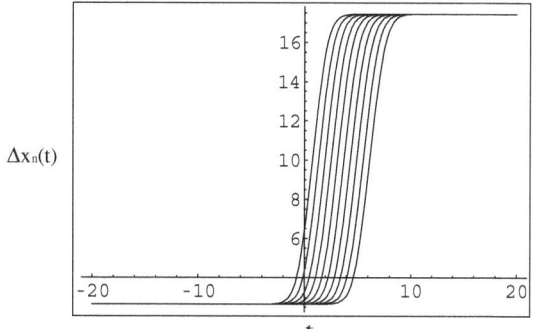

This is another shock wave solution with velocity $U = \frac{2b}{a}$ which represents a traffic jam propagating backwards, see Fig.12.4. Apart from the above type of solutions one can also construct explicit elliptic function waves propagating with velocity $U = \frac{1}{2\tau}$ and as a limiting form a kink like solution can also be obtained. For details see for example [2, 4].

12.5 Other Developments

Modeling of vehicular traffic is a complex dynamical problem, which has been attracting the attraction of scientists for more than half a century (for a review, see [7]). Yet the precise mechanism for generation and propagation of traffic jams is not fully understood. In this chapter, we have presented a couple of simple car following models which possess exact shock like solutions representing jams propagating backwards. More general models require detailed numerical analysis.

In the case of sparse traffic, it is well known that there exists a uniform flow equilibrium where vehicles follow each other with the same velocity while oscillations may arise when the traffic becomes more dense. One of the typical oscillations [8] is a stop-and-go-wave, where the velocity breaks down and vehicles become densely packed on a section of the highway and the congestion propagates upstream as a density wave with a characteristic wave speed. Several models exist to reproduce some aspects of such a flow. More general models than the simple car following model (12.1) start with the dynamical equation for the acceleration of the ith vehicle,

$$\dot{v}_i(t) = f\left(h_i(t - \tau_1), \dot{h}_i(t - \tau_2), v_i(t - \tau_3)\right). \tag{12.30}$$

Here v_i is the velocity of the ith vehicle, while h_i is the bumper to bumper distance between the ith and $(i + 1)$th vehicles called the headway. The reaction time delays $\tau_1, \tau_2, \tau_3 (\geq 0)$ are generally different, but often assumed to be same for simplicity. Here

$$\dot{h}_i(t) = x_{i+1}(t) - x_i(t) - l, \tag{12.31}$$

where x_i is the position of the front bumper of the ith vehicle and l is the length of the vehicle. The time derivative then gives

$$\dot{h}_i(t) = v_{i+1}(t) - v_i(t), \tag{12.32}$$

With suitable functional forms of the function f and appropriate boundary conditions, one may numerically analyze Eq. (12.30) to get detailed microscopic dynamics and macroscopic properties of traffic flow. For details see [8–10] for example.

In all the above models time-delay plays a crucial role.

References

1. G.B. Whitham, *Linear and Nonlinear Waves* (Wiley, New York, 1974)
2. K. Hasabe, A. Nakayama, Y. Sugiyama, Phys. Lett. A **259**, 135 (1999)
3. Y. Igarashi, K. Itoh, K. Nakanishi, J. Phys. Soc. Jpn. **68**, 791 (1999)
4. Y. Tutiya, M. Kanai, J. Phys. Soc. Jpn. **76**, 083001 (2007)
5. R. Hirota, *The Discrete Method in Soliton Theory* (Cambridge University Press, Cambridge, 2004)
6. M. Lakshmanan, S. Rajasekar, *Nonlinear Dynamics: Integrability, Chaos and Patterns* (Springer, Berlin, 2003)
7. D. Helbing, Rev. Mod. Phys. **73**, 1067 (2001)
8. G. Orosz, R.E. Wilson, R. Szalai, G. Stepan, Phys. Rev. E **80**, 046205 (2009)
9. R.E. Wilson, Philos. Trans. R. Soc. Lond. Ser. A **366**, 2017 (2008)
10. D. Helbing, M. Moussid, Eurphys. J. **B69**, 571 (2009)

Appendix A
Computing Lyapunov Exponents for Time-Delay Systems

A.1 Introduction

The hall mark property of a chaotic attractor, namely sensitive dependence on initial condition, has been associated by the Lyapunov exponents to characterize the degree of exponential divergence/convergence of trajectories arising from nearby initial conditions. At first, we will describe briefly the concept of Lyapunov exponent and the procedure for computing Lyapunov exponents of the flow of a dynamical system described by n-dimensional ordinary differential equations (ODEs), which is then extended to scalar delay differential equations (DDEs), which are essentially an infinite-dimensional systems. An important step in computing Lyapunov exponents of DDEs is that it is necessary to approximate the continuous evolution of an infinite-dimensional system by a finite-dimensional (appreciably large) iterated mapping. Then the Lyapunov exponents of the finite-dimensional map can be calculated by computing simultaneously the reference trajectories from the original map and the trajectories from their linearized equations of motion. Alternatively, it can also be calculated by computing the evolution of infinitesimal volume element formed by a set of infinitesimal separation vectors corresponding to the trajectories starting from nearby initial conditions.

A.2 Lyapunov Exponents of an n-Dimensional Dynamical System

Consider an n-dimensional dynamical system described by the system of first order coupled ordinary differential equation [1–3]

$$\dot{\mathbf{X}} = \mathbf{F}(\mathbf{X}), \tag{A.1}$$

where $\mathbf{X}(t) = (x_1(t), x_2(t), ..., x_n(t))$. We consider two trajectories in the n-dimensional phase space starting from two nearby initial conditions $\mathbf{X}_0$ and $\mathbf{X}'_0 = \mathbf{X}_0 + \delta\mathbf{X}_0$. They evolve with time yielding the vectors $\mathbf{X}(t)$ and $\mathbf{X}'(t) = \mathbf{X}(t) + \delta\mathbf{X}(t)$, respectively, with the Euclidean norm

M. Lakshmanan, D.V. Senthilkumar, *Dynamics of Nonlinear Time-Delay Systems*,
Springer Series in Synergetics, DOI 10.1007/978-3-642-14938-2,
© Springer-Verlag Berlin Heidelberg 2010

$$d\left(\mathbf{X}_0, t\right) = ||\delta\mathbf{X}\left(\mathbf{X}_0, t\right)|| \equiv \sqrt{\delta x_1^2 + \delta x_2^2 + \dots + \delta x_n^2} . \tag{A.2}$$

Here $d(\mathbf{X}_0, t)$ is simply a measure of the distance between the two trajectories $\mathbf{X}(t)$ and $\mathbf{X}'(t)$. The time evolution of $\delta\mathbf{X}$ is found by linearizing (A.1) to obtain

$$\delta\dot{\mathbf{X}} = M(\mathbf{X}(t)) \cdot \delta\mathbf{X} , \tag{A.3}$$

where $M = \partial\mathbf{F}/\partial\mathbf{X}|_{\mathbf{X}=\mathbf{X}_0}$ is the Jacobian matrix of $\mathbf{F}$. Then the mean rate of divergence of two close trajectories is given by

$$\lambda\left(\mathbf{X}_0, \delta\mathbf{X}\right) = \lim_{t\to\infty} \frac{1}{t} \log\left(\frac{d\left(\mathbf{X}_0, t\right)}{d\left(\mathbf{X}_0, 0\right)}\right) . \tag{A.4}$$

Furthermore, there are n-orthonormal vectors $\mathbf{e}_i$ of $\delta\mathbf{X}$, $i = 1, 2, \dots, n$, such that

$$\delta\dot{\mathbf{e}}_i = M\left(\mathbf{X}_0\right)\mathbf{e}_i , \quad M = \mathrm{diag}\left(\lambda_1, \lambda_2, \dots, \lambda_n\right) . \tag{A.5}$$

That is, there are n-Lyapunov exponents given by

$$\lambda_i\left(\mathbf{X}_0\right) = \lambda_i\left(\mathbf{X}_0, \mathbf{e}_i\right) , \quad i = 1, 2, \dots, n . \tag{A.6}$$

These can be ordered as $\lambda_1 \geq \lambda_2 \geq \dots \geq \lambda_n$. From (A.4) and (A.6) we may write

$$d_i\left(\mathbf{X}_0, t\right) \approx d_i\left(\mathbf{X}_0, 0\right) e^{\lambda_i t} , \quad i = 1, 2, \dots, n . \tag{A.7}$$

To identify whether the motion is periodic or chaotic it is sufficient to consider the largest nonzero Lyapunov exponent λ_m among the n Lyapunov exponents of the n-dimensional dynamical system.

A.2.1 Computation of Lyapunov Exponents

To compute the n-Lyapunov exponents of the n-dimensional dynamical system (A.1), a reference trajectory is created by integrating the nonlinear equations of motion (A.1). Simultaneously the linearized equations of motion (A.3) are integrated for n-different initial conditions defining an arbitrarily oriented frame of n-orthonormal vectors $(\Delta\mathbf{X}_1, \Delta\mathbf{X}_2, \dots, \Delta\mathbf{X}_n)$. There are two technical problems [4] in evaluating the Lyapunov exponents directly using (A.4), namely the variational equations have at least one exponentially diverging solution for chaotic dynamical systems leading to a storage problem in the computer memory. Further, the orthonormal vectors evolve in time and tend to fall along the local direction of most rapid growth. Due to the finite precision of computer calculations the collapse toward a common direction causes the tangent space orientation of all the vectors to become indistinguishable. Both the problems can be overcome by a repeated use of what is known as Gram-Schmidt reorthonormalization (GSR) procedure [5] which is well known in the theory of linear vector spaces. We apply GSR after τ time steps which orthonormalize the evolved vectors to give a new set $\{\mathbf{u}_1, \mathbf{u}_2, ..., \mathbf{u}_n\}$:

$$\mathbf{v}_1 = \Delta\mathbf{X}_1 , \tag{A.8}$$

$$\mathbf{u}_1 = \mathbf{v}_1/||\mathbf{v}_1|| , \tag{A.9}$$

$$\mathbf{v}_i = \Delta\mathbf{X}_i - \sum_{j=1}^{i-1}\langle\Delta\mathbf{X}_i , \mathbf{u}_j\rangle \mathbf{u}_j , \quad i = 2, 3, ..., n \tag{A.10}$$

$$\mathbf{u}_i = \mathbf{v}_i/||\mathbf{v}_i|| , \tag{A.11}$$

where $\langle , \rangle$ denotes inner product. In this way the rate of growth of evolved vectors can be updated by the repeated use of GSR. Then, after the N-th stage, for N large enough, the one-dimensional Lyapunov exponents are given by

$$\lambda_i = \frac{1}{N\tau} \sum_{k=1}^{N} \log ||\mathbf{v}_i^{(k)}|| . \tag{A.12}$$

For a given dynamical system, τ and N are chosen appropriately so that the convergence of Lyapunov exponents is assured. A fortran code algorithm implementing the above scheme can be found in [4].

A.3 Lyapunov Exponents of a DDE

As described in the Sect. 1.2.2 of Chap. 1, a DDE of the form

$$\dot{X} = F(t, X(t), X(t - \tau)), \tag{A.13}$$

can be approximated as an N-dimensional iterated map [6], $X(k + 1) = G(X(k))$, (k labels the kth iteration and $k + 1$ to its next iteration). Now, the Lyapunov exponents of the N-dimensional map can be calculated by computing simultaneously a reference trajectory and the trajectories that are separated from the reference trajectory by a small amount, corresponding to N-different initial conditions defining an arbitrarily oriented frame of N-orthonormal vectors as described above.

Alternatively, it can also be calculated by computing the evolution of infinitesimal volume element, formed by a set of infinitesimal separation vectors δx, which evolves according to

$$\delta x(k + 1) = \sum_{i=1}^{N} \frac{\partial G(x(k))}{\partial x_i(k)} \delta x_i(k). \tag{A.14}$$

Computational problems associated with computing adjacent trajectories can be avoided by calculating the evolution of infinitesimal separations directly from the above equation. The evolution equation of the infinitesimal volume element corresponding to the continuous DDE (A.13) can be written as

$$\frac{d\delta x}{dt} = \frac{\partial F(x, x_\tau)}{\partial x}\delta x + \frac{\partial F(x, x_\tau)}{\partial x_\tau}\delta x_\tau. \tag{A.15}$$

This equation can be solved using any convenient integration scheme. The small separations δx represents separation between two infinite-dimensional vectors. There are N such separations for every coordinate of the N-dimensional system corresponding to N Lyapunov exponents. Let $\delta \tilde{x}^i(k)$ denote the collection of all separations of ith coordinate during kth iteration, then its Lyapunov exponents can be given as

$$\lambda_i = \frac{1}{L\tau}\sum_{k=1}^{L} \log \frac{||\delta \tilde{x}^i(k)||}{||\delta \tilde{x}^i(k-1)||}. \tag{A.16}$$

For computing each exponent λ_i, arbitrarily select an initial separation $\delta \tilde{x}^i(0)$ and integrate for a time τ. Renormalize $\delta \tilde{x}^1(\tau)$ to have unit length. Using GSR procedure, orthonormalize the second separation function relative to the first, the third relative to the second, and so on. Repeat this procedure for L iterations. For sufficiently large L, it is numerically shown that the values of λ_i converge [6].

References

1. M. Lakshmanan, S. Rajasekar, *Nonlinear Dynamics: Integrability, Chaos and Patterns* (Springer, Berlin, 2003)
2. J.P. Eckmann, D. Ruelle, Rev. Mod. Phys. **57**, 617 (1985)
3. H.G. Schüster, *Deterministic Chaos* (Physik Verlag, Weinheim, 1984)
4. A. Wolf, J.B. Swift, H.L. Swinney, J. A. Vastano, Physica D **16**, 285 (1985)
5. C.R. Wylie, L.C. Barrett, *Advanced Engineering Mathematics* (McGraw-Hill, New York, 1995)
6. J.D. Farmer, Physica D **4**, 366 (1982)

Appendix B
A Brief Introduction to Synchronization in Chaotic Dynamical Systems

B.1 Introduction

Synchronization phenomenon is abundant in nature and can be realized in very many problems of science, engineering, and social life. Systems as diverse as clocks, singing crickets, cardiac pacemakers, firing neurons, and applauding audiences exhibit a tendency to operate in synchrony. The underlying phenomenon is universal and can be understood within a common framework based on modern nonlinear dynamics.

The history of synchronization goes back to the seventeenth century when the Dutch physicist Christiaan Huygens reported on his observation of phase synchronization of two pendulum clocks [1, 2]. Huygens briefly, but extremely precisely, described his observation of synchronization as follows.

> ... It is quite worth noting that when we suspended two clocks so constructed from two hooks imbedded in the same wooden beam, the motions of each pendulum in opposite swings were so much in agreement that they never receded the least bit from each other and the sound of each was always heard simultaneously. Further, if this agreement was disturbed by some interference, it reestablished itself in a short time. For a long time I was amazed at this unexpected result, but after a careful examination finally found that the cause of this is due to the motion of the beam, even though this is hardly perceptible. The cause is that the oscillations of the pendula, in proportion to their weight, communicate some motion to the clocks. This motion, impressed onto the beam, necessarily has the effect of making the pendula come to a state of exactly contrary swings if it happened that they moved otherwise at first, and from this finally the motion of the beam completely ceases. But this cause is not sufficiently powerful unless the opposite motions of the clocks are exactly equal and uniform.

Despite being the oldest scientifically studied nonlinear effects, synchronization was understood only in the 1920s when Edward Appleton [3] and Balthasar van der Pol [4] theoretically and experimentally studied synchronization of triode oscillators. Considering the simplest case, they showed that the frequency of a generator can be entrained, or synchronized, by a weak external signal of a slightly different frequency. These studies were of great practical importance because triode generators became the basic elements of radio communication systems. The synchronization phenomenon was used to stabilize the frequency of a powerful generator with the help of one which was weak but very precise.

Even though the notion of *synchronization* was identified well before the concept of chaos was realized, it was believed that *chaotic synchronization* was not feasible because of the hallmark property of chaos which is the extreme sensitivity to initial conditions. The latter property implies that two trajectories emerging from two different close by initial conditions separate exponentially in the course of time. As a result, chaotic systems intrinsically defy synchronization because even two identical systems starting from very slightly different initial conditions would evolve in time in an unsynchronized manner (the differences in the system states would grow exponentially). This is a relevant practical problem, insofar as experimental initial conditions are never known perfectly. Nevertheless, it has been shown that it is possible to synchronize chaotic systems, to make them evolve on the same chaotic trajectory, by introducing appropriate coupling between them due to the works of Pecora and Carroll and the earlier works of Fujisaka and Yamada [5–10]. Since the identification of synchronization in chaotic oscillators, the phenomenon has attracted considerable research activity in different areas of science and technology and several generalizations and interesting applications have been developed. The phenomenon of chaotic synchronization is of interest not only from a theoretical point of view but also has potential applications in diverse subjects such as as biological, neurological, laser, chemical, electrical and fluid mechanical systems as well as in secure communication, cryptography, system reconstruction, parameter estimation, controlling chaos, long term prediction of chaotic systems and so on [2, 11–21].

Chaotic synchronization, in general, can be defined as a process wherein two (or many) chaotic systems (either equivalent or nonequivalent) adjust a given property of their motion to a common behavior, due to coupling or forcing. This ranges from complete agreement of trajectories to locking of phases [11].

The first point we note here is that there is a great difference in the process leading to synchronized states, depending upon the particular coupling configuration, namely one should distinguish two main cases: unidirectional coupling and bidirectional coupling. When the evolution of one of the coupled systems is unaltered by the coupling, the resulting configuration is called *unidirectional coupling* or *drive-response coupling*. As a result, the response system is slaved to follow the dynamics of the drive system, which, instead, purely acts as an external but chaotic forcing for the response system. In such a case external synchronization is produced. Typical examples are communication using chaos. On the contrary, when both the systems are connected in such a way that they mutually influence each other's behavior then the corresponding configuration is called *bidirectional coupling*. Here both the systems are coupled with each other, and the coupling factor induces an adjustment of the rhythms onto a common synchronized manifold, thus inducing a mutual synchronization behavior. This situation typically occurs in physiology, e.g. between cardiac and respiratory systems or between neurons. These two processes are very different not only from a philosophical point of view; up to now no way has been discovered to reduce one process to another, or to link formally the two cases. Inside this classification, the appearance and robustness of synchronization states have been established by means of several different coupling schemes, such as the Pecora and Carrol method [8, 10, 21], the negative feedback [14], the sporadic driving [22], the active-passive decomposition [23, 24], the diffusive coupling and some other hybrid methods [25]. A description and analysis of some of these coupling schemes is given in [26] in a single mathematical framework. In the following studies we will consider only the so called *unidirectional coupling* or *drive-response coupling* configuration.

Chaos synchronization has been receiving a great deal of interest for more than two decades in view of its potential applications in various fields of science and engineering [5, 6, 8, 27–29]. Since the identification of chaotic synchronization, different kinds of synchronization have been proposed in interacting chaotic systems, which have all been identified both theoretically and experimentally. These include

1. complete or identical synchronization (CS) [5–8, 27],
2. phase synchronization (PS) [30–32],
3. lag synchronization (LS) [33–35],
4. anticipatory synchronization (AS) [36–38],
5. generalized synchronization (GS) [39–41],
6. intermittent lag synchronization (ILS) [33, 42–44],
7. intermittent anticipatory synchronization (IAS) [45],
8. intermittent generalized synchronization (IGS) [46],
9. imperfect or intermittent phase synchronization (IPS) [47–50],
10. almost synchronization (AS) [51],
11. time scale synchronization (TSS) [52] and
12. episodic synchronization (ES) [53].

Transition from one kind of synchronization to the other, coexistence of different kinds of synchronization in time series and also the nature of transitions have also been studied extensively [33–35, 54, 55] in coupled chaotic systems. There are also attempts to find a unifying framework for defining the overall class of chaotic synchronizations [56–58]. Before presenting the details of important types of aforesaid synchronization phenomena, we will discuss about the characterization for identifying the existence of synchronization in coupled chaotic systems.

B.2 Characterization of Synchronization

The existence of synchronization, in particular CS, is also characterized by quantitative measures in addition to qualitative pictures such as combined phase space plots of state variables, time trajectory of error variable, etc. Such quantitative measures are usually addressed in terms of a stability problem, that is, stability of the synchronized motion, and many criteria have been established in the literature to cope with it. One of the most popular and widely used criteria is the use of the

Lyapunov exponents as average measurements of expansion or shrinkage of small displacements along the synchronized trajectory.

Let us consider a set of two unidirectionally coupled identical chaotic systems whose temporal evolution is given by the system of coupled first order ODEs

$$\dot{\mathbf{X}} = F(\mathbf{X}), \qquad\qquad \left(\dot{} = \frac{d}{dt} \right) \qquad\qquad \text{(B.1a)}$$

$$\dot{\mathbf{Y}} = F(\mathbf{Y}, \mathbf{S}(t)), \qquad\qquad\qquad \text{(B.1b)}$$

where $\mathbf{X} = (x_1, x_2, ..., x_n)$ and $\mathbf{Y} = (y_1, y_2, ..., y_n)$ are n-dimensional state vectors corresponding to the drive and response systems, respectively, with F defining a vector field $F : R^n \rightarrow R^n$ and $\mathbf{S}(t)$ is some function of $\mathbf{X}(t)$, corresponding to the drive signal. The stability problem of identical coupled systems can be formulated in a very general way by addressing the question of the stability of the CS manifold $\mathbf{X} \equiv \mathbf{Y}$, or equivalently by studying the temporal evolution of the synchronization error $\mathbf{e} \equiv \mathbf{Y} - \mathbf{X}$. The evolution of $\mathbf{e}$ is given by

$$\dot{\mathbf{e}} = F(\mathbf{X}) - F(\mathbf{Y}, \mathbf{S}(t)). \qquad\qquad \text{(B.2)}$$

A CS regime exists when the synchronization manifold is asymptotically stable for all possible trajectories $\mathbf{S}(t)$ of the driving system within the chaotic attractor. This property can be proved by carrying out a stability analysis of the linearized system for small $\mathbf{e}$,

$$\dot{\mathbf{e}} = \mathbf{D}_X(\mathbf{S}(t))\mathbf{e}, \qquad\qquad \text{(B.3)}$$

where $\mathbf{D}_X$ is the Jacobian of the vector field $\mathbf{F}$ evaluated onto the driving trajectory $\mathbf{S}(t)$. Normally, when the driving trajectory $\mathbf{S}(t)$ is constant (fixed point) or periodic (limit cycle), the stability problem can be studied by evaluating the eigenvalues of $\mathbf{D}_X$ or the Floquet multipliers [59, 60]. However, if the response systems is driven by a chaotic signal, this method will not work.

A possible way out is to calculate the Lyapunov exponents of the system (B.3). In the context of drive-response coupling schemes, these exponents are usually called *conditional Lyapunov exponents* (CLEs) because they are the Lyapunov exponents of the response system under the explicit constraint that they must be calculated on the trajectory $\mathbf{S}(t)$ [10, 23]. Alternatively, they are called *transverse Lyapunov exponents* (TLEs) because they correspond to directions which are transverse to the synchronization manifold $\mathbf{X} \equiv \mathbf{Y}$ [25, 61]. These exponents may be defined, for an initial condition of the driver signal $\mathbf{S}_0$ and initial orientation of the infinitesimal displacement $\mathbf{U}_0 = \mathbf{e}(0)/|\mathbf{e}(0)|$, as

$$h(\mathbf{S}_0, \mathbf{U}_0) \equiv \lim_{t \to \infty} \frac{1}{t} ln \left(\frac{|\mathbf{e}(t)|}{|\mathbf{e}(0)|} \right) = \lim_{t \to \infty} \frac{1}{t} ln |\mathbf{Z}(\mathbf{S}_0, t).\mathbf{U}_0|, \qquad \text{(B.4)}$$

where $\mathbf{Z}(\mathbf{S}_0, t)$ is the matrix solution of the linearized equation,

$$dZ/dt = \mathbf{D}_X(\mathbf{S}(t))\mathbf{Z}, \tag{B.5}$$

subject to the initial condition $\mathbf{Z}(0) = I$. The synchronization error $\mathbf{e}$ evolves according to $\mathbf{e(t)} = \mathbf{Z}(\mathbf{S}_0, t)\mathbf{e_0}$ and then the matrix $\mathbf{Z}$ determines whether this error shrinks or grows in a particular direction. In most cases, however, the calculation cannot be made analytically, and therefore numerical algorithms should be used [62–64].

It is very important to emphasize that the negativity of the conditional Lyapunov exponents is only a necessary condition for the stability of the synchronized state. The conditional Lyapunov exponents are obtained from a temporal average, and therefore they characterize the global stability over the whole chaotic attractor. Relevant cases exist where these exponents are negative and nevertheless the systems are not perfectly synchronized, thus indicating that additional conditions should be fulfilled to warrant synchronization in a necessary and sufficient way [65].

The stability of a CS manifold can also be studied by the use of the Lyapunov function $L(\mathbf{e})$. It can be defined as a continuously differentiable real valued function with the following properties:

(a) $L(\mathbf{e}) > 0$ for all $\mathbf{e} \neq 0$ and $L(\mathbf{e}) = 0$ for $\mathbf{e} = 0$.
(b) $dL/dt < 0$ for all $\mathbf{e} \neq 0$.

If for a given coupled system one can find a Lyapunov function, then the CS manifold is globally stable. For illustrative examples one may refer to [13, 23, 28, 66]. Unfortunately, whether such functions exist and how one should construct them is known only in a very limited number of cases, whereas a general procedure to obtain these functions is not yet available.

At this stage, let us summarize the validity of the stability criteria discussed above. In general, only Lyapunov functions give a sufficient condition for the stability of the synchronization manifold, whereas the negativity of the conditional Lyapunov exponents provides a necessary condition. While the Lyapunov function criterion gives a local condition for stability, the other two (CLEs/TLEs) involve temporal averages over chaotic trajectories of the driving signal, and therefore they establish conditions for global stability. As a consequence, none of these latter criteria prevents from local desynchronization events that could occur within the CS manifold. This point is discussed in [61], where the synchronized behavior of two chaotic circuits coupled in a drive-response configuration is studied. The appearance of these local desynchronized states, despite Lyapunov exponents being negative, is also related with a small parameter mismatch between the coupled systems and low levels of noise, which are unavoidable effects in experimental devices and in numerical integration.

We have pointed in the above that the characterization of synchronization in coupled identical systems can be done using the stability of synchronized motion by referring to the stability of the CS manifold. When we deal with nonidentical coupled systems, similar stability criteria can be formulated, but additional problem will appear due to the more complicated structure of the synchronization manifold. Also, the other kinds of synchronization have their own characterizations, which we will discuss in the following sections.

B.2.1 Complete Synchronization

When one deals with coupled identical chaotic systems, synchronization appears as the equality of the state variables while evolving in time. Complete synchronization (CS) was the first discovered and simplest form of synchronization in chaotic systems. It is characterized by a perfect locking of the chaotic trajectories of two identical nonlinear systems which is achieved by means of a suitable coupling in such a way that the two trajectories remain in step with each other in the course of time, that is, $X(t) \equiv Y(t)$, where X and Y are n-dimensional state variables whose evolution is represented by (B.1), individually. This mechanism was first shown to occur when two identical chaotic systems are coupled unidirectionally, provided the conditional Lyapunov exponents of the subsystem (response) to be synchronized are all negative [8]. Complete synchronization is also called conventional synchronization or identical synchronization in the literature [67].

As an illustrative example for CS, we will consider a Pecora and Caroll drive-response configuration with a drive system given by the Lorenz system [68],

$$\dot{x}_1 = \sigma(y_1 - x_1), \tag{B.6a}$$
$$\dot{y}_1 = -x_1 z_1 + r x_1 - y_1, \tag{B.6b}$$
$$\dot{z}_1 = x_1 y_1 - b z_1, \tag{B.6c}$$

and with a response system given by the subspace containing the (y, z) variables, where x_1 acts as the driving signal for the response system,

$$\dot{y}_2 = -x_1 z_2 + r x_1 - y_2, \tag{B.7a}$$
$$\dot{z}_2 = x_1 y_2 - b z_2. \tag{B.7b}$$

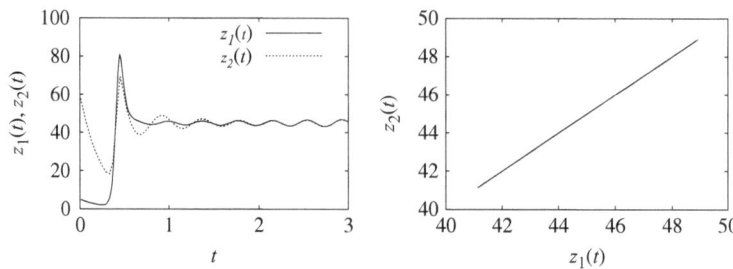

Fig. B.1 Complete synchronization between two coupled Lorenz systems using Pecora and Caroll method as represented by Eqs. (B.6) and (B.7). (**a**) Time trajectory plot and (**b**) Phase space plot

Here the control parameters σ, r and b are fixed as $\sigma = 16, r = 45.92$ and $b = 4$ so that Eqs. (B.6) give rise to chaotic dynamics. With this particular choice of the driving, CS sets in rather quickly as shown in Fig. B.1. Figure B.1a is a time trajectory plot of $z_1(t)$ and $z_2(t)$ showing complete synchronization and diagonal line in Fig. B.1b confirms the CS between $z_1(t)$ and $z_2(t)$. Note that the above configuration is also called a *homogeneous driving* configuration.

B.2.2 Phase Synchronization

Definition of chaotic phase synchronization (CPS) in coupled chaotic systems is derived from the classical definition of phase synchronization in periodic oscillators. Interacting chaotic systems are said to be in phase synchronized state when there exists entrainment between phases of the systems, $n\phi_1 - m\phi_2 =$const, while their amplitudes may remain chaotic and uncorrelated (In the presence of noise, a weaker condition for phase locking, $|n\phi_1 - m\phi_2| <$const, should be used instead). In other words, CPS exists when their respective frequencies and phases are locked [2, 11, 69]. To study CPS, one has to identify a well defined phase variable in both the coupled systems. If the flow of the chaotic oscillator has a proper rotation around a certain reference point, the phase can be defined in a straightforward way. For example, for the Rössler system [30] with standard parameters the projection of the chaotic attractor onto the (x, y) plane looks like a smeared limit cycle. In this and similar cases one can define the phase [2, 11] as

$$\phi(t) = \arctan(y(t)/x(t)). \tag{B.8}$$

A more general approach to define the phase in chaotic oscillators is the analytic signal approach [2, 11] introduced in [70]. The analytic signal $\chi(t)$ is given by

$$\chi(t) = s(t) + i\tilde{s}(t) = A(t) \exp^{i\Phi(t)}, \tag{B.9}$$

where $\tilde{s}(t)$ denotes the Hilbert transform of the observed scalar time series $s(t)$

$$\tilde{s}(t) = \frac{1}{\pi} P.V. \int_{-\infty}^{\infty} \frac{s(t')}{t - t'} dt', \tag{B.10}$$

where P.V. stands for the Cauchy principle value of the integral and this method is especially useful for experimental applications [2, 11] .

The phase of a chaotic attractor can also be defined based on an appropriate Poincaré surface of section which the chaotic trajectory crosses once for each rotation. Each crossing of the orbit with the Poincaré section corresponds to an increment of 2π of the phase, and the phase in between two crossings is linearly interpolated [2, 11],

$$\Phi(t) = 2\pi k + 2\pi \frac{t - t_k}{t_{k+1} - t_k}, \qquad (t_k < t < t_{k+1}) \qquad (B.11)$$

where t_k is the time of kth crossing of the flow with the Poincaré section. For the phase coherent chaotic oscillators, that is, for flows which have a proper rotation around a certain reference point, the phases calculated by these three different ways are in good agreement [2, 11].

As the simplest example of chaotic phase synchronization, we will consider two coupled Rössler systems [30, 71],

$$\dot{x}_{1,2} = -\omega_{1,2} y_{1,2} - z_{1,2} + C(x_{2,1} - x_{1,2}), \qquad (B.12a)$$

$$\dot{y}_{1,2} = \omega_{1,2} x_{1,2} + a y_{1,2}, \qquad (B.12b)$$

$$\dot{z}_{1,2} = 0.2 + z_{1,2}(x_{1,2} - 10), \qquad (B.12c)$$

where the parameters $\omega_{1,2} = 1 \pm \Delta\omega$ govern the frequency mismatch and C is the strength of coupling. As the coupling is increased for a fixed mismatch $\Delta\omega$, one can observe a transition from a regime, where the phases rotate with different velocities $\phi_1 - \phi_2 \sim \Delta\Omega t$, to a synchronous state, where the phase difference does not grow with time $|\phi_1 - \phi_2| < \text{const}$; $\Delta\Omega = 0$. This transition is illustrated in Fig. B.2a. Moreover, the correlation between the amplitudes of x_1 and x_2 is quite small (Fig. B.2b), although the phases are completely locked. In this example, it is shown that transition of one of the zero Lyapunov exponents to negative value as shown in Fig. B.3 corresponds to the critical point at which the phases become locked. It is known that in the absence of coupling each oscillator has one positive, one negative and one zero Lyapunov exponents. The zero Lyapunov exponents correspond to the transition along the trajectory. As the coupling strength is increased the interaction between the oscillators increases such that the phase difference $\phi_1 - \phi_2$ decreases and phases become locked eventually. Thus one of the zero exponents becomes negative to account for the phase locking phenomenon.

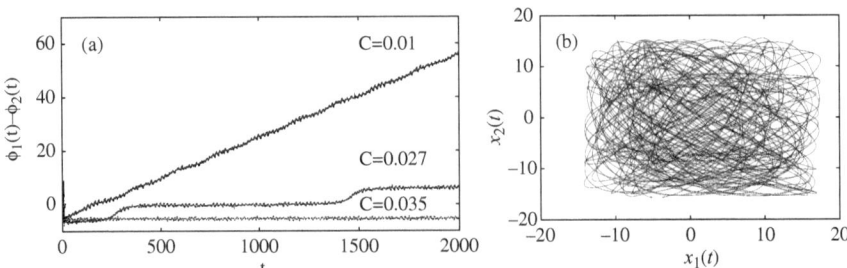

Fig. B.2 (**a**) Phase difference of two coupled Rössler systems (B.12) versus time for nonsynchronous ($C = 0.01$), nearly synchronous ($C = 0.027$) and synchronous ($C = 0.035$) states and (**b**) Amplitudes of (B.12) that remain uncorrelated for phase synchronous case. The frequency mismatch is $\Delta\omega = 0.015$ and the value of the parameter $a = 0.15$

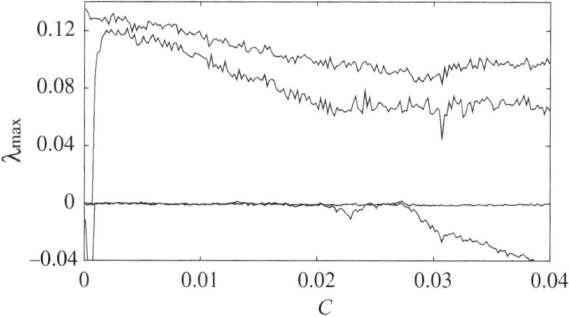

Fig. B.3 The four largest Lyapunov exponents of coupled coupled Rössler systems (B.12) as a function of the coupling strength C

B.2.3 Lag Synchronization

It has been shown in the previous section that when nonidentical chaotic oscillators are weakly coupled, the phases can be locked while the amplitudes remain highly uncorrelated. On further increase of the coupling strength, a relationship between the amplitudes may be established. Indeed, it has been demonstrated that there exists a regime of *lag synchronization* [33] where the states of the two oscillators are nearly identical, but one system lags in time with the other, that is, $Y(t) = X(t - \tau)$, $\tau > 0$.

To characterize lag synchronization quantitatively, Rosenbulm et al. [33] have introduced the notion of similarity function $S_l(\tau)$ as a time averaged difference between the variables x_1 and x_2 (with mean values being subtracted) taken with the time shift τ,

$$S_l^2(\tau) = \frac{\langle [x_2(t + \tau) - x_1(t)]^2 \rangle}{[\langle x_1^2(t) \rangle \langle x_2^2(t) \rangle]^{1/2}}, \tag{B.13}$$

where $\langle x \rangle$ means time average over the variable x, and $x_1(t)$ and $x_2(t)$ are the state variables of the drive and response systems, respectively. If the signals $x_1(t)$ and $x_2(t)$ are independent, the difference between them is of the same order as the signals themselves. If $x_1(t) = x_2(t)$, as in the case of complete synchronization, the similarity function reaches a minimum so that $S(\tau) = 0$ for $\tau = 0$. But for the case of nonzero positive value of time shift τ, if $S_l(\tau) = 0$, then there exists a time shift τ between the two signals $x_1(t)$ and $x_2(t)$ such that $x_2(t) = x_1(t - \tau)$, demonstrating lag synchronization.

We will consider the coupled Rössler systems (B.12) again for illustrative purpose with the same parameters as in the previous section except that the frequency mismatch now is given by $\omega_{1,2} = 0.97 \pm \Delta\omega$ with $\Delta\omega = 0.02$ [33] and the value of the parameter a is chosen as $a = 0.165$. It was noted in the previous section that as the coupling is increased from zero there exists entrainment of phases of the coupled systems in the weak coupling limit. As the coupling strength is increased further one can expect a stronger correlation in the amplitude resulting in the onset

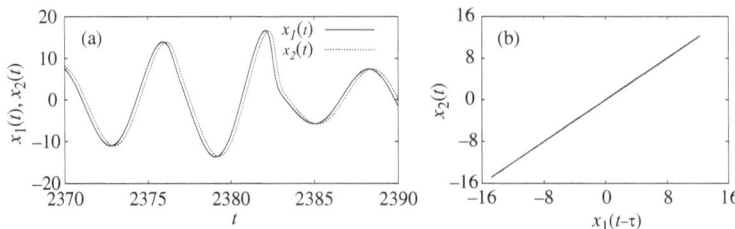

Fig. B.4 (**a**) Time series plot of the state variables $x_{1,2}$ showing the state of one of the systems evolving with a time lag $\tau = 0.21$ to the state of the other variable for the value of the coupling strength $C = 0.2$ and (**b**) Projection of the attractor of the coupled system on the delayed-coordinates, plot of $x_1(t - \tau)$ Vs $x_2(t)$, demonstrating that the state of one of the oscillators is delayed in time with respect to the other for the above values of the parameters

of lag synchronization for an appropriate value of the coupling strength. In fact, one finds that for $C = 0.2$, the state of one of the oscillators, x_2, lags in time to that of the other, x_1, with a lag time $\tau = 0.21$ which is illustrated in Fig. B.4a. Projection of the attractor of the coupled system (B.12) on the delayed-coordinate $x_1(t - \tau)$ Vs $x_2(t)$ is shown in Fig. B.4b.

B.2.4 Anticipatory Synchronization

It has also been shown that certain kinds of coupled chaotic systems may synchronize so that the response "anticipates" the driver, $Y(t) = X(t+\tau)$, by synchronizing with the future states. In [36] different unidirectional coupling schemes are considered such as a nonlinear time-delayed feedback either in the driver or in both the coupled systems. The results confirm that the anticipating synchronization can be globally stable due to the interplay between delayed feedback and dissipation, for any relatively small value of the lag time between response and driver. In addition, it has been shown that it is possible to achieve anticipation times larger than the characteristic time scales of the system dynamics, thus introducing a novel way of reducing the unpredictability of chaotic dynamics [37].

Anticipatory synchronization can also be characterized using the same similarity function $S_l(\tau)$ but with a negative time shift $\tau < 0$ instead of the positive time shift $\tau > 0$ in Eq. (B.13). In other words, one may define the similarity function for anticipatory synchronization as

$$S_a^2(\tau) = \frac{\langle [x_1(t - \tau) - x_2(t)]^2 \rangle}{\left[\langle x_1^2(t) \rangle \langle x_2^2(t) \rangle \right]^{1/2}}, \ \tau < 0 \qquad (B.14)$$

Then the minimum of $S_a(\tau)$, that is $S_a(\tau) = 0$, indicates that there exists a time shift $-\tau$ between the two signals $x_1(t)$ and $x_2(t)$ such that $x_2(t) = x_1(t - \tau)$, $\tau < 0$, demonstrating anticipatory synchronization.

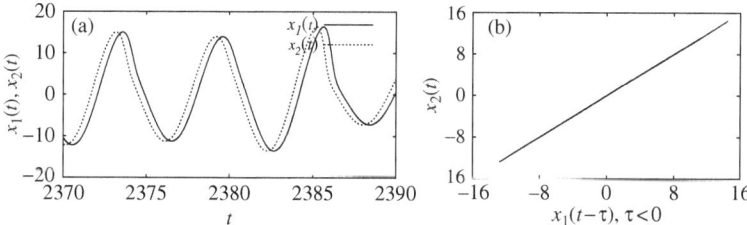

Fig. B.5 (a) Time series plot of the state variables $x_{1,2}$ showing that the drive $x_2(t)$ anticipates the state of the response system $x_1(t)$ with an anticipating time $|\tau| = 0.4$ for the value of the coupling strength $C = 1.0$ and (b) Projection of the attractor of the coupled system on the delayed-coordinates, $x_1(t - \tau)$ Vs $x_2(t)$, $\tau < 0$, demonstrating the existence of anticipating synchronization between the drive $x_1(t)$ and response $x_2(t)$ variables

As an illustrative example, we will consider the following unidirectionally coupled Rössler systems [36],

$$\dot{x}_1 = -y_1 - z_1, \tag{B.15a}$$
$$\dot{y}_1 = x_1 + ay_1, \tag{B.15b}$$
$$\dot{z}_1 = 0.2 + z_1(x_1 - 10), \tag{B.15c}$$
$$\dot{x}_2 = -y_2 - z_2 + C(x_1 - x_{2,\tau}), \tag{B.15d}$$
$$\dot{y}_2 = x_2 + ay_2, \tag{B.15e}$$
$$\dot{z}_2 = 0.2 + z_2(x_2 - 10), \tag{B.15f}$$

where the parameter a is fixed as 0.15. It can be easily checked that the above coupled systems exhibit anticipatory synchronization for small values of delay τ upon increasing the coupling strength. Figure B.5a illustrates that the response $x_2(t)$ anticipates the state of the drive $x_1(t)$ with anticipating time $\tau = 0.4$ for the value of the coupling strength $C = 1.0$ and the projection of the attractor of the coupled system (B.15) on the delayed-coordinates, $x_1(t)$ Vs $x_2(t - \tau)$, is shown in Fig. B.4b.

B.2.5 Generalized Synchronization

In general, completely identical synchronization may not be expected in nonidentical systems because there does not exist an invariant manifold $x = y$. In such cases where there exists an essential difference between the coupled systems, there is no hope to have a trivial manifold in the phase space attracting the system trajectories, and therefore it is not clear at a first glance whether nonidentical chaotic systems can synchronize. However, many works have shown that it is possible to generalize the concept of synchronization to include nonidenticity between the coupled systems and this phenomenon is called *generalized synchronization* [7, 39–41].

In order to define *generalized synchronization* (GS), let us consider the following coupled system

$$\dot{X} = F(X), \tag{B.16a}$$
$$\dot{Y} = G(Y, H_\mu(X)), \tag{B.16b}$$

where X is the n-dimensional state vector of the driver and Y is the m-dimensional state vector of the response. F and G are vector fields, $F : R^n \rightarrow R^n$, and $G : R^m \rightarrow R^m$. The coupling between the response and the driver is provided by the vector filed $H_\mu(X) : R^n \rightarrow R^m$, where the dependence of this function upon the parameters μ is explicitly considered. When $\mu = 0$, the response system evolves independently of the driver, and we assume that both systems are chaotic.

Some differences in the definition of GS exists in the literature. However, we will discuss here a more general definition given in [39, 40, 72]. When $\mu \neq 0$, the chaotic trajectories of the two systems are said to be synchronized in a generalized sense if there exists a transformation $\psi : X \rightarrow Y$ which is able to map asymptotically the trajectories of the driver attractor into the ones of the response attractor $Y(t) = \psi(X(t))$, regardless of the initial condition in the basin of the synchronization manifold $M = (X, Y) : Y = \psi(X)$.

The difference between various definitions of GS is based on the mathematical properties required for the map ψ. Reference [67] distinguishes between two types of GS, namely the so-called *weak synchronization* and *strong synchronization*. The latter corresponds to the case of a map ψ which is smooth, in the sense of being differentiable; on the other hand the former corresponds to the case of a map ψ which is non-smooth, in the sense of being not differentiable. Even a stronger version of strong synchronization is considered in [73], called *differentiable generalized synchronization*, requiring continuous differentiability of ψ. All of these different approaches have relevant consequences when one looks for the existence of GS in experimental situations. The stability of the manifold M of GS can be determined as in the case of CS, that is, by the negativity of conditional Lyapunov exponents [67] and the use of Lyapunov functions [40].

As an example, we consider the system studied in [74] where the drive system is described by

$$\mu \dot{x}_1 = y_1, \tag{B.17a}$$
$$\mu \dot{y}_1 = -x_1 - \delta y_1 + z_1, \tag{B.17b}$$
$$\mu \dot{z}_1 = \gamma(\alpha_1 f(x_1) - z_1) - \sigma y_1, \tag{B.17c}$$

which is realized in experiments with electrical chaotic circuits [75]. The response system equations are

$$\dot{x}_2 = y_2, \tag{B.18a}$$
$$\dot{y}_2 = -x_2 - \delta y_2 + z_2, \tag{B.18b}$$
$$\dot{z}_2 = \gamma(\alpha_2 f(x_2) - z_2 + gx_1) - \sigma y_2, \tag{B.18c}$$

where g is the coupling strength, and $\gamma = 0.294, \sigma = 1.52, \delta = 0.534$, and $\alpha_2 = 16.7$ are fixed system parameters. The nonlinear function $f(x)$ models the

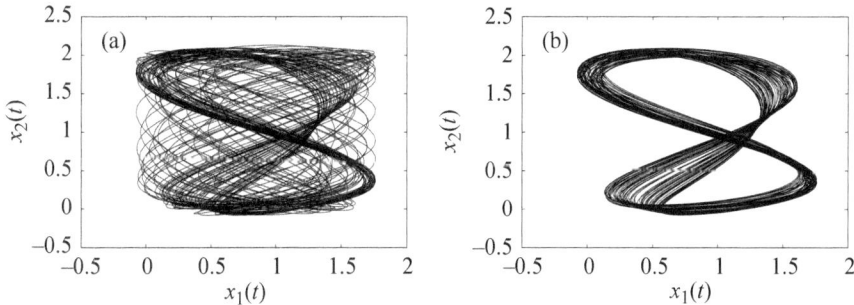

Fig. B.6 Projection of attractor constructed from the drive (B.17) and response attractors (B.18) and plotted for (x_1, x_2). (**a**) For $\alpha_1 = 15.94$ showing desynchronized state and (**b**) For $\alpha_1 = 15.93$ showing generalized synchronized state

input-output characteristics of a nonlinear converter in the circuit [74, 75]. The parameter μ in the drive system equations is the time scaling parameter that is used to select the desired frequency ratio of the synchronization. For the parameter values $g = 3.0$, $\mu = 0.498$ and $\alpha_1 = 15.94$ the above systems are in asynchronous state which is shown in Fig. B.6a and as the value of α is decreased to $\alpha = 15.93$ the above systems display generalized synchronization as illustrated in Fig. B.6b.

References

1. C. Huygens, *Horologium Oscillatorium* (Apud F. Muguet, France, 1673)
2. A.S. Pikovsky, M.G. Rosenblum, J. Kurths, *Synchronization – A Unified Approach to Nonlinear Science* (Cambridge University Press, Cambridge, 2001)
3. E.V. Appleton, Proc. Cambridge Philos. Soc. (Math. Phys. Sci.) **21**, 231 (1922)
4. B. van der Pol, J. van der Mark, Nature **120**, 363 (1927)
5. H. Fujisaka, T. Yamada, Prog. Theor. Phys. **69**, 32 (1983)
6. H. Fujisaka, T. Yamada, Prog. Theor. Phys. **70**, 1240 (1983)
7. V.S. Afraimovich, N.N. Verichev, M.I. Rabinovich, Izvestiya Vysshikh Uchebnykh Zavedenii Radiofizika **29**, 3050 (1986)
8. L.M. Pecora, T.L. Carroll, Phys. Rev. Lett. **64**, 821 (1990)
9. A.S. Pikovsky, Z. Phys. B **55**, 149 (1984)
10. L.M. Pecora, T.L. Carroll, Phys. Rev. A **44**, 2374 (1991)
11. S. Boccaletti, J. Kurths, G. Osipov, D.L. Valladares, C.S. Zhou, Phys. Rep. **366**, 1 (2002)
12. S. Hayes, C. Grebogy, E. Ott, Phys. Rev. Lett. **70**, 3031 (1993)
13. K.M. Cuomo, A.V. Oppenheim, Phys. Rev. Lett. **71**, 65 (1993)
14. T. Kapitaniak, Phys. Rev. E **50**, 1642 (1994)
15. G. Pérez, H.A. Cerdeira, Phys. Rev. Lett. **74**, 1970 (1995)
16. J.H. Peng, E.J. Ding, M. Ding, W. Yang, Phys. Rev. Lett. **76**, 904 (1996)
17. K. Pyragas, Phys. Lett. A **181**, 203 (1993)
18. A. Kittle, K. Pyragas, R. Richter, Phys. Rev. E **50**, 262 (1994)
19. R. Brown, N.F. Rulkov, E.R. Tracy, Phys. Rev. E **49**, 3784 (1994)
20. U. Parlitz, Phys. Rev. Lett. **76**, 1232 (1996)
21. R. He, P.V. Vaidya, Phys. Rev. A **46**, 7387 (1992)
22. R.E. Amritkar, N. Gupte, Phys. Rev. E **47**, 3889 (1993)

23. L. Kocarev, U. Parlitz, Phys. Rev. Lett. **74**, 5028 (1995)
24. U. Parlitz, L. Kocarev, T. Stojanovski, H. Preckel, Phys. Rev. E **53**, 4351 (1996)
25. J. Guemez, M.A. Matias, Phys. Rev. E **52**, R2145 (1995)
26. C.W. Wu, L.O. Chua, Int. J. Bifurcat. Chaos **4**, 979 (1994)
27. E. Ott, C. Grebogi, J.A. Yorke, Phys. Rev. Lett. **64**, 1196 (1990)
28. M. Lakshmanan, K. Murali, *Chaos in Nonlinear Oscillators: Controlling and Synchronization* (World Scientific, Singapore, 1996)
29. M. Lakshmanan, S. Rajasekar, *Nonlinear Dynamics: Integrability, Chaos and Patterns* (Springer, Berlin, 2003)
30. M.G. Rosenblum, A.S. Pikovsky, J. Kurths, Phys. Rev. Lett. **76**, 1804 (1996)
31. T. Yalcinkaya, Y.C. Lai, Phys. Rev. Lett. **79**, 3885 (1997)
32. E.R. Rosa, E. Ott, M.H. Hess, Phys. Rev. Lett. **80**, 1642 (1998)
33. M.G. Rosenblum, A.S. Pikovsky, J. Kurths, Phys. Rev. Lett. **78**, 4193 (1997)
34. S. Rim, I. Kim, P. Kang, Y.J. Park, C.M. Kim, Phys. Rev. E **66**, 015205(R) (2002)
35. M. Zhan, G.W. Wei, C.H. Lai, Phys. Rev. E **65**, 036202 (2002)
36. H.U. Voss, Phys. Rev. E **61**, 5115 (2002)
37. H.U. Voss, Phys. Rev. Lett. **87**, 014102 (2001)
38. C. Masoller, Phys. Rev. Lett. **86**, 2782 (2001)
39. N.F. Rulkov, M.M. Sushchik, L.S. Tsimring, H.D.I. Abarbanel, Phys. Rev. E **51**, 980 (1995)
40. L. Kocarev, U. Parlitz, Phys. Rev. Lett. **76**, 1816 (1996)
41. R. Brown, Phys. Rev. Lett. **81**, 4835 (1998)
42. S. Boccaletti, D.L. Valladares, Phys. Rev. E **62**, 7497 (2000)
43. S. Taherion, Y.C. Lai, Phys. Rev. E **59**, R6247 (1999)
44. D.L. Valladares, S. Boccaletti, Int. J. Bifurcat. Chaos **11**, 2699 (2001)
45. D.V. Senthilkumar, M. Lakshmanan, Chaos **17**, 013112 (2007)
46. A.E. Hramov, A.A. Koronovskii, Europhys. Lett. **79**, 169 (2005)
47. A. Pikovsky, G. Osipov, M. Rosenblum, M. Zaks, J. Kurths, Phys. Rev. Lett. **79**, 47 (1997)
48. A. Pikovsky, M. Zaks, M. Rosenblum, G. Osipov, J. Kurths, Chaos **7**, 680 (1997)
49. K.J. Lee, Y. Kwak, T.K. Lim, Phys. Rev. Lett. **81**, 321 (1998)
50. M.A. Zaks, E.-H. Park, M.G. Rosenblum, J. Kurths, Phys. Rev. Lett. **82**, 4228 (1999)
51. R. Femat, G. Solis-Perales, Phys. Lett. A **262**, 50 (1999)
52. A.E. Hramov, A.A. Koronovskii, Physica D **206**, 252 (2005)
53. I. Fischer, R. Vicente, J. M. Buldu, M. Peil, C.R. Mirasso, M.C. Torrent, J. Garcia-Ojalvo, Phys. Rev. Lett. **97**, 123902 (2006)
54. M. Zhan, Y. Wang, X. Gang, G.W. Wei, C.H. Lai, Phys. Rev. E **68**, 036208 (2003)
55. A. Locquet, F. Rogister, M. Sciamanna, P. Megret, M. Blandel, Phys. Rev. E **64**, 045203(R) (2001)
56. R. Brown, L. Kocarev, Chaos **10**, 344 (2000)
57. S. Boccaletti, L.M. Pecora, A. Pelaez, Phys. Rev. E **63**, 066219 (2001)
58. A.E. Hramov, A.A. Koronovskii, Chaos **14**, 603 (2004)
59. F. Verhulst, *Nonlinear Differential Equations and Dynamical Systems* (Springer, Berlin, Heidelberg, 1990)
60. L. Yu, E. Ott, Q. Chen, Phys. Rev. Lett. **65**, 2935 (1990)
61. D.J. Gauthier, J.C. Bienfang, Phys. Rev. Lett. **77**, 1751 (1996)
62. G. Benettin, C. Froeschlé, H.P. Scheidecker, Phys. Rev. A **19**, 454 (1976)
63. I. Shimada, T. Nagashima, Prog. Theor. Phys. **61**, 1605 (1979)
64. A. Wolf, J.B. Swift, H.L. Swinney, J.A. Vastano, Physica D **16**, 285 (1985)
65. J.L. Willems, *Stability Theory of Dynamical Systems* (Wiley, New York, 1970)
66. K. Murali, M. Lakshmanan, Phys. Rev. E **49**, 4882 (1994)
67. K. Pyragas, Phys. Rev. E **54**, R4508 (1996)
68. C. Sparrow, *The Lorenz Equations, Bifurcations, Chaos, and Strange Attractors*, (Springer, New York, 1982)
69. G.V. Osipov, A.S. Pikovsky, M.G. Rosenblum, J. Kurths, Phys. Rev. E **55**, 2353 (1997)

70. D. Gabor, J. IEE London **93**, 429 (1946)
71. O.E. Rössler, Phys. Let. A **57**, 397 (1976)
72. H.D.I. Abarbanel, N.F. Rulkov, M.M. Sushchik, Phys. Rev. E **53**, 4528 (1996)
73. B.R. Hunt, E. Ott, J.A. Yorke, Phys. Rev. E **55**, 4029 (1997)
74. N.F. Rulkov, C.T. Lewis, Phys. Rev. E **63**, 065204(R) (2001)
75. N.F. Rulkov, Chaos **6**, 262 (1996)

Appendix C
Recurrence Analysis

C.1 Introduction

The concept of recurrence dates back to Poincaré [1], who proved that after a sufficiently long time the trajectory of a chaotic system in phase space will return arbitrarily close to any former point of its path with probability one. However, the concept of recurrence within the framework of chaotic systems was not considered until the sixties, when the now famous Lorenz equation was derived by E. Lorenz as a simplified equation of convection rolls [2, 3]. Later in 1987, Eckmann et al. introduced the method of recurrence plots (RPs), a technique that visualizes the recurrences of a dynamical system and gives information about the behavior of its trajectory in phase space [4]. This technique has become popular in the last decade because of its applicability to rather short and non-stationary time series. Further, cross recurrence plots (CRPs) (a bivariate extension of the RP) was introduced by Zbilut et al. [5] and Marwan and Kurths [6] to analyse the dependencies between two different systems by comparing their states [5, 6]. As an extension of CRPs to analyse physically different systems with different phase space dimensions, joint recurrence plots (JRPs) were introduced. Also, in order to go beyond the visual inspection of RPs, several measures of complexity which quantify the small scale structures in RPs have been proposed [7–10] and are known as recurrence quantification analysis (RQA). These measures are based on the recurrence point density and the diagonal and vertical line structures of the RP. Furthermore, a more theoretical study of the relationship between RPs and the properties of dynamical systems has also been addressed [10–15]. The concept of recurrence plots and its measures have been applied in numerous fields of research including astrophysics [16, 17], earth sciences [18–20], engineering [21, 22], biology [23, 24] and cardiology/neuroscience [25–28]. In the following, we describe briefly the concept of recurrence plots along with CRP and JRP. We will also discuss the various quantification measures introduced to characterize synchronization transitions in coupled chaotic systems.

C.2 Recurrence Plots and Their Variants

In this section, we will describe briefly the concept of recurrence plots and their variants such as cross recurrence plots and joint recurrence plots to analyse the data of different physical systems of same or even different dimensions along with suitable illustrations.

C.2.1 Recurrence Plots

As mentioned in the introduction, RPs provide a visual impression of the trajectory of a dynamical system in phase space. Suppose that the time series $\{X_i\}_{i=1}^N$ representing the trajectory of a system in phase space is given, with $X_i \in \mathbb{R}^d$. The RP efficiently visualises recurrences and can be formally expressed by the matrix

$$\mathbb{R}_{i,j} = \Theta(\varepsilon - ||X_i - X_j||), \qquad i, j = 1, \cdots, N, \qquad (C.1)$$

where N is the number of measured points X_i, ε is a predefined threshold, Θ is the Heaviside function (i.e. $\Theta(x) = 0$, if $x < 0$, and $\Theta(x) = 1$ otherwise) and $||.||$ is the Euclidean norm. For ε-recurrent states, that is for states which are in an ε-neighbourhood, we have the following notion:

$$X_i \approx X_j \iff \mathbb{R}_{i,j} \equiv 1. \qquad (C.2)$$

The graphical representation of the matrix $\mathbb{R}_{i,j}$ is called recurrence plot (RP). The RP is obtained by plotting the recurrence matrix, Eq. (C.1), using different colors for its binary entries, for example by marking a black dot at the coordinates (i, j), if $\mathbb{R}_{i,j} = 1$, and a white dot, if $\mathbb{R}_{i,j} = 0$. Since $\mathbb{R}_{i,i} \equiv 1 \mid_{i=1}^N$ by definition, the RP has always a black main diagonal line. Furthermore, the RP is symmetric by definition with respect to the main diagonal, that is $\mathbb{R}_{i,j} \equiv \mathbb{R}_{j,i}$.

A crucial parameter of an RP is the threshold ε. Therefore, special attention has to be required for its choice. If ε is chosen too small, there may be almost no recurrence points and we cannot learn anything about the recurrence structure of the underlying system. On the other hand, if ε is chosen too large, almost every point is a neighbour of every other point, which leads to a lot of artefacts. A too large ε includes also points into the neighbourhood which are simple consecutive points on the trajectory. Hence, one has to find an appropriate value for ε. Moreover, the influence of noise can entail choosing a larger threshold, because noise would distort any existing structure in the RP [10].

Several methods have been advocated in the literature to estimate the value of threshold ε with their own advantages and disadvantages which has been discussed in [10]. Among them, we use the approach that preserves the fixed recurrence point density. In order to find an ε which corresponds to a fixed recurrence point density or recurrence rate (RR) defined as

$$RR(\varepsilon) = \frac{1}{N^2} \sum_{i,j=1}^{N} \mathbb{R}_{i,j}(\varepsilon), \tag{C.3}$$

the cumulative distribution of the N^2 distances between each pair of vectors can be used. The RRth percentile is then the required ε. An alternative is to fix the number of neighbours for every point of the trajectory. In this case, the threshold is actually different for each point of the trajectory. The advantage of these two methods is that both of them preserve the recurrence point density and allow one to compare RPs of different systems without the necessity of normalising the time series beforehand. Nevertheless, the choice of ε depends strongly on the system under study.

For illustration, we will show the RPs of three different motions, namely (i) of a periodic motion on a circle (Fig. C.1a), (ii) of a chaotic attractor of Rössler system (Fig. C.1b) and (iii) of a Gaussian white noise (Fig. C.1c). In all our simulation, we have chosen the threshold value for ε as $\varepsilon = 0.03RR$ and the sampling interval to be $\Delta t = 0.1$. The RP of the purely periodic oscillation shown in Fig. C.1a consists of uninterrupted diagonal lines separated by the distance T, where T is the period of the oscillation. This is due to the fact that the position of the system in the phase space recurs exactly at the same point after a cycle and hence one has identical recurrence. The RP of Gaussian white noise depicted in Fig. C.1c is rather homogeneous, consisting of mainly single points, indicating the randomness of its behavior. The RP of chaotic attractor of Rössler system is illustrated in Fig. C.1b, which shows that the predominant structures are intermediate between that of periodic oscillations and that of purely stochastic motions. The RP of Rössler attractor also shows diagonal lines which are shorter (interrupted) and the vertical distance between the diagonal lines is not constant because of the multiple time scales of the chaotic system. The interrupted diagonal lines are due to the exponential divergence of nearby trajectories (sensitive to slightly different initial conditions). However, on the upper right of Fig. C.1b, there is a small rectangular patch which rather looks like the RP of the periodic motion and this structure corresponds to an unstable periodic orbit of the Rössler attractor [10]. It is also conjectured that shorter the diagonals in the RP, the less the predictability of the system [29], and indeed it was suggested that the inverse of the longest diagonal (except the main diagonal for which $i = j$) is proportional to the largest Lyapunov exponent of the system by Eckmann et al. [4].

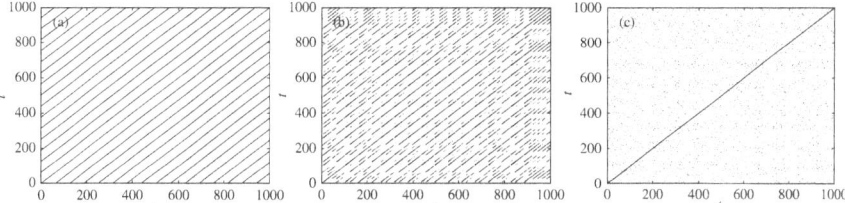

Fig. C.1 Recurrence Plots of (**a**) a periodic oscillation, (**b**) a chaotic attractor of Rössler system and (**c**) a Gaussian white noise

C.2.2 Cross Recurrence Plots (CRP)

As mentioned in the introduction, CRP is a bivariate extension of the RP and was introduced to analyse the difference between two different systems [5, 6]. CRPs can be regarded as a generalisation of the linear cross-correlation function [10]. The cross recurrence matrix, analogous to RP, of two dynamical systems represented by the trajectories X and Y in a d-dimensional phase space is defined by

$$CR_{i,j}^{X,Y} = \Theta(\varepsilon - ||X_i - Y_j||), \qquad i = 1, \cdots, N, \qquad j = 1, \cdots, M, \quad (C.4)$$

where N and M are the lengths of the trajectories X and Y, respectively. Note that N may not be equal to M and hence the matrix CR is not necessarily a square matrix. As a CRP is plotted for those times when a state of the first system recurs to that of the other system, both the systems are represented in the same phase space. The components of X_i and Y_i are usually normalised before computing the cross recurrence matrix, while the other possibilities are to use the fixed amount of neighbours for each X_i in which case the components need not be normalised. It has been shown that the latter choice of the fixed neighborhood has the additional advantage of suitability for slowly changing trajectories [10].

As an illustration, the CRP of the coupled Rössler systems (Eq. (B.12)) for the same value of the parameters as in Sect. B.2.2 and for the value of the coupling strength $C = 0.01$ is shown in Fig. C.2. As the values of the main diagonal $CR_{i,i}$ are not necessarily unity, CRPs do not have a black main diagonal line as in RPs as in Fig. C.2. It has been shown that measures based on the length of the diagonally oriented lines are used to find the nonlinear interactions between two systems, which cannot be detected by the common cross-correlation function [6, 10]. An important property of CRPs is that they reveal the local difference of the dynamical evolution of close trajectory segments, represented by bowed lines. A time dilation or time compression of one of the trajectories causes a distortion of the diagonal lines. For

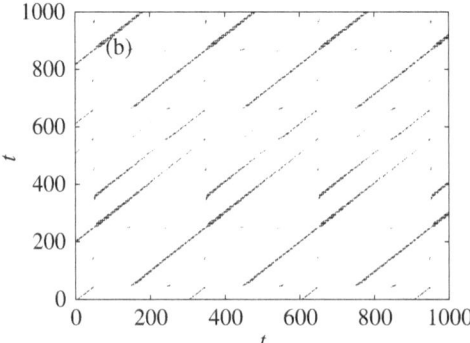

Fig. C.2 Cross recurrence plot of the coupled Rössler systems (Eq. (B.12)) for the same value of the parameters as in Sect. B.2.2 and for the value of the coupling strength $C = 0.01$

two identical trajectories, the CRP is the RP of a single trajectory and contains the main black diagonal line.

C.2.3 *Joint Recurrence Plots (JRP)*

We have seen above that CRP can be used to analyse the interrelation between two different systems. However, CRP cannot be used to analyse two physically different systems because the two different physical units or different phase space dimensions do not make sense in computing CRP. A different possibility to compare the states of different systems is to consider the recurrences of their trajectories in their corresponding phase spaces separately and then look for the times when both of them recur simultaneously, that is when joint recurrence occurs. The individual phase spaces are preserved by this approach and different thresholds for each system ε^X and ε^Y are considered, in respect of the natural measure of both the systems. Joint recurrence matrix for two systems X and Y can be defined as

$$JR_{i,j}^{X,Y}(\varepsilon^X, \varepsilon^Y) = \Theta(\varepsilon^X - ||X_i - X_j||)\Theta(\varepsilon^Y - ||Y_i - Y_j||), \qquad i, j = 1, \cdots, N. \tag{C.5}$$

JRP of the coupled Rössler systems (Eq. (B.12)) for the same value of the parameters as in Sect. B.2.2 and for the value of the coupling strength $C = 0.01$ is shown in Fig. C.3.

The bivariate joint recurrence plot can be generalized to analyse n systems $(X_{(1)}, X_{(2)}, ..., X_{(n)})$ by using multivariate joint recurrence matrix, which can be represented using Eq. (C.1) as

$$JR_{i,j}^{X_{(1,2,...,n)}}(\varepsilon^{X_{(1)}}, ..., \varepsilon^{X_{(n)}}) = \prod_{k=1}^{n} \mathbb{R}_{i,j}^{X_{(k)}}(\varepsilon^{X_{(k)}}), i, j = 1, \cdots, N. \tag{C.6}$$

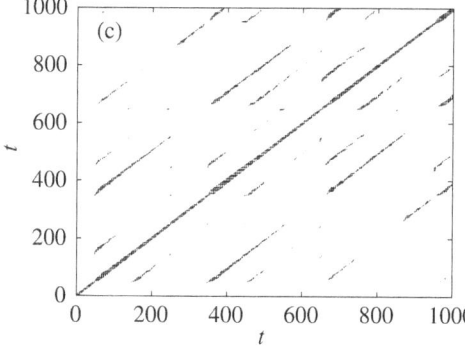

Fig. C.3 Joint recurrence plot of the coupled Rössler systems (Eq. (B.12)) for the same value of the parameters as in Sect. B.2.2 and for the value of the coupling strength $C = 0.01$

In addition, a delayed version of the joint recurrence matrix can also be introduced as

$$JR_{i,j}^{X,Y}(\varepsilon^X, \varepsilon^Y, \tau) = \mathbb{R}_{i,j}^X(\varepsilon^X)\mathbb{R}_{i+\tau,j+\tau}^Y(\varepsilon^Y), \qquad i, j = 1, \cdots, N - \tau, \quad (C.7)$$

to analyse the interacting delayed systems [10]. JRP is invariant under permutation of the coordinates in one or more of the systems. It can also be computed using a fixed amount of nearest neighbours. In this case, each RPs which contributes to the JRP is computed using the same number of nearest neighbours. These RPs obtained from CRP, JRP and their variants are exploited in quantifying several dynamical properties and their transitions using recurrence quantification analysis as discussed in the next section.

C.3 Recurrence Quantification Analysis (RQA)

Several measures of complexity which quantify the small scale structures in RPs have been proposed and are known as recurrence quantification analysis. These measures are based on the recurrence point density, the diagonal and vertical line structures of the RP. Studies based on RQA measures show that they are able to identify bifurcation points, including chaos-order and chaos-chaos transitions [10]. Several recurrence quantification measures have been introduced for different requirements. Some of the most important measures include Recurrence Rate (RR), Determinism (DET), Divergence (DIV), Entropy ($ENTR$), Trend ($TREND$), Ratio ($RATIO$), Linearity (LAM), Trapping Time (TT), Maximal vertical length (V_{max}), etc. It has also been shown that several dynamical invariants such as correlation entropy, correlation dimension, generalized mutual information, etc can also be calculated using RQA. Detailed discussion on all of the above RQAs can be found in [10] and, all of the methods and procedure discribed in this appendix are available in the CRP toolbox for Matlab (Provided by TOCSY: http://tocsy.agnld.uni-potsdam.de). However, in the following, we will focus our discussion on some of the RQAs that have been introduced to characterize and to identify different kinds of synchronization transitions in coupled chaotic systems.

C.3.1 Generalized Autocorrelation Function, P(t)

Generalized autocorrelation function $P(t)$ has been defined as [10, 30]

$$P(t) = \frac{1}{N - t} \sum_{i=1}^{N-t} \Theta(\varepsilon - ||X_i - X_{i+t}||). \qquad (C.8)$$

If any two coupled oscillators are in phase synchronization (PS), then the distances between the diagonal lines in their respective RPs coincide as their phases, and hence their time scales are locked to each other. As PS is characterized by entrainment in the phases of the interacting systems while their amplitudes remain uncorrelated, their respective RPs remain non-identical. However, if the probability that the first oscillator recurs after t time steps is high, then the probability that the second oscillator recurs after the same time interval is also high, and vice versa. Therefore, looking at the probability $P(t)$ that the system recurs to the ε neighborhood of a former point of the trajectory X after t time steps and comparing $P(t)$ of both the system allows to detect and quantify PS.

Generalized autocorrelation function $P(t)$ can be considered as a statistical measure about how often the phase ϕ has increased by 2π or multiples of 2π within the time t in the original space. If two systems are in a phase synchronized state, their phases increase on the average by $K.2\pi$, where K is a natural number, within the same time interval t. The value of K corresponds to the number of cycles when $||X(t+T) - X(t)|| \sim 0$, or equivalently when $||X(t+T) - X(t)|| < \varepsilon$, where T is the period of the system. Hence, looking at the coincidence of the positions of the maxima of $P(t)$ for both the systems, one can qualitatively identify CPS. It is to be noted that the heights of the local maxima are in general different for both systems if they are only in PS.

C.3.2 Correlation of Probability of Recurrence (CPR)

A criterion to quantify phase synchronization between two systems is the cross correlation coefficient between $P_1(t)$ and $P_2(t)$ ($P_1(t)$ represents the probability of recurrence of the first system and $P_2(t)$ that of the second system) which can be defined as Correlation of Probability of Recurrence (CPR)

$$CPR = \langle \bar{P}_1(t)\bar{P}_2(t)\rangle/\sigma_1\sigma_2, \tag{C.9}$$

where $\bar{P}_{1,2}$ means that the mean value has been subtracted and $\sigma_{1,2}$ are the standard deviations of $P_1(t)$ and $P_2(t)$, respectively. If both systems are in CPS, the probability of recurrence is maximal at the same time t and CPR ≈ 1. If they are not in CPS, the maxima do not occur simultaneously and hence one can expect a drift in both the probability of recurrences and low values of CPR.

It has also shown that this method is highly efficient even for non-phase coherent oscillators and it is able to detect PS even in time series which are strongly corrupted by noise. One of the most important applications of this method is that it can also be applied to experimental time series with noise.

C.3.3 Joint Probability of Recurrence (JPR)

Joint probability of recurrence to quantify the existence of generalized synchronization (GS) between two systems is defined as

$$JPR = \frac{S - RR}{1 - RR}, \tag{C.10}$$

where, $S = \frac{RR_{1,2}}{RR}$, $RR_{1,2}$ is the recurrence rate of the JRP of both the systems and $RR_1 = RR_2 = RR$ is the recurrence rate of the individual systems.

C.3.4 Similarity of Probability of Recurrence (SPR)

As the recurrence matrix contains only information about the neighborhood of each point of a time series, the RPs of systems in GS must be almost identical. Hence, it follows that their respective probabilities of recurrence must coincide and this suggests the similarity coefficient between $P_1(t)$ and $P_2(t)$ represented as

$$SPR = 1 - \langle (\bar{P}_1(t) - \bar{P}_2(t))^2 \rangle / \sigma_1 \sigma_2, \tag{C.11}$$

is of order 1 if both systems are in GS and approximately zero or negative if they evolve independently.

C.4 Synchronization and Recurrences

In this section, we will investigate the onset, existence and transition among different kinds of synchronizations by using recurrence plots and recurrence quantification analysis discussed above. It may be noted that these indices based on the recurrence are of considerable importance in synchronization analysis of experimental systems and, in particular, in the case of very small available data set. With these indices, one can quantify the degree of synchronization in complex interacting systems, specifically in the case of non-coherent attractors. These methods are more appropriate for non-stationary data. In the following, we will analyse (i) phase synchronization in mutually coupled Rössler systems [31] and (ii) transition from phase to lag synchronization again in mutually coupled Rössler systems [32] but in slightly different parameter regimes using recurrence plots and recurrence indices discussed above.

C.4.1 PS in Mutually Coupled Rössler Systems

Phase synchronization has already been discussed in detail in Sect. B.2.2 and it has been illustrated using mutually coupled Rössler systems [31]. Now, we will discuss about the structure of recurrence plots, the nature of generalized autocorrelation function, $P(t)$, and correlation of probability of recurrence, CPR, for two different values of the coupling strength corresponding to non-synchronized and phase synchronized state in these systems. It is well known that PS is characterized by entrainment in the phase of the interacting systems while their amplitudes remain uncorrelated. During PS, the phases get locked and so also the frequencies. Therefore, the recurrence plots of both the systems have the same distance (vertical) between the diagonal lines, which corresponds to the period of oscillation, while their respective RPs remain nonidentical.

Recurrence plot of both of the mutually coupled Rössler systems (Eq. (B.12)) for the same values of the parameters as in Sect. B.2.2 are shown in Fig. C.4a, b, respectively, for the value of coupling strength $C = 0.01$ in the non-synchronized regime. The generalized autocorrelation functions, $P_{1,2}(t)$ of both the systems are shown in Fig. C.4c, which indicates that the positions of local maxima are not in coincidence and there exists a drift between them indicating non-synchronized state. The value of correlation of probability of recurrence, $CPR = 0.022$, is rather low confirming the non-synchronized state. Similarly, RPs of both the systems are shown in Fig. C.5a, b, respectively, for the value of coupling strength $C = 0.035$ corresponding to PS regime. Now both $P_1(t)$ and $P_2(t)$ are in perfect coincidence in their positions of local maxima indicating PS (Fig. C.5c). In addition, the value of the correlation coefficient $CPR = 0.91$ which is rather high, indicating a high degree of PS.

The transition from non-synchronized state to PS and the onset of PS can also be clearly revealed by the index CPR. It has been demonstrated [31] that the onset of PS occurs at the value of coupling strength $C = 0.027$ and PS exists for values $C > 0.027$ as indicated by the Lyapunov exponents shown in Fig. C.6a in the range of coupling strength $C \in (0, 0.04)$. The onset of PS at this value is also clearly revealed

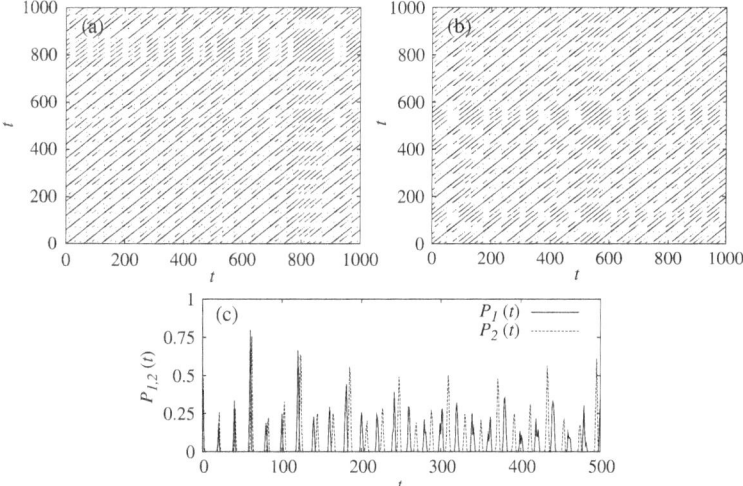

Fig. C.4 Recurrence plots of the coupled Rössler systems (Eq. (B.12)) for the same value of the parameters as in Sect. B.2.2 but for the value of the coupling strength $C = 0.01$ in the non-synchronized state. (**a**) First system, (**b**) Second system and (**c**) Generalized autocorrelation functions, $P_{1,2}(t)$, of both the systems

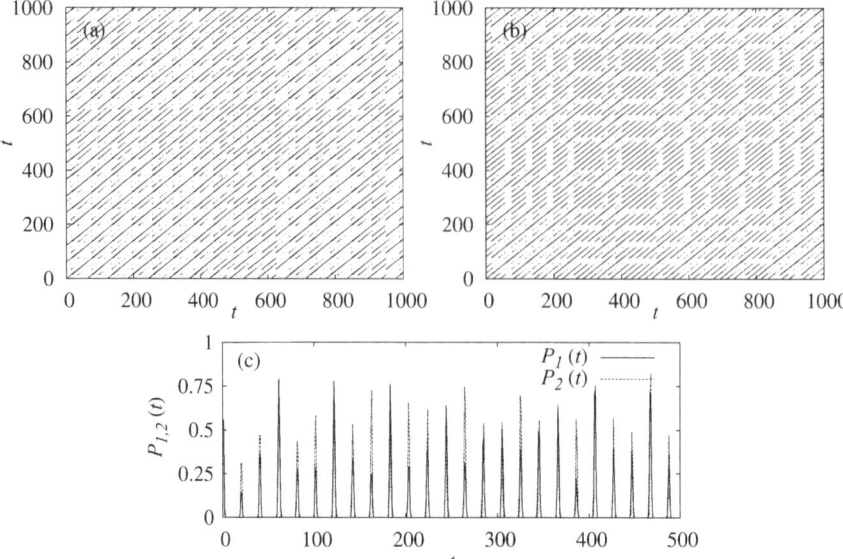

Fig. C.5 Recurrence plots of the coupled Rössler systems (Eq. (B.12)) for the same value of the parameters as in Sect. B.2.2 but for the value of the coupling strength $C = 0.035$ in the PS state. (**a**) First system, (**b**) Second system and (**c**) Generalized autocorrelation functions, $P_{1,2}(t)$, of both the systems

by the index *CPR* shown in Fig. C.6b in the same range of the coupling strength C of the mutually coupled Rössler systems (Eq. (B.12)). The value of the *CPR* shows a sudden increase in its value at $C = 0.027$ and above this value of coupling strength *CPR* fluctuates near to but less than unity characterizing the degree of PS.

C.4.2 Phase to Lag Synchronization

Lag synchronization (LS) has also been already discussed in Sect. B.2.3, along with an illustration as demonstrated in [32]. With the same values of parameters as discussed in Sect. B.2.3 for mutually coupled Rössler systems (Eq. (B.12)), we will characterize the transition from non-synchronized state to PS and then to an LS state using the recurrence indices. As LS is a special case of generalized synchronization (GS) all the discussion for LS will also hold for GS.

Since RPs and generalized autocorrelation functions for both the coupled systems are already shown in the non synchronized and PS regimes, we concentrate here on LS only. RPs and $P_{1,2}(t)$ of the mutually coupled Rössler systems (Eq. (B.12)) for

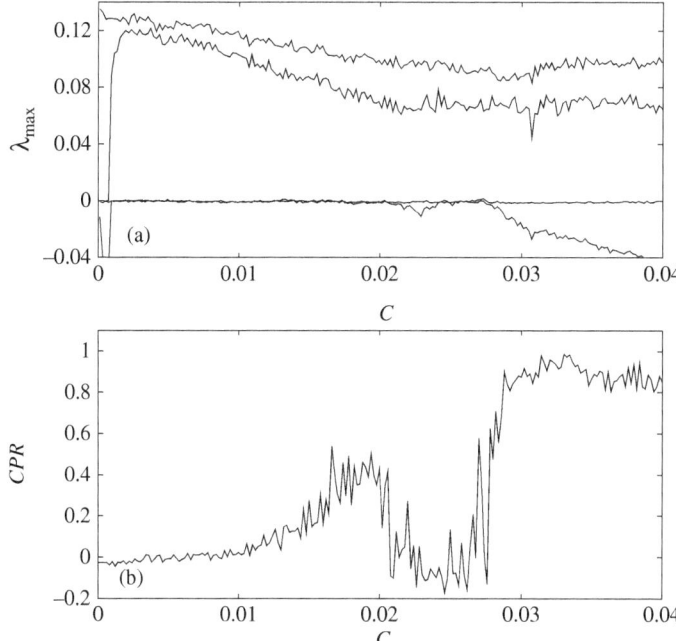

Fig. C.6 (**a**) Four largest Lyapunov exponents in the range of coupling strength $C \in (0, 0.04)$ of the mutually coupled Rössler systems (Eq. (B.12)) studied in Sect. B.2.2 and (**b**) Correlation of probability of recurrence, *CPR*, in the same range of the coupling strength characterizing the onset of PS

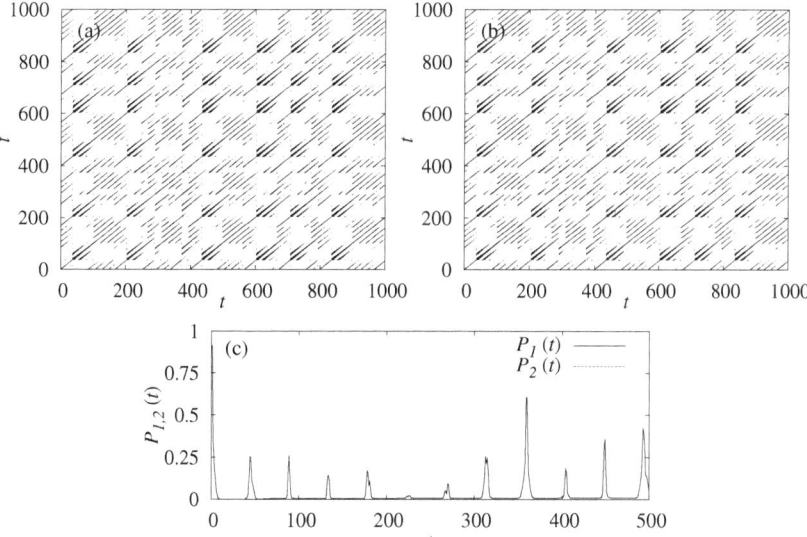

Fig. C.7 Recurrence plots of the coupled Rössler systems (Eq. (B.12)) for the same value of the parameters as in Sect. B.2.3 and for the value of the coupling strength $C = 0.2$ in the LS state. (**a**) First system, (**b**) Second system and (**c**) Generalized autocorrelation functions, $P_{1,2}(t)$, of both the systems

the value of coupling strength $C = 0.2$ is shown in Fig. C.7, where both the systems are in LS. It is evident that the RPs of both the systems are identical confirming the existence of lag (generalized) synchronization between the coupled systems. Furthermore, the generalized autocorrelation functions, $P_{1,2}(t)$, are also in perfect coincidence both with their positions and with their amplitudes confirming the existence of lag (generalized) synchronization. Correspondingly, the value of the indices $CPR = 0.881$ and $SPR = 0.999$ are rather high attributing to the degree of LS.

Transition from the non-synchronized state to PS and then from PS to LS in mutually coupled Rössler systems has been demonstrated in [32]. It has been shown that the onset of PS occurs at the critical value of the coupling strength $C_p = 0.036$ and that of LS occurs at $C_l = 0.14$ as indicated by the largest Lyapunov exponents of the coupled Rössler systems shown in Fig. C.8a. Indices CPR, JPR, SPR are depicted in Fig. C.8b in the range of coupling strength $C \in (0, 0.2)$. Indices CPR and SPR indicate the onset of PS at the critical value of the coupling strength $C_p = 0.036$ as indicated by the Lyapunov exponents, by a sudden increase in their values. The onset of LS in the coupled Rössler systems is also indicated by the indices JPR and SPR exactly at the same critical value of the coupling strength $C_l = 0.14$ by saturation in their amplitudes at high values near to unity.

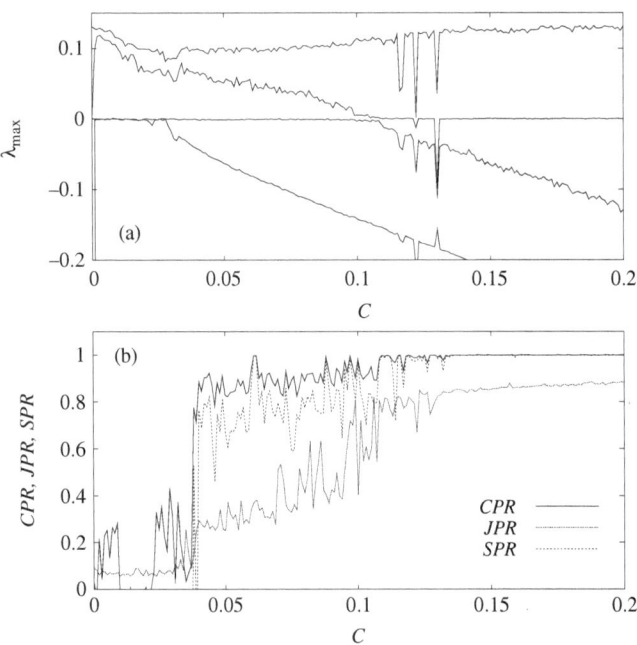

Fig. C.8 (**a**) Four largest Lyapunov exponents in the range of coupling strength $C \in (0, 0.2)$ of the mutually coupled Rössler systems (Eq. (B.12)) studied in Sect. B.2.3 and (**b**) Indices, CPR, JPR, SPR, in the same range of the coupling strength characterizing the onset of PS, LS and transition among them

References

1. H. Poincaré, Acta Mathematica **13**, 1 (1890)
2. E.N. Lorenz, J. Atmos. Sci. **20**, 130 (1963)
3. E.N. Lorenz, J. Atmos. Sci. **26**, 636 (1969)
4. J.-P. Eckmann, S.O. Kamphorst, D. Ruelle, Europhys. Lett. **5**, 973 (1987)
5. J.P. Zbilut, A. Giuliani, C.L. Webber Jr., Phys. Lett. A **246**, 122 (1998)
6. N. Marwan, J. Kurths, Phys. Lett. A **302**, 299 (2002)
7. J.P. Zbilut, C.L. Webber Jr., Phys. Lett. A **171**, 199 (1992)
8. C.L. Webber Jr., J.P. Zbilut, J. Appl. Physiol. **76**, 965 (1994)
9. N. Marwan, N. Wessel, U. Meyerfeldt, A. Schirdewan, J. Kurths, Phys. Rev. E **66**, 026702 (2002)
10. N. Marwan, M.C. Romano, M. Thiel, J. Kurths, Phys. Rep. **438**, 237 (2007)
11. M.C. Casdagli, Physica D **108**, 12 (1997)
12. P. Faure, H. Korn, Physica D **122**, 265 (1998)
13. J. Gao, H. Cai, Phys. Lett. A **270**, 75 (1999)
14. M. Thiel, M.C. Romano, J. Kurths, R. Meucci, E. Allaria, T. Arecchi, Physica D **171**, 138 (2002)
15. M. Thiel, M.C. Romano, P. Read, J. Kurths, Chaos **14**, 234 (2004)
16. J. Kurths, U. Schwarz, C.P. Sonett, U. Parlitz, Nonlinear Processes Geophys. **1**, 72 (1994)
17. N.V. Zolotova, D.I. Ponyavin, Astron. Astrophys. **499**, L1 (2006)
18. N. Marwan, M. Thiel, N.R. Nowaczyk, Nonlinear Processes Geophys. **9**, 325 (2002)
19. N. Marwan, M.H. Taruth, M. Vuille, J. Kurths, Clim. Dyn. **21**, 317 (2003)
20. T.K. March, S.C. Chapmann, R.O. Dendy, Physica D **200**, 171 (2005)
21. A.S. Elwakil, A.M. Soliman, Chaos Solit. Fract. **10**, 1399 (1999)
22. J.M. Nichols, S.T. Trickey, M. Seaver, Mech. Syst. Signal Process. **20**, 421 (2006)
23. A. Giuliani, C. Manetti, Phys. Rev. E **53**, 6336 (1996)
24. C. Manetti, A. Giuliani, M.A. Ceruso, C. L. Webber Jr., J. P. Zbilut, Phys. Lett. A **281**, 317 (2001)
25. J.E. Naschitz, I. Rosner, M. Rozenbaum, M. Fields H. Isseroff, J.P. Babich, E. Zuckerman, N. Elias, D. Yeshurun, S. Naschitz, E. Sabo , QJM:Int. J. Med. **97**, 141 (2004)
26. N. Thomasson, T.J. Hoeppner, C.L. Webber Jr., J.P. Zbilut, Phys. Lett. A **279**, 94 (2001)
27. N. Marwan, A. Meinke, Int. J. Bifurcat. Chaos **14**, 761 (2004)
28. U.R. Acharya, O. Faustand, N. Kannathal, T.L. Chua, S. Laxminarayan, Comput. Meth. Prog. Biomed. **80**, 37 (2005)
29. F.M. Atay, Y. Altintas, Phys. Rev. E **59**, 6 (1999)
30. M.C. Romano, M. Thiel, J. Kurths, I.Z. Kiss, J.L. Hudson, Europhys. Lett. **71**, 466 (2005)
31. M.G. Rosenblum, A.S. Pikovsky, J. Kurths, Phys. Rev. Lett. **76**, 1804 (1996)
32. M.G. Rosenblum, A.S. Pikovsky, J. Kurths, Phys. Rev. Lett. **78**, 4193 (1997)

Appendix D
Some More Examples of DDEs

D.1 Introduction

In addition to the examples of different kinds of DDEs presented in Chap. 1 and other chapters, we will describe briefly some of the available DDEs of various forms that have been used in the literature in different areas of science and technology.

D.2 DDEs with Constant Delay

DDEs with constant delays have been discussed in Sect. 1.1.1 of Chap. 1 along with some of the instances where they appear. In the following we will present few more of them briefly.

D.2.1 Hutchinson's Equation/Delayed Logistic Equation

Hutchinson [1, 2] proposed a more realistic logistic delay equation for single species dynamics by assuming egg formation to occur τ time units before hatching represented as follows,

$$\frac{dx}{dt} = rx(t)\left[1 - \frac{x(t-\tau)}{K}\right], \tag{D.1}$$

where $x(t)$ denotes the population size at time t, $r > 0$ is the intrinsic growth rate and $K > 0$ is the carrying capacity of the population. This equation is often referred to as the *Hutchinson's equation or delayed logistic equation.*

D.2.2 Gopalsamy and Ladas Population Model

Gopalsamy and Ladas [3] proposed a single species population model exhibiting the Allee effect in which the per capita growth rate is a quadratic function of the density and is subject to more than one identical time-delay terms represented as

$$\frac{dx}{dt} = x(t)\left[a + bx(t - \tau) - cx^2(t - \tau)\right], \tag{D.2}$$

where $a > 0, c > 0, \tau > 0$ and b are real constants. In this model, when the density of the population is not small, the positive feedback effects of aggregation and cooperation are dominated by density-dependent stabilizing negative feedback effects due to intraspecific competition. In other words, intraspecific mutualism dominates at low densities and intraspecific competition dominates at higher densities [2, 3].

D.2.3 Stem-Cell Model

The dynamics of pluripotential stem-cell population is governed by the pair of coupled DDEs [4, 5]

$$\frac{dx}{dt} = -\gamma x(t) + \beta x(t)^2 - exp(-\gamma\tau)\beta y_\tau^2, \tag{D.3}$$

$$\frac{dy}{dt} = -\left[\beta y(t) + \delta\right]y(t) + 2exp(-\gamma\tau)\beta y_\tau^2, \quad y_\tau = y(t - \tau), \tag{D.4}$$

where τ is the time required for a cell to traverse the proliferative phase and β is the resting to proliferative phase feedback rate. Further details can be found in [4, 5].

D.2.4 Pupil Cycling Model

Pupil cycling is described by the following DDE with piecewise constant negative feedback

$$\frac{dx}{dt} = y(t), \tag{D.5}$$

$$\frac{dy}{dt} = f(x_\tau), \quad x_\tau = x(t - \tau), \tag{D.6}$$

where the piecewise constant negative feedback is given as

$$f(x) = \begin{cases} a, & x > \theta \\ b, & x \le \theta. \end{cases} \tag{D.7}$$

Here $x(t)$ is the pupil area at time t, τ is the time-delay, a, b describe retinal illumination $(a > b)$ and θ is a threshold area [6, 7].

D.3 DDEs with Discrete Delays

Given the general form of DDEs with discrete delays as in Sect. 1.1.2 along with suitable examples, we will describe here some other examples of discrete/multiple delays with their explicit equations and their details.

D.3.1 Australian Blowfly Model

Braddock and van den Driessche [2, 3, 8] proposed a logistic equation with two different delays to mimic the population $x(t)$ of the Australian blowfly *Lucila cuprina*, which is represented as follows:

$$\frac{dx}{dt} = rx(t)\left[1 + ax(t - \tau_1) - bx(t - \tau_2)\right], \tag{D.8}$$

where $r > 0$, $a > 0$ and $b > 0$ are real constants, $\tau_1 > 0$ and $\tau_2 > 0$ corresponds to regeneration and reproductive delays, respectively.

D.3.2 Wilson and Cowan Model

Wilson and Cowan [9, 10] model describes the evolution of a network of synaptically interacting neural populations, typically one being excitatory and the other inhibitory, in the presence of two different delays represented as

$$\frac{dx}{dt} = -x(t) + f\left[\theta_x + ax(t - \tau_1) + by(t - \tau_2)\right], \tag{D.9}$$

$$\frac{dy}{dt} = \alpha\left(-y(t) + f\left[\theta_y + cx(t - \tau_2) + dy(t - \tau_1)\right]\right), \tag{D.10}$$

where $x(t)$ and $y(t)$ represent the synaptic activity of the two populations with a relative time scale for the response set by α^{-1}. The architecture of the network is fixed by the weights a, b, c, d, while $\theta_{x,y}$ describe background drives and f is the common firing rate function.

D.3.3 Human Respiratory Model

A simple model of the respiratory control mechanism in humans is represented as [11]

$$\frac{dx}{dt} = p - \alpha W \left[x(t - \tau_1), y(t - \tau_2) \right] (x(t) - x_I), \quad (D.11)$$

$$\frac{dy}{dt} = -\sigma + \beta W \left[x(t - \tau_1), y(t - \tau_2) \right] (y(t) - y_I), \quad (D.12)$$

where $x(t)$ and $y(t)$ denote the arterial CO_2 and O_2 concentrations, respectively. $W(\cdot, \cdot)$ is the ventilation function (the volume of gas moved by the respiratory systems), $\tau_{1,2}$ are transport delays, x_I and y_I are inspired CO_2 and O_2 concentrations, p is the CO_2 production rate, σ is the O_2 consumption and α, β are positive constants referring to the diffusibility of CO_2 and O_2, respectively.

D.4 DDEs with Distributed Delay

In the following we will present a few examples for DDEs with distributed delay in addition to the details presented in Sect. 1.1.3.

D.4.1 Volterra's Logistic Equation

The Hutchinson's equation (D.1) assumed that the regulatory effect depends on the population at a fixed earlier time $t - \tau$. However, in a more realistic model the delay effect should be an average over past populations and this requires an equation with a distributed delay. Volterra [2, 12] suggested the first model of logistic equation with distributed delay and he used a distributed delay term to examine a cumulative effect in the death rate of a species, depending on the population at all times, represented as

$$\frac{dx}{dt} = rx(t) \left[1 + \frac{1}{K} \int_{-\infty}^{t} G(t - s) x(s) ds \right], \quad (D.13)$$

where $G(t)$ is the delay kernel, corresponding to a weighting factor which indicates how much emphasis should be given to the size of the population at earlier times to determine the present effect on resource availability.

D.4.2 Neural Network with Distributed Delay

Hopfield neural networks with distributed delays are considered [13] to take into account the distribution of conduction velocities along parallel pathways with a variety of axon sizes and lengths as [13]

$$\frac{dx}{dt} = -x(t) + a \tanh \left[x(t) - b \int_{0}^{\infty} x(t - s) k(s) ds - c \right], \quad (D.14)$$

where $x(t)$ is the state of neuron, a, b and c are non-negative constants.

D.4.3 Chemostat Model

A chemostat model of a single species feeding on a limiting nutrient supplied at constant rate is proposed as [14]

$$\frac{dS}{dt} = \left(S^0 - S(t)\right) D - ax(t)p\left(S(t)\right), \tag{D.15}$$

$$\frac{dx}{dt} = x(t)\left[-D_1 + \int_{-\infty}^{t} F(t-s)p(S(t))ds\right], \tag{D.16}$$

where $S(t)$ and $x(t)$ denote the concentration of the nutrient and the population of microorganism at t. S^0 denotes the input concentration of nutrient, D is referred to as the dilution rate and D_1 denotes the sum of the dilution rate and the death rate of the population of microorganism. The function $p(S)$ describes the species specific growth rate and a^{-1} is referred to as the growth yield constant.

D.5 DDEs with State-Dependent Delay

We will discuss some of the DDEs with state-dependent delay that have been used in the literature in some detail. General discussion and some other examples are presented in Sect. 1.1.4,

D.5.1 Population Model

Considering the birth rate as population density dependent rather than age dependent certain population dynamics is also modeled with delay equations with state dependent delay. Assuming the lifespan L of individuals in the population as a function of the current population size, $x(t)$, and taking into account the crowding effect, a DDE with state dependent delay for population dynamics is suggested [15, 2],

$$\frac{dx}{dt} = \frac{bx(t) - bx(t - L[x(t)])}{1 - L'[x(t)]bx(t - L[x(t)])}. \tag{D.17}$$

D.5.2 Logistic Model with State Dependent Delay

Logistic model with a state dependent delay has also been proposed [2],

$$\frac{dx}{dt} = rx(t)\left[1 - \frac{x(t - \tau(x(t)))}{K}\right]. \tag{D.18}$$

D.5.3 Mechanical Model for Machine Tool Chatter

Turning process in machine tool chatter is governed by the following state dependent DDE [16],

$$m\ddot{x}(t) + c_x \dot{x}(t) + k_x x(t) = K_x w \left[v\tau(x_t) + y(t - \tau(x_t)) - y(t) \right]^q, \quad (D.19)$$

$$m\ddot{y}(t) + c_y \dot{y}(t) + k_y y(t) = K_y w \left[v\tau(x_t) + y(t - \tau(x_t)) - y(t) \right]^q, \quad (D.20)$$

where m, c_x, c_y, k_x and k_y are the modal mass, the damping and the stiffness parameters in the x and y directions, respectively. $K_{x,y}$ are the cutting coefficients, w is the depth of cut, q is an exponent and v is the speed of feed. More details on the system and its stability analysis can be found in [16].

D.6 DDEs with Time-Dependent Delay

In the following we will present explicit equations for two models described by DDEs with time-dependent delays.

D.6.1 Stem-Cell Equation

The stem cell equation can be put in the form [17]

$$\dot{S}(t) = 2M(t - \tau_m(t))S(t - \tau_m(t)) - S(t)\left[M(t) + \omega\right], \quad (D.21)$$

where $S(t)$ is the available stem cell population. The rate $M(t)S(t)$ at which stem cells enter the mitotic channel is controlled by the mitotic operator, $M(t)$, acting on the stem cell population and the rate at which they return after dividing is $2M(t - \tau_m(t))S(t - \tau_m(t))$, assuming that there are no losses. $\tau_m(t)$ represents the delay between cells leaving the stem cell population to enter the mitotic cycle and the return of two daughter cells.

D.6.2 Neural Network Model

A neural network model with time-varying delay is represented as [18]

$$\frac{dX}{dt} = -DX(t) + AG(X(t)) + BF(X(t - \tau(t))) + I(t), \quad (D.22)$$

where $X(t) = [x_1(t), x_2(t), \cdots, x_n(t)]$ is the state vector of the network at time t, $D = \text{diag}[d_1, d_2, \cdots, d_n]$ with $d_i > 0$ denotes the rate with which the cell i resets its potential to the resting state when isolated from other cells and inputs,

$A = (a_{kl})_{n \times n}$, $B = (b_{kl})_{n \times n} \in \mathbb{R}^{n \times n}$ represent the connection weight matrix and the delayed connection weight matrix, respectively. a_{kl}, b_{kl} denote the strengths of connectivity between the cell k and l at time t and $t - \tau(t)$, respectively. $F(X) = \left[f_1(x_1(t)), \cdots, f_n(x_n(t)) \right]$, $G(X) = \left[g_1(x_1(t)), \cdots, g_n(x_n(t)) \right]$ are activation functions.

Further details on all these examples can be found in their respective references.

References

1. G.E. Hutchinson, Ann. N.Y. Acad. Sci. **50**, 221 (1948)
2. S. Ruan, Delay differential equations in single species dynamics, in *Delay Differential Equations and Applications*, ed. by O. Arino et al. (Springer, Berlin, 2006), pp. 477–517
3. K. Gopalsamy, G. Ladas, Quart. Appl. Math. **48**, 433 (1990)
4. M.C. Mackey, Blood **51**, 941 (1978)
5. M.C. Mackey, Dynamic haematological disorders of stem cell origin, in *Biophysical and Biochemical Information Transfer in Recognition*, ed. by J.G. Vassileva-Popova, E.V. Jensen (Plenum Press, New York, 1979), pp. 373–409
6. J.G. Milton, A. Longtin, Vision Res. **30**, 515 (1990)
7. J.G. Milton, A. Longtin, T.H. Krikham, G.S. Francis, Am. J. Ophthalmol. **105**, 402 (1988)
8. R.D. Braddock, P. van den Driessche, J. Austral. Math. Soc. Ser. B **24**, 292 (1983)
9. H.R. Wilson, J.D. Cowan, Biophys. J. **12**, 1 (1972)
10. S. Coombes, C. Laing, Philos. Trans. R. Soc. A **367**, 1117 (2009)
11. K.L. Cooke, J. Turi, J. Math. Biol. **32**, 535 (1994)
12. V. Volterra, C.R. Acad. Sci. **199**, 1684 (1934)
13. F.Y. Zhang, H.F. Huo, Discrete Dyn. Nat. Soc. **2006**, 27941 (2006)
14. S. Ruan, G.S.K. Wolkowicz, J. Math. Anal. Appl. **204**, 786 (1996)
15. J. Bélair, Population models with state-dependent delay, in *Mathematical Populations Dynamics*, ed. by O. Arino, D.E. Axelrod, M. Kimmel (Marcel Dekker, New York, 1991), pp. 165–176.
16. T. Insperger, D.A.W. Barton, G. Stépán, Int. J. Non-linear Mech. **43**, 140 (2008)
17. J. Kirk, T.E. Wheldon, W.M. Gray, J.S. Orr, Bio-Med. Comput. **1**, 291(1970)
18. W. Wang, J. Cao, Physica A **366**, 197 (2006)

Glossary

Amplitude death The phenomenon of suppression of oscillations in dynamical systems mainly due to time-delay feedback or time-delay coupling is termed as amplitude death.

Analog simulation circuit An electronic circuit designed to mimic the dynamics of a system modelled by a linear/nonlinear evolution equation.

Analytic signal approach It is one of the approaches to calculate the phase of a non-phase-coherent chaotic/hyperchaotic attractor. The complex analytical signal $\chi(t)$ is constructed from a scalar time series $s(t)$ via Hilbert transform (HT).

Anticipatory synchronization Anticipatory synchronization is a special kind of generalized synchronization (see below), where one (receiver) of the coupled systems anticipates the state of the other (transmitter) with finite anticipating time.

Attractor It is a bounded region of phase space of a dynamical system towards which nearby trajectories asymptotically approach. The attractor may be a point or a closed curve or an unclosed but bounded orbit.

Autonomous system A system with no explicit time-dependent term in its equation of motion.

Band merging bifurcation Merging of two or more bands of a m-band chaotic attractor at a critical value of a control parameter.

Bifurcation A sudden/abrupt qualitative change in the dynamics of a system at a critical value of a control parameter when it is varied smoothly.

Bifurcation diagram A plot illustrating qualitative changes in the dynamical behavior of a system as a function of a control parameter.

Bifurcation route The nature of sudden/abrupt qualitative changes in the dynamical behavior of a system as a function of a control parameter indicating the mechanism responsible for the change.

Chaos A phenomenon or process of occurrence of bounded nonperiodic evolution in deterministic nonlinear systems with high sensitive dependence on initial conditions. A consequence is that nearby chaotic orbits diverge exponentially (in a time average sense) in phase space. A measure of quantification of the degree of divergence is the set of Lyapunov exponents. Chaotic motion is characterized by at least one positive Lyapunov exponent.

Chimera state The coexistence of coherent (synchronized) and incoherent (desynchronized) states in coupled identical oscillators is called a chimera state.

Chua's circuit A simple, third-order, autonomous electronic circuit consisting of two linear capacitors, a linear inductor, a linear resistor, and only one nonlinear element, namely Chua's diode, having a piecewise linear characteristic.

Chemostat model A chemostat (from *Chem*ical environment is *stat*ic) is a bioreactor to which fresh medium is continuously added, while culture liquid is continuously removed to keep the culture volume constant. By changing the rate with which the medium is added to the bioreactor the growth rate of the microorganism can be easily controlled.

Complete synchronization Complete synchronization (CS) refers to the identical evolution of the trajectories of two identical linear/nonlinear systems which is achieved by means of a suitable coupling in such a way that the two trajectories remain in step with each other in the course of time.

Complex network In the context of network theory, a complex network is a network (graph) with non-trivial topological features (Examples: scale-free networks and small-world networks) that do not occur in simple networks such as lattices or random graphs. The study of complex networks is an active area of scientific research of this decade inspired largely by the empirical study of real-world networks such as computer networks and social networks.

Connection delay Delay caused due to the finite time required for the propagation of signals from output to the receiver end or among the interconnected dynamical systems.

Correlation dimension A quantitative measure used to describe geometric and probabilistic features of attractors. It is an integer for regular attractors such as a fixed point, a limit cycle or a quasiperiodic orbit. It is non-integer for a strange (chaotic) attractor.

Correlation function A statistical measure used to characterize regular and chaotic motions. For periodic motion it oscillates while for chaotic motion it decays to zero.

Correlation of probability of recurrence (CPR) A cross correlation coefficient between the generalized autocorrelation functions of two systems, $P_{1,2}(t)$, is defined as correlation of probability of recurrence (CPR).

Cross recurrence plot (CRP) A cross recurrence plot (CRP) is a bivariate extension of the recurrence plot and was introduced to analyse the difference between two different systems. CRP can be regarded as a generalisation of the linear cross-correlation function.

Delay differential equation (DDE) A delay-differential equation (DDE) comprises of an unknown function and certain of its derivatives, evaluated at arguments that differ by fixed numerical values. For example, $\dot{x}(t) = F(t, x(t), x(t - \tau))$ is a retarded DDE for $\tau > 0$. DDEs (also called functional differential equations or retarded differential-difference equations) generalize the concept of differential equations by allowing the state of the system to depend on states different from the present one. DDEs can also be of neutral and advanced types.

Delay time modulation (DTM) Delay time modulation refers to the case of time varying delay $\tau(t)$, where the time-delay τ evolve in time or even it can be a function of state variable $\tau(x)$ (in which case it is referred to as state dependent delay).

El Niño-Southern oscillation The El Niño-Southern Oscillation is often abbreviated as ENSO and in popular usage is called simply El Niño. It is defined by sustained differences in the Pacific ocean surface temperatures when compared with the average value. The accepted definition is a warming or cooling of at least $0.5\,°C(0.9\,°F)$ averaged over the east-central tropical Pacific ocean. When this happens for less than 5 months, it is classified as El Niño or La Niña conditions; if the anomaly persists for 5 months or longer, it is called an El Niño or La Niña "episode". Typically, this happens at irregular intervals of 27 years and lasts 9 months to 2 years.

Embedding theorem Delay embedding theorem gives the conditions under which a chaotic dynamical system can be reconstructed from a sequence of observations of the state of a dynamical system. The reconstruction preserves the properties of the dynamical system that do not change under smooth coordinate changes, but it does not preserve the geometric shape of structures in phase space.

Epidemiology It is a branch of science dealing with spreading of diseases in human population. The model proposed to study the nature of spreading and to identify the measures to control a specific disease is called an epidemic model. The term "epidemics" is derived from Greek *epi-* upon + *demos* people. An epizootic is the analogous circumstance within an animal population.

Equilibrium point An admissible solution of $F(X) = 0$ for a dynamical system $\dot{X} = F(X)$, $X = (x_1, x_2, \cdots, x_n)^T$. It is also called fixed point or singular point of the system.

Error feedback Error feedback refers to the feedback given as a linear/nonlinear function of the difference of the state variables of the coupled systems.

Feedback delay Finite time taken by a signal that is fed back into the system causes the feedback delay. For instance, in semiconductor lasers the coherent light is converted into chaotic signal due to the feedback of the light through a cavity and the round trip time results in the feedback delay. Feedback delay can give rise to plethora of new behaviors in dynamical systems (see Chaps. 5 and 6).

FitzHugh-Nagumo oscillator FitzHugh-Nagumo model $\dot{x} = x - x^3/3 - y + I$, $\dot{y} = 0.08(x + 0.7 - 0.8y)$ is a two-dimensional simplification of the Hodgkin-Huxley model of spike generation in squid giant axons. It is used to isolate conceptually the essentially mathematical properties of excitation and propagation from the electro-chemical properties of sodium and potassium ion flow.

Generalized autocorrelation function In recurrence quantification analysis, gen-eralized autocorrelation function, $P(\varepsilon, t)$, can be considered as the probability that the system recurs to the ε-neighbourhood of a former point $\mathbf{x}_i$ of the trajectory after t time steps. Comparing $P(\varepsilon, t)$ of two systems, one can characterize quan-titatively and qualitatively the existence of phase synchronization between the two systems.

Generalized synchronization Synchronization can be achieved even in the case of coupled non-identical systems and in this case, it is termed as generalized syn-chronization where there exists some functional relationship between the variables of the coupled systems.

Globally coupled chimera (GCC) state The coexistence of chimera states in a system of identical oscillators with (sub) populations with time-delay coupling is termed as globally coupled chimera states. It is demonstrated that coupling delay can induce globally clustered chimera (GCC) states in systems having more than one coupled identical oscillator (sub) populations. By GCC one refers to the state of a system, which has more than one (sub) population, that splits into two different groups, one synchronized and the other desynchronized, each group comprising of oscillators from both the populations.

Hopf bifurcation It corresponds to the birth of a limit cycle from an equilibrium point when a control parameter is varied. If the limit cycle is stable (unstable) then the bifurcation is called supercritical (subcritical).

Hyperchaos It represents chaotic motion with more than one positive Lyapunov exponents. It has at least two exponentially diverging directions in its orthonormal phase space.

Ikeda system Ikeda system was introduced to describe the dynamics of an opti-cal bistable resonator, which is specified by the state equation $\frac{dx}{dt} = -\alpha x(t) - \beta \sin x(t - \tau)$. Physically $x(t)$ is the phase lag of the electric field across the res-onator and thus may clearly assume both positive and negative values, α is the relaxation coefficient, β is the laser intensity injected into the system and τ is the round-trip time of the light in the resonator.

Intermittency route to chaos A route to chaos where regular orbital behavior is intermittently interrupted by short time irregular bursts. As the control parameter is varied, the durations of the bursts increase, leading to full scale chaos.

Inverse period doubling It denotes the bifurcation sequence of a nonlinear dynam-ical system which is inverse to the period doubling bifurcation as a control parameter is varied.

Invertible map A map is invertible when its inverse exists and is unique for each point in the phase space.

Jacobian matrix Jacobian matrix is the matrix of all first-order partial derivatives of a vector-valued function. The Jacobian determinant (often simply called the Jacobian) is the determinant of the Jacobian matrix. These concepts are named after the mathematician Carl Gustav Jacob Jacobi.

Joint recurrence plot (JRP) A joint recurrence plot is introduced to compare the states of different systems by estimating the recurrences of their trajectories in their corresponding phase spaces separately and then look for the times when both of them recur simultaneously, that is when joint recurrence occurs.

Kelvin waves A Kelvin wave is a wave in the ocean or atmosphere that balances the earth's Coriolis force against a topographic boundary such as a coastline, or a waveguide such as the equator. A feature of a Kelvin wave is that it is non-dispersive, i.e., the phase speed of the wave crests is equal to the group speed of the wave energy for all frequencies. This means that it retains its shape in the alongshore direction over time.

Krasovskii-Lyapunov theory Krasovskii-Lyapunov theory is the direct extension of Lyapunov second theorem on stability, which states that if a positive definite function $V(x) : R^n \rightarrow R$ exists such that $V(x) \geq 0$ with equality if and only if $x = 0$ and $\dot{V}(x) \leq 0$ with equality if and only if $x = 0$ (negative definite), then the equilibrium state is Lyapunov stable.

Kuramoto oscillators It is a mathematical model used to describe synchronization. More specifically, it is a model for the behavior of a large set of coupled oscillators. Its formulation was motivated by the behavior of systems of chemical and biological oscillators, and it has found widespread applications. The most popular form of the model has the following governing equations: $\frac{d\theta_i}{dt} = \omega_i + \frac{K}{N} \sum_{j=1}^{N} \sin(\theta_j - \theta_i)$, $i = 1 \ldots N$, where the system is composed of N limit-cycle oscillators.

Lag synchronization Lag synchronization is a special case of generalized synchronization, where one of the coupled systems always evolve in lag with respect to the other with a finite lag time.

Limit cycle An isolated closed orbit in the phase space associated with a dynamical system.

Linear superposition principle A property associated with linear differential equations. The property is that if u_1 and u_2 are two linearly independent solutions of a linear homogeneous differential equation then $u = au_1 + bu_2$ is also a solution of it, where a and b are arbitrary (complex) constants.

Localized set It refers to the sets obtained by observing one of the coupled systems whenever a defined event occurs in the other system and viceversa. The concept of localized sets has been introduced recently as a new framework to identify phase synchronization in chaotic/hyperchaotic attractors without explicitly calculating the phase variable.

Logistic map A discrete map analog of the logistic equation for population growth. The map is represented as $x_{n+1} = ax_n(1 - x_n)$, where a is a parameter with $0 \leq a \leq 4$ and $0 < x < 1$.

Lorenz equation The paradigmic nonlinear chaotic system originally introduced by E. Lorenz in 1963 in connection with atmospheric convection, represented by a set of three coupled ordinary differential equations $\frac{dx}{dt} = \sigma(y-x)$, $\frac{dy}{dt} = x(\rho-z) - y$, $\frac{dz}{dt} = xy - \beta z$, where σ is called the Prandtl number and ρ is called the Rayleigh number.

Lyapunov exponent Lyapunov exponent of a dynamical system is a quantity (a number) that characterizes the rate of separation of infinitesimally close trajectories. Different types of orbits can be distinguished depending on the value of its Lyapunov exponents. All negative exponents represent regular and periodic orbits, while at least one positive exponent indicates the presence of chaotic motion. More than one positive exponent indicate the presence of hyperchaotic motion.

Lyapunov function Lyapunov function is a function which can be used to prove the stability of a certain fixed point in a dynamical system or autonomous differential equation.

Mackey-Glass system The Mackey-Glass system, which was originally deduced as a model for blood production in patients with leukemia, can be represented by the first order nonlinear DDE $\dot{x} = -bx(t) + \frac{ax(t-\tau)}{(1.0+x(t-\tau)^c)}$, where a, b and c are positive constants. Here, $x(t)$ represents the concentration of blood at time t (density of mature cells in bloodstreams), when it is produced, $x(t - \tau)$ is the concentration when the "request" for more blood is made and τ is the time-delay between the production of immature cells in the bone marrow and their maturation for release in circulating bloodstreams.

Noise In common use, the word noise means any unwanted sound. In both analog and digital electronics, noise is an unwanted perturbation to a wanted signal. In signal processing or computing it can be considered unwanted data without meaning.

Nonautonomous system A system with at least one explicit time-dependent term in its equation of motion.

Non-invertible map Maps that are not invertible are non-invertible maps, that is, one for which inverse does not exist.

Non-phase-coherent attractor If the flow of a dynamical system does not have a proper center of rotation around a fixed reference point, then the corresponding attractor is termed as a non-phase-coherent attractor. For instance, the funnel Rössler attractor for the parameter values $a = 0.25$, $b = 0.2$, and $c = 8.5$ shown in Fig. 10.1b of Chap. 10, is an example of non-phase-coherent attractor.

Orthonormalization A form of orthogonalization in which the resulting vectors are all unit vectors. Gram-Schmidt orthogonalization, also called the Gram-Schmidt process, is a procedure which takes a nonorthogonal set of linearly independent functions and constructs an orthogonal basis over an arbitrary interval with respect to an arbitrary weighting function $w(x)$.

Phase-coherent attractor If the flow of a dynamical system has a proper rotation around a fixed reference point as its center, then the corresponding attractor is called a phase-coherent attractor. For instance, Rössler attractor for the standard parameter values $a = 0.15, b = 0.2$, and $c = 8.5$ shown in Fig. 10.1a of Chap. 10, is an example of phase-coherent attractor.

Period doubling It denotes the bifurcation sequence of periodic motions for a non-linear dynamical system in which the period doubles at each bifurcation as a control parameter is varied. Beyond a critical accumulation parameter value, chaotic motion occurs. It is also referred as subharmonic bifurcation or flip bifurcation.

Phase flip bifurcation It denotes the abrupt change in the relative phase of the coupled oscillators from zero to π as a function of the delay time.

Phase point A point in the phase space representing the state of a system at any instant of time.

Phase space As abstract space where each of the variables needed to specify the dynamical state of a system represents an orthogonal coordinate.

Phase synchronization Phase synchronization can be defined as perfect locking of the phase/frequency of the coupled systems, while their amplitudes remain uncorrelated and often chaotic in the case of coupled chaotic systems.

Piecewise linear system A piecewise linear system is usually referred to a non-linear dynamical system, whose nonlinear function $f(x)$ is composed of piecewise linear segments.

Poincaré section Any suitable hyperplane of the phase space is a Poincaré section (or surface of section). The relation between the successive intersections of the phase trajectories with this section in a single direction constitutes the Poincaré map.

Propagation delay See connection delay.

Pseudospace Any additional phase space created by embedding technique is referred to as pseudospace.

PSPICE simulation PSPICE, is an acronym for Personal Simulation Program with Integrated Circuit Emphasis, is a SPICE (Simulation Program with Integrated Circuit Emphasis) analog circuit and digital logic simulation software that runs on personal computers.

Recurrence analysis Recurrence analysis is a powerful technique that visualizes the recurrences of a dynamical system and gives information about the behavior of its trajectory in the phase space.

Recurrence plot (RP) A recurrence plot (RP) is the graphical representation of a binary symmetric square matrix which encodes the times when two states are in close proximity (neighbours), that is the time of recurrence in the phase space.

Recurrence quantification analysis (RQA) Several measures of complexity which quantify the small scale structures in recurrence plots have been proposed and are known as recurrence quantification analysis (RQA).

Rossby waves Rossby waves are giant meanders in high-altitude winds that are a major influence on weather. Their emergence is due to shear in rotating fluids so that the Coriolis force changes along the sheared coordinate. In planetary atmospheres, they are due to the variation in the Coriolis effect with latitude. The waves were first identified in the Earth's atmosphere in 1939 by Carl-Gustaf Arvid Rossby who went on to explain their motion. Rossby waves are a subset of inertial waves.

Rössler system Otto Rössler designed the Rössler attractor in 1976, but the theoretical equations were later found to be useful in modeling equilibrium in chemical reactions. The defining equations are: $\frac{dx}{dt} = -y - z$, $\frac{dy}{dt} = x + ay$, $\frac{dz}{dt} = b + z(x - c)$. Rössler studied the chaotic attractor with $a = 0.2$, $b = 0.2$, and $c = 5.7$, though properties of $a = 0.1$, $b = 0.1$, and $c = 14$ have been more commonly used since.

Runge-Kutta method In numerical analysis, the Runge-Kutta methods are an important family of implicit and explicit iterative methods for the approximation of solutions of ordinary differential equations. These techniques were developed by the German mathematicians C. Runge and M.W. Kutta.

Stochastic process A stochastic process is the counterpart to a deterministic process (or deterministic system). Instead of dealing with only one possible "reality" of how the process might evolve under time (as is the case, for example, for solutions of an ordinary differential equation), in a stochastic or random process there is some indeterminacy in its future evolution described by probability distributions. This means that even if the initial condition (or starting point) is known, there are many possibilities the process might go to, but some paths are more probable and others less.

Synchronization The word *synchronous* is derived from Greek terminology *chronous* means time and *syn* means common. Put together synchronous/synchronization has its direct meaning "share the common time" or "occurring in the common time". Technically, it can be defined as "entrainment of a dynamical property/share a common property of motion" or "as degree of correlation" between the interacting dynamical systems.

Time series The measured values of a physical variable of a dynamical system at regular intervals of time.

Transient motion An initial time evolution of a system before getting settled into its steady state behavior.

Unstable periodic orbit An unstable period-1, or 2, $\cdots$, or n fixed point or limit cycle. A chaotic orbit is regarded as a pool of unstable periodic orbits.

Index